METHODS IN MOLECULAR BIOLOGY

For further volumes:
http://www.springer.com/series/7651

For over 35 years, biological scientists have come to rely on the research protocols and methodologies in the critically acclaimed Methods in Molecular Biology series. The series was the first to introduce the step-by-step protocols approach that has become the standard in all biomedical protocol publishing. Each protocol is provided in readily-reproducible step-by step fashion, opening with an introductory overview, a list of the materials and reagents needed to complete the experiment, and followed by a detailed procedure that is supported with a helpful notes section offering tips and tricks of the trade as well as troubleshooting advice. These hallmark features were introduced by series editor Dr. John Walker and constitute the key ingredient in each and every volume of the Methods in Molecular Biology series. Tested and trusted, comprehensive and reliable, all protocols from the series are indexed in PubMed.

Manuscript Instructions https://drive.google.com/file/d/1K4Az5vpAFVzYadJ3jKx2-QSo7Dqa_SNA/view?usp=sharing

Allergic Disease Models

Methods and Translational Protocols

Edited by

José de la Fuente

SaBio, Instituto de Investigación en Recursos Cinegéticos (IREC), Ciudad Real, Spain

Alejandro Cabezas-Cruz

microbioTick Lab, UMR BIPAR, INRAE, Anses, Maisons-Alfort, France

Humana Press

Editors
José de la Fuente
SaBio
Instituto de Investigación en Recursos Cinegéticos (IREC)
Ciudad Real, Spain

Alejandro Cabezas-Cruz
microbioTick Lab
UMR BIPAR, INRAE, Anses
Maisons-Alfort, France

ISSN 1064-3745 ISSN 1940-6029 (electronic)
Methods in Molecular Biology
ISBN 978-1-0716-5634-1 ISBN 978-1-0716-5635-8 (eBook)
https://doi.org/10.1007/978-1-0716-5635-8

This Humana imprint is published by the registered company Springer Science+Business Media, LLC, part of Springer Nature.
The registered company address is: 1 New York Plaza, New York, NY 10004, U.S.A.

Preface

Allergic diseases such as Alpha-Gal Syndrome (AGS) affect human and animal health worldwide. This *Methods in Molecular Biology* volume offers the first comprehensive and protocol-driven resource covering the entire spectrum of experimental models and translational applications for allergic diseases. The book integrates in vivo animal models such as ticks, mice, and zebrafish, novel omics methodologies for the study of tick-borne allergies and AGS, microbiota-based protocols, multiomics integration workflows, bioinformatics pipelines, with AI and data integration, translational and participatory models, and clinical diagnostic methods with highly methodical, reproducible, and translational protocols. The volume uniquely includes citizen science approaches and is designed to provide step-by-step, validated methods to accelerate research, clinical application, and biomarker discovery in allergic disease research. These methods provide tools to advance in the study of allergic diseases and the development of diagnostic, preventive, and therapeutic interventions.

Ciudad Real, Spain *José de la Fuente*
Maisons-Alfort, France *Alejandro Cabezas-Cruz*

Contents

Contributors

NÚRIA AMIGÓ • *Lipid RDI Cluster, Biosfer Teslab, Reus, Spain; Institut d'Investigació Sanitària Pere Virgili, Reus, Spain; Department of Basic Medical Sciences, Universitat Rovira i Virgili, Reus, Spain; Centro de Investigación Biomédica en Red de Diabetes y Enfermedades Metabólicas Asociadas (CIBERDEM), Instituto de Salud Carlos III (ISCIII), Madrid, Spain*

JOSÉ MIGUEL URRA ARDANAZ • *Immunology, Hospital General Universitario de, Ciudad Real, Spain*

SARA ARTIGAS-JERÓNIMO • *SaBio, Instituto de Investigación en Recursos Cinegéticos, IREC (CSIC, UCLM, JCCM), Ciudad Real, Spain*

PAVLE BANOVIĆ • *Clinic for Lyme Borreliosis and Other Tick-Borne Diseases, Department of Prevention of Rabies and Other Infectious Diseases, Pasteur Institute Novi Sad, Novi Sad, Serbia; Department of Microbiology with Parasitology and Immunology, Faculty of Medicine, University of Novi Sad, Novi Sad, Serbia; Diagnostics and Laboratory Research Task Force, Balkan Association for Vector-Borne Diseases, Novi Sad, Serbia*

LUIS G. BERMÚDEZ-HUMARÁN • *INRAE, Université Paris-Saclay, AgroParisTech, Micalis Institute, Jouy-en-Josas, France*

FRANCISCO FEO BRITO • *Faculty of Medicine, University of Castilla-La Mancha, Ciudad Real, Spain*

ALEJANDRO CABEZAS-CRUZ • *UMR BIPAR (INRAE, Anses, EnvA), Maisons-Alfort, France; ANSES, INRAE, Ecole Nationale Vétérinaire d'Alfort, UMR BIPAR, Laboratoire de Santé Animale, Maisons-Alfort, France;*

MARINELA CONTRERAS • *SaBio,Instituto de Investigación en Recursos Cinegéticos,IREC (CSIC, UCLM, JCCM), Ciudad Real, Spain*

JOSÉ DE LA FUENTE • *SaBio, Instituto de Investigación en Recursos Cinegéticos, IREC (CSIC, UCLM, JCCM), Ciudad Real, Spain; Department of Veterinary Pathobiology, Center for Veterinary Health Sciences, Oklahoma State University, Stillwater, OK, USA*

VINCENT C. DURU • *Institute for Parasitology and Tropical Veterinary Medicine, Freie Universität Berlin, Berlin, Germany; Veterinary Centre for Resistance Research, Freie Universität Berlin, Berlin, Germany; Melbourne Veterinary School, Faculty of Science, The University of Melbourne, Werribee, Victoria, Australia*

AGUSTÍN ESTRADA-PEÑA • *Faculty of Veterinary Medicine, University of Zaragoza, Zaragoza, Spain; Ministry of Human Health, CSAI-Foundation, Madrid, Spain*

DAVID FERNÁNDEZ-CASTELLANOS • *SaBio, Instituto de Investigación en Recursos Cinegéticos, IREC (CSIC, UCLM, JCCM), Ciudad Real, Spain*

EMILIA FORNAL • *Department of Bioanalytics, Faculty of Medical Sciences, Medical University of Lublin, Lublin, Poland*

ABDUL GHAFAR • *Melbourne Veterinary School, Faculty of Science, The University of Melbourne, Werribee, Victoria, Australia*

ALMUDENA GONZÁLEZ-GARCÍA • *SaBio, Instituto de Investigación en Recursos Cinegéticos, IREC (CSIC, UCLM, JCCM), Ciudad Real, Spain*

JASMINA GRUJIĆ • *Department of Transfusion Medicine, Faculty of Medicine, University of Novi Sad, Novi Sad, Serbia; Blood Transfusion Institute of Vojvodina, NoviSad, Serbia*
ABDUL JABBAR • *Melbourne Veterinary School, Faculty of Science, The University of Melbourne, Werribee, Victoria, Australia*
MARIELA JIMÉNEZ • *Departamento de Microbiología, Universidad Autónoma de Aguascalientes, Aguascalientes, México*
ANDRIJA JOVIĆ • *Faculty of Medicine Niš, University of Niš, Nis, Serbia; Clinic for Dermatovenerology, University Clinical Center Niš, Nis, Serbia; Clinical Medicine Task Force, Balkan Association for Vector-Borne Diseases, Novi Sad, Serbia*
ANNA KOZUB-PĘDRAK • *Department of Bioanalytics, Faculty of Medical Sciences, Medical University of Lublin, Lublin, Poland*
JOANNA KULISZ • *Department of Biology and Parasitology, Faculty of Health Sciences, Medical University of Lublin, Lublin, Poland*
MARKO LILIĆ • *School of Medicine, University of Mostar, Mostar, Bosnia and Herzegovina*
KINGA LIS • *Department of Allergology, Clinical Immunology and Internal Medicine, Ludwik Rydygier Collegium Medicum in Bydgoszcz, Nicolaus Copernicus University in Torun, Bydgoszcz, Poland*
APOLLINE MAITRE • *ANSES, INRAE, Ecole Nationale Vétérinaire d'Alfort, UMR BIPAR, Laboratoire de Santé Animale, Maisons-Alfort, France*
LOURDES MATEOS-HERNÁNDEZ • *ANSES, INRAE, Ecole Nationale Vétérinaire d'Alfort, UMR BIPAR, Laboratoire de Santé Animale, Maisons-Alfort, France*
LORENA MAZUECOS • *Biochemistry Section, Faculty of Chemical Sciences and Technologies, University of Castilla-La Mancha (UCLM), Ciudad Real, Spain; DOE Research Group,Institute of Biomedicine, Instituto de Investigación Sanitaria de Castilla-La Mancha (IDISCAM), University of Castilla-La Mancha (UCLM), Ciudad Real, Spain*
ANA MARÍA BURGOS MONTERO • *Allergy Department, Hospital General La Mancha Centro, Alcázar de San Juan, Spain*
BAHAR E. MUSTAFA • *Melbourne Veterinary School, Faculty of Science, The University of Melbourne, Werribee, Victoria, Australia*
ARD M. NIJHOF • *Institute for Parasitology and Tropical Veterinary Medicine, Freie Universität Berlin, Berlin, Germany; Veterinary Centre for Resistance Research, Freie Universität Berlin, Berlin, Germany*
HANNA NIKOLAICHUK • *Department of Bioanalytics, Faculty of Medical Sciences, Medical University of Lublin, Lublin, Poland*
ELIANNE PILOTO-SARDIÑAS • *Direction of Animal Health, National Center for Animal and Plant Health, Carretera de Tapaste y Autopista Nacional, San José de las Lajas, Mayabeque, Cuba*
DIANA REYES-PAVÓN • *Departamento de Microbiología, Universidad Autónoma de Aguascalientes, Aguascalientes, México*
RAÚL GUZMÁN RODRÍGUEZ • *Allergy Department, Hospital Universitario de Toledo, Toledo, Spain*
EVA SALINAS • *Departamento de Microbiología, Universidad Autónoma de Aguascalientes, Aguascalientes, México*
SARA SAMINO • *Lipid RDI Cluster, Biosfer Teslab, Reus, Spain*
ISIDRO SOBRINO • *SaBio, Instituto de Investigación en Recursos Cinegéticos, IREC (CSIC, UCLM, JCCM), Ciudad Real, Spain*

RITA VAZ-RODRIGUES • *SaBio, Instituto de Investigación en Recursos Cinegéticos, IREC (CSIC, UCLM, JCCM), Ciudad Real, Spain*
MARGARITA VILLAR • *SaBio, Instituto de Investigación en Recursos Cinegéticos, IREC (CSIC, UCLM, JCCM), Ciudad Real, Spain; Biochemistry Section, Faculty of Science and Chemical Technologies, University of Castilla- La Mancha, Ciudad Real, Spain*
ANETA WOŹNIAK • *Department of Biology and Parasitology, Faculty of Health Sciences, Medical University of Lublin, Lublin, Poland*
ALEJANDRA WU-CHUANG • *ANSES, INRAE, Ecole Nationale Vétérinaire d'Alfort, UMR BIPAR, Laboratoire de Santé Animale, Maisons-Alfort, France; Facultad de Ciencias Químicas, Universidad Nacional de Asunción, San Lorenzo, Paraguay*
ZBIGNIEW ZAJĄC • *Department of Biology and Parasitology, Faculty of Health Sciences, Medical University of Lublin, Lublin, Poland*

Part I

Foundations and Framework

Check for updates

Chapter 1

Introduction to Allergy Modeling Across Species with Ethical and Regulatory Considerations

José de la Fuente and Alejandro Cabezas-Cruz

Abstract

Allergic diseases affect the health and quality of life of humans and domestic animals. The prevalence of allergic diseases is increasing and likely associated with changes in climate and lifestyle conditions with decreased exposure to beneficial microorganisms. Animal models include multiple species (e.g., mice, rats, zebrafish, guinea pigs, dogs, cats, sheep, swine, monkeys) that are essential to study the immunological mechanisms of allergic diseases and to evaluate new therapies that can be beneficial for both humans and animals. The animal models contribute to overcoming ethical limitations in human research and have been scientifically validated with the required approval of ethical committees for animal experimentation. The animal models guide allergy modeling with different methods and translational protocols disclosed in the Methods in Molecular Biology book, "Allergic Disease Models: Methods and Translational Protocols".

Keywords Alpha-Gal Syndrome, Allergy, Artificial Intelligence, Citizen science, Data integration, Disease, Microbiota, Model

1 Introduction

Allergy modeling across species uses different animal models such as mice, rats, guinea pigs, and other rodents, zebrafish, dogs, cats, and larger animals like sheep and swine, among others [1–5]. In these animal models, it is possible to simulate human type I hypersensitivity to study the allergy-mediated immunological mechanisms and to design and evaluate new interventions. These animal models have been validated and are essential to overcome ethical limitations in human research and require approval of ethical committees for animal experimentation. Nevertheless, variations between animal species, protocols, and allergens used should be considered to select the most appropriate animal models [5–9].

For example, mice are commonly used to model respiratory inflammatory diseases such as asthma, dermatitis, and food allergies associated with or not with the Alpha-Gal Syndrome (AGS). For food allergies and control interventions associated with the AGS,

José de la Fuente and Alejandro Cabezas-Cruz (eds.), *Allergic Disease Models*, Methods in Molecular Biology, vol. 3092,
https://doi.org/10.1007/978-1-0716-5635-8_1,

zebrafish and humanized mice with inactivated Alpha-Gal synthesis to reproduce human conditions are used as model animals [8, 10, 11]. Rats such as the Brown Norway model have been used in the study of various immune-mediated diseases and food allergies due to their high IgE responses [12]. Dogs are used as a natural model for environmental allergies such as human atopic dermatitis [13]. Large animals such as sheep, swine, cattle, horses, and monkeys show physiological similarity to humans and are used as models for chronic diseases such as asthma and long-term evaluation [14].

2 Allergy Modeling

Allergic models include food allergy, AGS, passive sensitization models, allergic airway diseases, and epicutaneous diseases [15–18]. Food allergy models use oral sensitization with allergens combined with adjuvants and immunostimulatory agents to induce allergic and immune responses that may result in anaphylaxis. The AGS model reproduces a pathognomonic immunoglobulin E (IgE)-mediated delayed anaphylaxis to consumption of foods such as mammalian meat or dairy products containing the oligosaccharide galactose-α-1,3-galactose (α-Gal). Passive sensitization models involve transferring IgE antibodies into naïve animal hosts such as mice to study the immune-mediated mechanisms causing rapid anaphylactic responses. Allergic airway disease models focus on asthma, with sensitization followed by aerosolized or intranasal challenge using compounds such as house dust mites to mimic eosinophilia and airway hyperresponsiveness in human asthma. Epicutaneous disease models are used to reproduce skin sensitization in atopic dermatitis and the mechanisms affecting the development of food allergy. Some of these allergic models are also used for developing and testing preventive and therapeutic interventions. Additionally, AI and computational approaches are used together with animal models to integrate multiomics datasets to model three-dimensional structures of allergens and predict protein allergenicity and cross-reactive epitopes.

3 Ethical and Regulatory Considerations

Despite recent advances in In Vitro and in silico models, it has not been possible to avoid experimental animal models to address scientific challenges, and current and future research will lead the development of valid alternatives to reduce the use of experimental animals.

Ethical considerations regarding animal experimentation focus on treating animals as sentient beings, which has led to Ethical Committees' adoption of the "four Rs" principles (Reduction, Refinement, Replacement and Responsibility) as a guide when evaluating scientific proposals to make decisions regarding animal experimentation [19]. The proposals must justify the need and appropriateness of animal use while minimizing pain and distress, ensuring proper housing, and balancing scientific benefits against animal suffering. Accordingly, ethical review, personnel training, and adherence to regulations are important to comply with ethical principles [19].

4 Methods and Translational Protocols

This Methods in Molecular Biology volume offers the first comprehensive and protocol-driven resource covering the entire spectrum of experimental models and translational applications for allergic diseases. The book integrates cutting-edge in vivo animal models, novel modeling for arthropod-borne allergies and Alpha-Gal Syndrome, microbiota-based protocols, multiomics integration workflows, bioinformatics pipelines, and clinical diagnostic methods, all highly methodical, reproducible, and translational. The volume includes citizen science approaches and is designed to provide step-by-step, validated methods to accelerate research, clinical application, and biomarker discovery in allergic disease research.

To approach these methods and translational protocols, the book covers the following topics:

1. Alpha-Gal Syndrome and arthropod allergy models.
2. Food allergy models.
3. Microbiota and host-microbiome interactions in allergy.
4. Multiomics approaches, AI and data integration.
5. Translational and participatory models.

Competing Interests

The authors have no conflicts of interest to declare that are relevant to the content of this chapter.

References

1. Feinberg SM, Feinberg AR (1970) Animal models of allergic diseases: Monkeys. Folia Allergol (Roma) 17(4):423–424. PMID: 4195743
2. Stechschulte D (1970) Animal models of allergic diseases. Rats, mice and guinea pigs. Folia Allergol (Roma) 17(4):420. PMID: 5457995
3. Herz U, Renz H, Wiedermann U (2004) Animal models of type I allergy using recombinant allergens. Methods 32(3):271–280. https://doi.org/10.1016/j.ymeth.2003.08.013
4. Jönsson F, Pazos-Castro D, Castell M, Johansen P (2024) Editorial: animal models of allergic diseases: from basic to preclinical research. Front Allergy 5:1526536. https://doi.org/10.3389/falgy.2024.1526536

5. Santoro D, Marsella R (2014) Animal models of allergic disease. Vet Sci 1:192–212. https://doi.org/10.3390/vetsci1030192
6. Aun MV, Bonamichi-Santos R, Arantes-Costa FM, Kalil J, Giavina-Bianchi P (2017) Animal models of asthma: utility and limitations. J Asthma Allergy 10:293–301. https://doi.org/10.2147/JAA.S121092
7. Helm RM (2002) Food allergy animal models: an overview. Ann N Y Acad Sci 964:139–150. https://doi.org/10.1111/j.1749-6632.2002.tb04139.x
8. Vaz-Rodrigues R, de la Fuente J (2025) Is Zebrafish a good mModel for the Alpha-Gal syndrome?. FASEB J 39(9):e70602. https://doi.org/10.1096/fj.202500687R
9. Larsen JM, Bøgh KL (2018) Animal models of allergen-specific immunotherapy in food allergy: overview and opportunities. Clin Exp Allergy 48(10):1255–1274. https://doi.org/10.1111/cea.13212
10. Kepley CL, Wang Y, Yelton A, Siebert ER, Iweala OI (2025) Ticked off: allergic effector cells in the pathogenesis of Alpha-Gal syndrome. Curr Allergy Asthma Rep 25(1):57. https://doi.org/10.1007/s11882-025-01237-2
11. Bamgbose T, Alberdi P, Abdullahi IO, Inabo HI, Bello M, Sinha S, Anvikar AR, Mateos-Hernandez L, Torres-Maravilla E, Bermúdez-Humarán LG, Cabezas-Cruz A, de la Fuente J (2022) Functional characterization of α-Gal producing lactic acid bacteria with potential probiotic properties. Sci Rep 12(1):7484. https://doi.org/10.1038/s41598-022-11632-8
12. Bernard I, Fournié GJ, Saoudi A (2010) Genomics studies of immune-mediated diseases using the BN-LEW rat model. Methods Mol Biol 597:389–402. https://doi.org/10.1007/978-1-60327-389-3_26
13. Tamamoto-Mochizuki C, Mishra SK (2025) Investigation of neutrophil infiltration in the acute canine atopic dermatitis model. Front Allergy 6:1716716. https://doi.org/10.3389/falgy.2025.1716716
14. Kirschvink N, Reinhold P (2008) Use of alternative animals as asthma models. Curr Drug Targets 9(6):470–484. https://doi.org/10.2174/138945008784533525
15. LLi L, Chen D, Zhou M, Zhang H, Zheng R, Su J, Chen Y, Xiong G, Luo X, Liang G, Zhang K, Dai M, Zhang P, Li Y, Huang X, Shi Z, Zhang Y, Gan Z, Tao J, Xu C, Chen Z, Feng P, Zhou Y, Li P, Wang D, Yang Q, Zhou Q (2025) Global research trends in chronic care for comorbid allergic diseases: a web of science bibliometric review (1999—2024). Curr Allergy Asthma Rep 25(1):58. https://doi.org/10.1007/s11882-025-01238-1
16. Benedé S, Berin MC (2021) Applications of mouse models to the study of food allergy. Methods Mol Biol 2223:1–17. https://doi.org/10.1007/978-1-0716-1001-5_1
17. Rao S, Rathore SS, Goyal RM, Narakornpichit A, Karkra R, Al-Khazraji A, Grover M (2026) Understanding Alpha-Gal syndrome: a systematic review and meta-analysis of clinical manifestations and outcomes. Neurogastroenterol Motil 38(1):e70245. https://doi.org/10.1111/nmo.70245
18. Vaz-Rodrigues R, Mazuecos L, de la Fuente J (2022) Current and future strategies for the diagnosis and treatment of the Alpha-Gal Syndrome (AGS). J Asthma Allergy 15:957–970. https://doi.org/10.2147/JAA.S265660.
19. Kiani AK, Pheby D, Henehan G, Brown R, Sieving P, Sykora P, Marks R, Falsini B, Capodicasa N, Miertus S, Lorusso L, Dondossola D, Tartaglia GM, Ergoren MC, Dundar M, Michelini S, Malacarne D, Bonetti G, Dautaj A, Donato K, Medori MC, Beccari T, Samaja M, Connelly ST, Martin D, Morresi A, Bacu A, Herbst KL, Kapustin M, Stuppia L, Lumer L, Farronato G, Bertelli M (2022) Ethical considerations regarding animal experimentation. J Prev Med Hyg 63(3):E255–E266. https://doi.org/10.15167/2421-4248/jpmh2022.63.2S3.2768

Check for updates

Chapter 2

Guidelines to Apply for Ethical Approval of Animal Experimentation and Use of Human Samples

David Fernández-Castellanos and José de la Fuente

Abstract

Animal experimentation is a fundamental tool in biomedical research for the study of diseases and the development of interventions such as drugs, vaccines, and probiotics for the treatment of human pathologies. Animal models enable analysis of pathophysiological processes in complex organisms with genetic and functional similarities to humans, which is essential for evaluating the safety and efficacy of new therapeutic interventions and surgical techniques before their clinical application. Among the most commonly used models are mice, rats, and fish, employed in the study of various diseases, including allergies like Alpha-Gal Syndrome and autoimmune disorders such as Guillain-Barré syndrome, characterized by progressive symptoms and potentially life-threatening conditions. At the same time, the use of human biological samples, such as blood, tissues, and biomolecules (e.g., DNA, RNA, proteins), plays a key role in the advancement of medical knowledge, especially in the identification of biomarkers associated with allergic diseases.

Keywords Allergy, Animal model, Ethics, Human samples

1 Introduction

Animal experimentation using models such as mice and zebrafish has become an essential approach for the study of diseases, primarily used in biomedical research [1–4]. These studies are based on the development of drugs, vaccines, or probiotics that may be effective against diseases, allergies, and pathologies in humans. Thanks to these animal models, diseases can be studied comprehensively in complex organisms similar to humans, something not achievable with cell cultures alone. It is important for testing the safety and efficacy of interventions such as treatments, vaccines, and surgical techniques before applying them to humans.

The animals most used in these models are mice, rats, and fish, which share genetic and physiological similarities with humans [1–4]. These studies contribute to understanding pathologies such as allergies like Alpha-Gal Syndrome (AGS) and autoimmune disorders

José de la Fuente and Alejandro Cabezas-Cruz (eds.), *Allergic Disease Models*, Methods in Molecular Biology, vol. 3092,
https://doi.org/10.1007/978-1-0716-5635-8_2,

such as Guillain-Barré syndrome (GBS), disorders that can cause life-threatening conditions and even death [1, 5].

Another aspect to consider today is the importance of using human samples in research. Specifically, human biological samples, such as blood, tissues, DNA, RNA, and proteins, among others, are crucial for advancing medical knowledge and biomedical research. For allergy studies in particular, obtaining human samples such as blood is important for identifying allergy-related markers.

This chapter describes the procedures for requesting approval for the facility where animal experiments will be conducted and for obtaining authorization to perform animal experimentation. This approval is governed by specific regulations and an ethical framework based on the "three Rs" principle (Replacement, Reduction, Refinement) to minimize its impact [6]. It also covers the specific regulations and application process for the collection and use of experimental samples in research. Experimental animal models and human samples contribute to the advancement of research to develop effective and safe treatments for allergic diseases affecting humans.

2 Methods

2.1 Animal Experimentation

2.1.1 Approval of the Facility for Animal Experimentation

1. To conduct animal experimentation, the facilities must be approved by the competent authority. Depending on the country, this will be the national government or, in the specific case of Spain, the regional authorities (Departments of Agriculture or Livestock). However, the procedures follow national regulations coordinated by the Ministry of Agriculture, Fisheries and Food (MAPA). In Spain, this approval must be requested from the Directorate General of Livestock or the Department of Rural Development/Livestock of the corresponding autonomous community to obtain the farm registration code (REGA). The regional administration will carry out a technical inspection upon receiving the application to verify that the facility complies with current regulations and does not require improvements or corrections before final approval.
2. The regulations governing facilities for animal experimentation include key aspects such as:
   Design and Installation: Facilities should be designed to maximize animal welfare, including rest areas, exercise areas, and ensuring that animals cannot escape. Noise, light, and ventilation, which must be appropriate for the animal species, should be controlled.

Safety and Hygiene: Eating, drinking, or smoking is prohibited in animal areas. Aerosol generation must be minimized, and personal protective equipment must be used.
Animal Facility Management: Cages and enclosures must be sufficiently large and appropriate for the animals, promoting environmental enrichment and group housing for social species.
Biosafety: It is mandatory to follow specific regulations, such as Real Decreto 664/1997 (*see* **Note 1**), to prevent occupational hazards and infections, with strict cleaning and disinfection of surfaces after use.
Personnel and Control: A person responsible for animal welfare, advised by a veterinarian, must be designated, and access must be restricted to unauthorized personnel to prevent quarantines or contamination.

3. Once approved, the facility will be granted an operating code. Depending on the species to be kept, the housing conditions and required technical specifications will vary. To complete this application, it must include:
Animal Welfare Officer (OEBA).
Trained and accredited personnel for animal handling.
Designated veterinarian for animal welfare advice.
4. The required documentation will be:
A detailed technical report of the facilities, including the location, photos, and plans of both the site and the facility. A description of the facilities, their characteristics, and the equipment included. In addition, a health report will be required, specifying the housing conditions for the species, its veterinary care, and waste disposal procedures.
5. Standard Operating Procedures (POEs in Spanish).
Emergency and contingency plan.
Technical documentation guaranteeing the suitability of the center for the animal species to be used (*see* **Note 2**).
6. In addition, European Union (EU) regulations regarding the specific requirements for each species must be considered, as detailed in Annex III of the Real Decreto 53/2013 (*see* **Note 3**).

2.1.2 Design of the Animal Experimentation

1. Several aspects must be considered before approaching an experimental design. The first is the species of animal to be used, followed by strategic planning to obtain valid scientific data while minimizing the use and suffering of animals, thus adhering to the 3Rs [7]:
Replacement: Use of alternative methods that do not involve animals (cell cultures, computer models, or species with lower pain perception) whenever possible.

Reduction: Use of the minimum number of animals necessary to obtain statistically significant results without compromising the integrity of the study.

Refinement: Adjustment of procedures and housing conditions to minimize pain and stress and improve animal welfare throughout their lives [8].

2. Once the 3Rs are established, animal experimentation is designed by applying for ethical evaluation of research projects, which must include the following information:

   Project title.

   Duration of the animal experimentation and the funding entity if it is related to a research project, or if there is no funding entity because the project will be carried out with the applicant research group's own funds.

   The personnel involved, including the principal investigator, the person responsible for designing the procedures with the animals, and all other individuals participating in the development of these procedures.

   Housing, zootechnical, and care conditions for the animals: where the animals will be housed, indicating the REGA code of the facility [9].

   Indicate whether express authorization is required, depending on the chosen species and its status.

   Type of research project: Type I, II, or III [10]. A summary of the research project should explain the context of the scientific, health (human/animal), or production problem that the study seeks to solve, why this research is necessary, and why the use of animals is required instead of alternative methods. The main purpose of the research should be explained, indicating the specific steps to achieve the objectives, which should also be described together with applied procedures.

   The objectives of the research project should be stated, indicating the main objective and the specific objectives, which must be related to the procedures performed.

   The methodology and experimental design should be described schematically, outlining the experimental groups, sample sizes, and explaining the experimental procedure that will be followed in each group. The survival time of the animals and a description of any invasive procedures should be clearly indicated.

   The severity classification of the procedures described should indicate the severity that the animals are expected to experience during the experiment: non-recovery, mild, moderate, or severe. The sample size of the experimental groups must be justified based on statistical, bibliographic, or previous experience criteria. Indicate whether the animals will receive anesthesia or analgesia during the procedures, whether any substances will be administered, whether blood samples will be taken, and whether biological agents, radioisotopes, or chemical substances will be used.

Provide a statement regarding replacement, reduction, refinement, and measures that have been taken to avoid severity of the procedures described above.

Explain how the animals will be monitored, including any anticipated suffering, and provide a monitoring protocol that outlines the parameters to be monitored, the frequency of monitoring, and the person responsible for it. Also, provide a list of endpoint criteria, detailing the symptoms the animals may experience, their assessment, and the actions that will determine whether or not humane euthanasia is necessary.

Explain the animals' final disposition, whether it be reuse, slaughter, or other options.

2.1.3 Approval of the Animal Experimentation

1. Once the experimental design has been completed through the application for ethical evaluation of research projects, it must be initially approved by the Animal Experimentation Ethics Committee (AEEC) to which your institution belongs. Once approved by the Ethics Committee, it must be submitted to the relevant state authority in each country. In the case of Spain, it must be approved by the regional authorities (Departments of Agriculture or Livestock) before it can be carried out (*see* **Note 4**). The application to the regional authority must include the favorable report from the AEEC, as well as the corresponding annexes requested by the regional administration (*see* **Note 5**), the resolution and report of the associated project (if any), and the training certificates of the personnel who will perform the procedures with the animals.

2.2 Use of Human Samples in Research

2.2.1 Approval of the Facility for Human Samples in Research

1. To conduct experiments with human samples, the facilities must be approved by the competent authority, which, depending on the country, will be the national government or, in the case of Spain, the regional health authority (Consejería de Sanidad). This is governed primarily by Law 14/2007 on Biomedical Research and Real Decreto 1716/2011 (*see* **Note 6**). The procedure depends on whether the facility will be established as a biobank or will manage sample collections for projects. In the case at hand, the collection of samples for research projects will be maintained as a collection linked to a specific area of research.
2. The request for administrative authorization will be made to the competent authority (in the case of Spain, to the Ministry of Health of the Regional Government of Castilla-La Mancha)

and is governed by Real Decreto 1277/2003 and the Carlos III Health Institute (ISCIII) (*see* **Note 7**).

3. The regional administration will carry out a technical inspection once the application is received to verify that the installation complies with current regulations (Real Decreto 53/2013) and does not require improvements or corrections before its final approval (*see* **Note 2**).
4. The facility requirements are as follows:
   Traceability: Labeling systems (barcodes/QR codes) that link the sample to the consent form without revealing the identity (pseudonymization).
   Physical Security: Storage equipment (−80 °C freezers or liquid nitrogen) with 24-h alarm systems and emergency power supply.
   Responsible Party: A Principal Investigator (PI) must be designated to act as the legal custodian of the sample collections.
5. Approval from the research ethics committee regarding the use of the samples and the research project protocol [11]. The ethics committee will be that of the workplace or the autonomous community to which the workplace belongs.
6. Informed Consent Form: This is the document signed by donors, specifying that the samples will be used only for the specific project or line of research (*see* **Note 8**).
7. Register the collections with the National Biobank Registry (e.g., at the ISCIII Carlos III Health Institute). This process involves submitting a report detailing the facilities and conservation equipment to the Health Department of the region where the research is conducted. Once authorized, the National Biobank and Collection Registry is notified electronically (in Spain, this procedure is carried out through the electronic platform of the ISCIII, where the authorization resolution details and collection characteristics must be provided). An annual report on the status and use of the samples must also be provided (*see* **Note 9**).
8. Once the facility is approved, its first use for biomedical research must be notified. This notification will be made by submitting the report to the competent regional authority, which will forward it to the National Biosafety Center (CNBV) in Spain. This entire process is mandatory if the human samples used in the biomedical research are considered Group 2 or 3 biological agents. This notification only needs to be submitted once for the first use and is not required for subsequent uses. This notification form will include the following content:
   Information about the entity and those responsible for the facility.
   Description of the facility and containment measures.

Description of activities and risk assessment.
Waste management and accident prevention plan.

9. Once the notification is submitted, the National Biosafety Center reviews it and may request a verification visit to check the containment measures. Once it verifies that everything is in order, it will issue the authorization for use.

2.2.2 Design of the Research Project Protocol and Annex Documents

1. When designing a research protocol using human samples, several aspects must be considered before proceeding. The first is the type of human sample to be used. The second aspect is strategic planning, which must clearly justify the ethical implications of sample use, including the drafting of informed consent and the handling of materials for sample collection, storage, anonymization, and final disposal. The third aspect is that the protocol design must be coherent and ethical, ensuring a favorable risk-benefit balance. All designs must adhere to the Declaration of Helsinki (*see* **Note 10**).
2. The main structure of the protocol should be as follows:
   Identification: Clear title, researchers, and center(s).
   Abstract and Justification: Research problem and its relevance.
   Objectives and Hypotheses: Primary and secondary objectives.
   Methodology and Design: Type of study (observational, clinical), study location.
   Subject Selection: Inclusion and exclusion criteria.
   Biological Sample Management (Critical Section): Origin (new collection or use from a biobank/collection).
   Detail the sample type (blood, tissues, fluids) and quantity.
   Procedure: Collection, transport, and processing.
   Anonymization/Coding: Ensure data confidentiality.
   Final Disposition: Destruction, return to the patient, or storage for future use.
   Ethical and Legal Considerations: Obtain informed consent, which is the written consent form from the patient or their legal representative. In addition to confidentiality, consider the protection of personal data (GDPR), and finally, whether there is any insurance or compensation for conducting the clinical study.
   Statistical Analysis: Sample size calculation.
   Timeline and Budget: Time and resources.
3. In addition, informed consent must be provided along with this protocol (*see* **Note 8**), which must have the following structure:
   of the procedure or study: Clear identification of what will be done.
   Identification of the parties: Full name and ID number of the patient/participant and the responsible professional.

Description of the objective: Explanation of the purpose of the treatment or research.
Procedures and duration: Details of the tests, visits, or steps to be followed.
Risks and benefits: Information on possible side effects and expected improvements.
Voluntariness and withdrawal: Explicit statement that participation is voluntary and can be withdrawn at any time without consequences.
Confidentiality: Guarantee of data processing in accordance with current regulations such as the GDPR (*see* **Note 11**).

4. In addition, the patient information sheet (*see* **Note 12**) must be provided. This is a mandatory ethical and legal document in biomedical research that ensures the donor understands the purpose, risks, and destination of their biological samples (blood, tissue, fluids). This document is accompanied by the informed consent form, where the patient signs their voluntary acceptance. The mandatory ethical and legal document form contains the following points:
Research objective: Clearly explain the intended use of the samples (specific study or sample bank/biobank).
Collection procedure: Sample type, quantity, and whether it involves an additional intervention or is surplus material from a diagnostic procedure.
Voluntariness: The patient must be informed that their participation is entirely voluntary.
Confidentiality: Ensure that personal data will be coded or anonymized to protect the donor's identity.
Right of withdrawal: The right to withdraw the sample or request its destruction at any time without explanation or affecting their medical care.
No financial compensation: Generally, donation is altruistic, although travel expenses may be covered.
Genetic research: If the samples will be used for genetic studies, this must be explicitly stated.
Contact: Name of the principal investigator and study managers to answer any questions.

5. This informed consent form must be signed by the patient, who will declare that they have read and understood the information, has been able to ask questions, consent to the use of the samples and associated data (medical history), and consent to their storage (or transfer to third parties) for future research. This document must consider the processing of samples, which is subject to data protection regulations (GDPR) and specific legislation on biomedical research and biobanks.

6. Finally, the PI curriculum vitae (CV) must be included, highlighting experience in project leadership, high-impact

publications, funding obtained, and academic qualifications, using a reverse chronological format (*see* **Note 13**). It should focus on achievements, patents, collaborative networks, and team management. Including a brief professional profile and using the Standardized Curriculum Vitae (CVN) format, if applicable, is essential. It should have the following structure:
Contact Information and Professional Profile: Name, current affiliation, contact information, and a five- to six-line summary highlighting specialization and years of experience.
Academic Background: PhD, master's degrees, and bachelor's degrees, indicating university and year of graduation.
Current Employment Status: Position, institution, and research area.
Research Areas: Brief description with keywords of current specialization.
Research Experience: Projects, clinical trials, internships, and techniques mastered, specifying the role (PI, co-investigator) and responsibilities.
Scientific Output: Publications (articles, books), notable patents, conferences, and congresses.
Funding and Grants: R&D projects, amounts, and funding agencies.

7. Once all this documentation is available, the ethical review request must be submitted, including the following information:
Project title
Principal Investigator's information
Research team
Duration of the experiment and the funding entity if it is related to a research project, or if there is no funding entity because the project will be carried out with the applicant research group's own funds.
Type of research project and type of information to be collected.
Group to which the participating population belongs
Project justification, explaining why the use of human samples is required instead of alternative methods. Explain the main purpose of the research, indicating the specific steps to achieve the objectives.
Research objectives
Methodology and experimental design, description of the type of study
Location where the research will be carried out and where the samples will be stored. Storage conditions and maintenance protocols for the samples.
Sample size
Participant selection criteria

Variables collected in the research and description of the interventions to be carried out (you must specify the bibliographic reference of the evaluation protocol, both the original and the translated version, if applicable. Remember to attach a copy of the evaluation protocol to this document. In the case of the use of biological samples, specify their destination)
Research timeline
Involvement of each team member in the research:
International implications of the study (indicate if the research participants or data are wholly or partially from other countries. In this case, attach information regarding the regulations in force in the country for that research and the permissions from the institutions or places where the research will be carried out or the data obtained)
Origin of the participants or the research data (attach authorization from the center from which the participants or data originate)
How the participants were identified, invited, and recruited (justify that the participants' freedom was not coerced)
Risks of intervention and potential benefits:
Whether or not the participants will receive any type of Compensation
Information and consent that will be requested by the research team.
Data protection and processing/confidentiality

8. Once the committee has reviewed the application and determined that no further information or modifications are required, it will grant approval for the experiment to proceed. This is subject to final approval by the competent state or regional authority, depending on the country.

2.2.3 Approval of the Use of Human Samples for Research

1. Once approved by the institutional Ethics Committee, the application must be submitted to the relevant state authority in each country. In the specific case of Spain, it must be approved by the regional health authorities (Health Departments) before the experiment can proceed. This application must include the favorable report from the Animal Experimentation Ethics Committee (AEEC), the corresponding annexes requested by the regional administration, the project resolution and its accompanying report, and the accreditation of the laboratory where the experiment will be conducted. This laboratory must be authorized and meet the necessary safety standards to carry out the entire research process.

3 Notes

1. The protection of workers against risks related to exposure to biological agents during work must be considered. In Spanish legislation, Real Decreto 664/1997. BOE-A-1997-11144 Real Decreto 664/1997, de 12 de mayo, sobre la protección de los trabajadores contra los riesgos relacionados con la exposición a agentes biológicos durante el trabajo.
2. The technical documentation for the suitability of an animal facility (breeding or experimentation facility) must ensure compliance with high standards of animal welfare, biosafety, and legal requirements. In the context of the European Union and Spain, this documentation is primarily governed by Directive 2010/63/EU and Real Decreto 53/2013. https://www.boe.es/buscar/act.php?id=BOE-A-2013-1337
3. Link del Real Decreto 53/2013, of February 1, which establishes the basic rules applicable to the protection of animals used in experimentation and other scientific purposes, including teaching. https://www.boe.es/buscar/act.php?id=BOE-A-2013-1337
4. Link to the University of Castilla La Mancha ethics committee website to request authorization for animal experimentation and to obtain information for the procedure.https://www.uclm.es/misiones/lauclm/organosdegobierno/consejogobierno/comisiones/comisiones_vipc/comiteeticaexperimentacionanimal
5. Link to the website of the Regional Government of Castilla-La Mancha to apply for the permit/authorization for animal experimentation. https://www.jccm.es/tramites/1003200
6. Real Decreto 1716/2011, of November 18, which establishes the basic requirements for the authorization and operation of biobanks for biomedical research purposes and the treatment of biological samples of human origin and regulates the operation and organization of the National Register of Biobanks for biomedical research. https://www.boe.es/buscar/doc.php?id=BOE-A-2011-18919
7. Link del Real Decreto 1277/2003, of October 10, which establishes the general bases on the authorization of health centers, services and establishments. https://www.boe.es/buscar/doc.php?id=BOE-A-2003-19572
8. Link to the University of Castilla La Mancha's informed consent model, which is the document signed by donors of human samples. https://share.google/u0SeMKtrDPiJt7gzS

9. Link to the website where the application for biobank registration is made. https://www.isciii.es/servicios/biobancos/registro-nacional-biobancos
10. Link to the Declaration of Helsinki, which governs the protocols for the design of research projects with human samples. https://www.wma.net/policies-post/wma-declaration-of-helsinki/
11. Link to Regulation (EU) 2016/679 of the European Parliament and of the Council of 27 April 2016 on the protection of natural persons with regard to the processing of personal data and on the free movement of such data and repealing Directive 95/46/EC (General Data Protection Regulation). https://www.boe.es/buscar/doc.php?id=DOUE-L-2016-80807
12. Link to the PDF of the patient information sheet from the Andalusian Regional Government, which serves as an example. https://share.google/gwrBnZ6FjDHrIKAz0
13. Link to the FECYT website where you can create a researcher's CV, which includes several templates for different applications and calls for proposals. https://cvn.fecyt.es/editor/#HOME
14. Link to the regulations and ethical principles in scientific research of the University of Castilla La Mancha (CIOMAB), the body in charge of approvals for the use of human samples in experiments. https://www.uclm.es/misiones/investigacion/serviciosinvestigacion/portaleticacientifica/ciomabb/normativa
15. Link to the information sheet for participants in the human sample experimentation project at the University of Castilla La Mancha (CEIS). https://share.google/SZuS0C0oKyB5PAdwR
16. Link to the MITECO website where all legislation, official forms, official notifications already evaluated, etc., regarding GMOs can be found. https://www.miteco.gob.es/es/calidadyevaluacionambiental/temas/biotecnologia/organismos-modificadosgeneticamenteomg-/

Competing Interests

The authors have no conflicts of interest to declare that are relevant to the content of this chapter.

References

1. Vaz-Rodrigues R, de la Fuente J (2025) Is zebrafish a good model for the alpha-gal syndrome? FASEB J 39(9):e70602. https://doi.org/10.1096/fj.202500687R
2. Sánchez-Morales L, Porras N, García-Seco T, Pérez-Sancho M, Cruz F, Chinchilla B, Barroso-Arévalo S, Diaz-Frutos M, Buendía A, Moreno I, Briones V, Risalde MLÁ, de la Fuente J, Juste R, Garrido J, Balseiro A, Gortázar C, Rodríguez-Bertos A, Domínguez M, Domínguez L (2024) Neuropathological lesions in intravenous BCG-stimulated K18-hACE2 mice challenged with SARS-CoV-2. Vet Res 55(1):71. https://doi.org/10.1186/s13567-024-01325-7
3. Rabie EC, González-García A, Contreras M, Yousif MD, de la Fuente J, Murdan S, Stutzer C, Maritz-Olivier C (2026) Evaluation of host immune responses following immunization with MSP4 antigen adsorbed onto poly(lactic-co-glycolic acid) microparticles in a BALB/c murine model. Ticks Tick Borne Dis 17(2):102619. https://doi.org/10.1016/j.ttbdis.2026.102619
4. Contreras M, González-García A, de la Fuente J (2022) Zebrafish animal model for the study of allergic reactions in response to tick saliva biomolecules. J Vis Exp 187(10):3791/64378. https://doi.org/10.3791/64378
5. Berciano J (2025) Microscopical anatomy of the peripheral nervous system: an essential notion for understanding the pathophysiology of very early classic Guillain-Barré syndrome. Neuropathology 45(2):85–99. https://doi.org/10.1111/neup.13006
6. Goetz C, Payros D, Knudsen C, Grundy MM, Lacroix-Lamandé S, Lalmanach AC, Wiedemann A, Piriou-Guzylack L, Archer F, Descamps D, Finot L, Beaumont M (2026) Review: in vitro and ex vivo models advancing the study of host-microbiota interactions across organ systems in farm animals. Animal. 101810. https://doi.org/10.1016/j.animal.2026.101810
7. Díaz L, Zambrano E, Flores ME et al. (2020) Ethical considerations in animal research: the principle of 3R's. Rev Invest Clin 73(4):199–209. https://doi.org/10.24875/RIC.20000380
8. Jar AM (2014) Bienestar animal y el uso de animales de laboratorio en la experimentación científica. [Animal welfare and the use of laboratory animals in scientific research]. Rev Argent Microbiol 46(2):77–79. https://doi.org/10.1016/S0325-7541(14)70051-3
9. American Association for Laboratory Animal Science (AALAS) (2000) Improving the scientific basis for regulation of animal care and use. Comp Med 50(5):481
10. Geraldi J, Söderlund J (2018) Project studies: what it is, where it is going. Int J Proj Manag 36(1):55–70. https://doi.org/10.1016/j.ijproman.2017.06.004
11. Slowther A, Johnston C, Goodall J, Hope T (2004) Development of clinical ethics committees. BMJ 328(7445):950–952. https://doi.org/10.1136/bmj.328.7445.950

Part II

Alpha-Gal Syndrome and Arthropod Allergy Models

Chapter 3

Artificial Feeding of Hard Ticks Using Silicone Membranes

Vincent C. Duru, Bahar E. Mustafa, Abdul Ghafar, Abdul Jabbar and Ard M. Nijhof

Abstract

Artificial tick feeding systems (ATFS) provide an ethical alternative to using experimental animals for feeding ticks. Silicone membrane-based systems enable controlled feeding of diverse Ixodid tick species and life stages while supporting the 3R principles to replace, reduce, and refine animal use. Here, we present a protocol for silicone membrane-based ATFS, outlining practical steps, essential components, and key troubleshooting considerations.

Keywords Artificial feeding, Ixodid ticks, Silicone membrane

1 Introduction

Ticks are hematophagous arthropods that require a vertebrate host to obtain a blood meal necessary for their development, survival, and reproduction. Laboratory research on tick biology, tick physiology, pathogen transmission, or acaricide efficacy has traditionally relied on the use of live vertebrate hosts, such as sheep, cattle, rabbits, and rodents, for colony maintenance and for feeding different life stages [1,2]. However, with growing interest and public demand for improved animal welfare, there has been a concerted shift towards alternatives to animal experimentation and implementation of the 3R principles (replacement, reduction, and refinement) in tick research [3].

Artificial tick feeding systems (ATFS) provide an in vitro platform that allows the feeding of ticks on blood through a membrane rather than on a live host. Beyond the clear ethical advantage, it offers controlled experimental conditions and a host-independent system that reduces the practical challenges associated with feeding ticks on animals [4, 5]. Since the first reported ATF experiments in 1912 [6] using rat diaphragm and skin membranes, the technique has evolved substantially and is now an important tool in tick biology

José de la Fuente and Alejandro Cabezas-Cruz (eds.), *Allergic Disease Models*, Methods in Molecular Biology, vol. 3092,
https://doi.org/10.1007/978-1-0716-5635-8_3,

and tick-borne disease research. ATFS has been applied to studies on both hard and soft tick biology and physiology, laboratory maintenance of soft tick colonies [7], pathogen transmission [8–14], and drug and vaccine discovery and development [15, 16].

Currently, ATFS approaches fall broadly into three main categories: (first) capillary tube-based feeding, (second) semi-artificial systems using natural animal skin membranes, and (third) artificial feeding systems employing silicone membranes. The latter reduces the reliance on animal-derived materials for the membrane itself, although all systems still require animal blood as the nutrient source. A typical silicone membrane-based ATFS comprises several essential components: a confined space to hold the ticks, a silicone membrane mimicking host skin, attachment stimuli, a blood meal supplemented with phagostimulants, a heating system to maintain blood at a physiological temperature, and an incubator to regulate appropriate temperature and relative humidity. In some setups, a heated water bath can substitute for the heating plate and incubator [5].

Here, we describe in detail a silicone-membrane-based ATF protocol routinely used in our laboratory. The foundations for this protocol were established by researchers at the University of Neuchâtel, Switzerland, who were the first to successfully feed all life stages of *Amblyomma hebraeum* [17] and later refined the method in subsequent studies [4,15]. We have since adapted and further developed the approach to support artificial feeding of all life stages of species such as *Hyalomma dromedarii*, *Hyalomma excavatum*, *Hyalomma scupense*, and *Ixodes ricinus* [18, 19]. Although silicone membrane–based ATFS provide a versatile framework, ATF performance can vary substantially between species and life stages, and the method is not universally applicable across all tick species or developmental stages. It should therefore be regarded as a general framework, with species-specific optimization likely required to achieve reliable attachment and feeding success.

2 Materials

2.1 Silicone Membranes

1. Membrane:
   a. Goldbeater's skin (also known as Baudruche membrane) of ~20 μm thickness (Altenburger Pergament and Trommelfell GmbH, Altenburg, Germany or Preservation Equipment Ltd, Diss, UK) for feeding ticks with mouthparts length <50 μm (*see* **Note 1**), or
   b. Lens cleaning paper of ~50 μm thickness (70 x 120 mm, Tiffen, Happauge, USA) for feeding ticks with mouthparts length >50 μm.
2. Adhesive tape.

3. Weighing balance.
4. A4-sized plastic laminating sheets or a polypropylene disposal bag cut along one long and one short side to obtain a flat sheet.
5. Flat glass sheet, or another smooth flat surface.
6. Elastosil E4 Silicone (Wacker Chemie, München, Germany).
7. Elastosil FL white color paste (optional) (Wacker Chemie, München, Germany) (*see* **Note 2**).
8. Silicone oil with a viscosity of ~200 mPa.s.
9. Hexane.
10. Disposable plastic container (>100 ml).
11. Wooden spatula.
12. Stainless steel scraper with a smooth edge (*see* **Note 3**).
13. Mechanical micrometer, measuring range 0–25 mm.

2.2 Hair Extract

1. 10–50 g of freshly collected animal hair from animals that were not previously treated with acaricides.
2. Chloroform.
3. Methanol.
4. Three glass beakers of 1000 mL.
5. Glass filter funnel equipped with appropriately sized, organic-solvent-compatible filter paper
6. Water bath.
7. Glass vials for storage.

2.3 Blood Meal

1. Defibrinated or heparinized aseptic bovine blood. Use 2–10 ml blood per feeding unit per day, depending on the size of the tick feeding unit and the type of blood receptacle.
2. 0.1 mol/L of adenosine triphosphate (ATP) in 0.9% NaCl is sterilized by filtration through a 0.2 μm membrane filter. Aliquot are in small volumes and stored at −20 °C.
3. 10 mg/ml concentration of Gentamicin is aliquoted into smaller volumes and stored at 4 °C.
4. Prepare a stock solution of 200 g per liter of D-glucose solution (*see* **Note 4**).
5. Water bath or bottle warmer set at 37 °C.

2.4 Vitamin B Solution (Adapted From [20])

1. Thiamine.
2. Riboflavin.
3. Nicotinic acid.
4. Pantothenic acid.

5. Pyridoxine.
6. Biotin.
7. Folic acid.
8. Cobalamin.
9. Choline chloride.
10. Meso-inositol.

2.5 Vitamin B Preparation

Supplementation of the blood meal with a vitamin B cocktail is optional but may be beneficial when the use of antibiotics is expected to negatively affect tick endosymbionts and thus the endogenous provision of B vitamins to the tick [19, 20].

1. Pantothenic acid: Dissolve 1.5 g of pantothenic acid in 15 ml of 0.9% NaCl. Mix thoroughly and vortex until fully dissolved. Store in a 15-ml Falcon tube at room temperature, protected from light.
2. Choline chloride: Dissolve 9.25 g of choline chloride in 50 ml of 0.9% NaCl. Filter-sterilize through a 0.2-μm membrane filter and store in a sterile glass container.
3. Prepare the remaining vitamin solutions:
   - Biotin: 20 mg in 20 ml of 0.9% NaCl (store at 4 °C).
   - Cobalamin: 20 mg in 20 ml of 0.9% NaCl.
   - Meso-Inositol: 236 mg in 2 ml of 0.9% NaCl.
   - Nicotinic acid: 200 mg in 2 ml of 0.9% NaCl.
   - Pyridoxine: 200 mg in 2 ml of 0.9% NaCl.
   - Thiamine: 200 mg in 2 ml of 0.9% NaCl.
   - Riboflavin: 40 mg in 2 ml of 0.9% NaCl.
   - Folic acid: 60 mg in 2 ml 1 N $NaHCO_3$
4. Combine and sterilize by transferring 1.5 ml from each vitamin solution into a sterile tube, mix thoroughly by vortexing, and filter-sterilize using a 0.45-μm membrane filter.
5. Aliquot the final vitamin cocktail into 2 ml Eppendorf tubes and store at 4 °C, protected from light.

2.6 Feeding Units and Other Components

1. Borosilicate glass tubes: inner diameter 28 mm, wall thickness 2.8 mm, height 65 mm (Neubert Glas, Geschwenda, Germany or Glastechnik Rahm GmbH, Muttenz, Switzerland). This size is suitable for use in six-well plates. Smaller glass tubes with an inner diameter of 15–16 mm, a wall thickness of 2.5 mm and a height of 40 mm may be used for feeding small juvenile ticks in a 12-well plate format.

2. Mite plug made from polyurethane foam with a diameter of 29 mm (K-TK e.K., Retzstadt, Germany).
3. Broad O-rings with an inner diameter of 30 mm, or an inner diameter of 18 mm for the smaller glass tubes (Emil Lux GmbH, Wermelskirchen, Germany).
4. 50 ml glass beakers, 6-well or 12-well cell culture plates.
5. 0.9% sodium chloride (NaCl) solution.
6. Sandpaper.
7. Small paintbrush.
8. Silicone glue; Elastosil E41(Wacker Chemie).
9. Scalpel blade.
10. 70% ethanol.
11. Stereomicroscope.
12. Falcon tubes, size depends on the amount of blood required for each blood change (15 ml or 50 ml).
13. Forceps (serrated).
14. Non-impregnated mosquito netting (optional), for adult feeding.
15. Animal hair (optional).
16. Nystatin solution, 10,000 units (U) per ml (optional).
17. 1× phosphate-buffered saline (PBS) (optional).
18. Heating plate (with a flat and large surface area, to accommodate 6- or 12-well plates) suitable for use inside an incubator under high-humidity conditions.
19. Incubator (ideally a CO_2 incubator; *see* **Note 5**). A water bath could be used alternatively to the heating plate and incubator.

3 Methods

3.1 Silicone Membrane Preparation

1. Measure the thickness of the plastic laminating sheets using a mechanical micrometer and record the value; this will later be subtracted from the combined measurement to determine the true membrane thickness.
2. Position the laminating sheet on a smooth, clean flat glass plate placed on a level surface. Secure the sheet at all four edges with adhesive tape.
3. Prepare the goldbeater's skin by cutting it into strips of approximately 15 × 12 cm.
4. Place the goldbeater's skin or lens-cleaning paper onto the laminating sheet, ensuring that the surface is smooth, taut,

and free of wrinkles. Secure both short edges with tape. Multiple pieces may be mounted on the same laminating sheet, provided they do not overlap.

5. Prepare the silicone mixture in a disposable plastic container as follows (work in a fume hood and use appropriate personal protective equipment due to hexane vapors):
   a. 16.5 g Elastosil E4 Silicone
   b. 0.165 g Elastosil FL white color paste (optional)
   c. 4.95 g silicone oil
   d. 3.19 g hexane
6. Mix thoroughly with a wooden spatula until a homogeneous mixture is obtained. This volume is sufficient to coat approximately eight pieces of lens-cleaning paper, yielding a final membrane thickness of ~70 μm.
7. Apply the silicone mixture by placing an appropriate amount of the mixture at one short end of the mounted membrane. The required amount depends on the desired final membrane thickness.
8. Spread the silicone evenly across the membrane surface using a stainless-steel scraper, applying consistent pressure and maintaining an even angle to ensure uniform thickness.
9. Leave the membrane to dry overnight at room temperature (preferably in a dust-free environment). For thicker membranes, extend drying time until the silicone is fully cured and no longer tacky.
10. Trim around the edges of the silicone membrane, leaving the laminating sheet attached.
11. Measure the membrane thickness at multiple locations, e.g., two peripheral areas and one central point, using a micrometer. Subtract the thickness of the laminating sheet measured under Step 1 to obtain the membrane thickness, calculate the mean value, and record it.
12. Store the membrane between paper towels in a dry, dust-free cupboard or in a sealed plastic bag until use. Membranes can be stored up to several months under these conditions.

3.2 Tick Feeding Unit Preparation

1. Prepare new glass tubes by lightly roughening the edge that will be glued to the membrane by rubbing it on sandpaper. This increases surface friction and improves adhesive bonding. When using previously used, washed, and autoclaved borosilicate glass tubes, check the tube edges for irregularities. Carefully remove any remaining silicone glue using a scalpel blade, followed by gentle smoothing with sandpaper. Clean

the tubes by wiping the prepared edge thoroughly with a paper towel soaked in 70% ethanol and allow it to air-dry.

2. Select an appropriate membrane (depending on tick life stage and hypostome length) and place it on a flat surface with the silicone-coated side facing upward.
3. Using a small paintbrush, apply a thin but continuous layer of Elastosil E41 silicone glue around the rim of the glass tube.
4. Gently place the glued rim of the glass tube onto the membrane and press lightly to ensure uniform contact. Carefully remove any excess silicone glue from the inner rim to prevent formation of a thick silicone layer, as many tick species and life stages prefer attaching at the corners and may be hindered by excess silicone. Apply pressure by placing a weight (~500 g per feeding unit) on top of each tube to maintain firm, even contact. Allow the assembly to set overnight at room temperature in a dust-free environment.
5. After the adhesive has set (at least 24 h), use a scalpel to cut around the perimeter of each glass tube, trimming as closely as possible so the feeding unit will fit into a six-well cell culture plate. Carefully remove the underlying laminating sheet.
6. To consider: Glue a 20 × 20 mm piece of mosquito netting to the center of the membrane using a small drop of Elastosil E41. Leave the edges of the netting unattached to provide space for ticks to crawl underneath. In our experience, the addition of mosquito netting can facilitate attachment and feeding in some tick species, for instance, for adult *Dermacentor reticulatus* ticks.
7. Test the feeding units for leakage by filling a small container, 6- or 12-well plate with 70% ethanol, and place the feeding units into the ethanol for approximately 10 min. After incubation, gently run a soft tissue along the inner wall of each feeding unit. Any moisture indicates leakage. Discard all leaking units.
8. If not used immediately, store the feeding units at room temperature in a sealed plastic bag.

3.3 Hair Extract Preparation

Hair extracts can be used as attachment stimuli during artificial feeding. In our experience, bovine hair extracts work well for most tick species, but hair from other host animals may be more effective for species with a pronounced host specificity. The choice of hair extract can therefore be adapted depending on the tick species and its natural host range. Perform all preparation steps in a fume hood, as methanol and chloroform are hazardous solvents.

1. Immerse 50 g of freshly collected hair in 240 ml of a chloroform: methanol mixture (2:1; 160 ml chloroform, 80 ml methanol) in a glass beaker for 2 h at room temperature. Ensure that

the hair is fully submerged; cover the beaker with aluminum foil. If using a smaller amount of hair, adjust the solvent volumes proportionally.

2. Decant and collect the supernatant in a clean glass beaker.
3. Repeat the extraction by immersing the hair in 240 ml of chloroform:methanol (1:1; 120 ml each) for 2 h.
4. Decant and pool the supernatant with the extract collected in Step 2.
5. Perform a third extraction step by immersing the hair in 240 ml of chloroform:methanol (1:2; 80 ml chloroform, 160 ml methanol) for 2 h at 45 °C in a water bath.
6. Decant and add the supernatant to the combined extracts from Steps 2 and 4.
7. Filter the combined solvents through a glass filter with an appropriately sized filter paper to remove particulate material.
8. Prepare aliquots in glass vials and store at −20 °C until use.

3.4 Preparation of the Blood Meal

The volume of blood required per feeding unit depends on the receptacle used. Approximately 5 ml is needed for 50 ml beakers, 3 ml for six-well plates, and 1 ml for 12-well plates. The recipe below is based on a 5 ml preparation; volumes can be scaled proportionally for other receptacle types.

1. Aliquot 5 ml of defibrinated or heparinized sterile bovine blood for each feeding unit into a Falcon tube.
2. Add 50 μl ATP solution, 50 μl glucose solution, 0.5 μl gentamicin (optional, *see* **Note 6**), and 50 μl vitamin B solution (optional) for every 5 ml of blood.
3. Gently swirl the tube to mix all components thoroughly. Place the tube in a water bath or bottle warmer at 37 °C for 5 min to pre-warm the blood meal.
4. Add 5 ml of the prepared blood meal to each beaker (or the appropriate scaled volume for other receptacles).

3.5 Initiation of Artificial Tick Feeding

1. Disinfect the feeding units by immersing them in 70% ethanol for 1 h or by exposing them to UV light for 15–20 min. If ethanol disinfection is used, allow the units to dry completely before proceeding.
2. Place the feeding unit, suspended with O-rings, into a beaker or well plate so that the membrane does not touch the bottom. Apply 75 μl of the prepared hair extract directly onto the membrane (30 μl for smaller feeding units) and allow it to dry for at least 2 h under a fume hood to ensure complete evaporation of the solvent.

3. Once the membrane is dry and you are ready to start the feeding, transfer the required volume of blood into Falcon tubes, add phagostimulants and antibiotics (if using) as described above, and pre-warm the mixture in a water bath or bottle warmer.
4. While the blood is warming, place autoclaved 50 ml glass beakers on hot plates set to 38–42 °C inside a flow cabinet. If using cell culture plates, heat them in the same way.
5. Once the blood is warmed, add the blood meal to the 50 ml beakers (5 ml per beaker) or to the cell culture plates (3 ml per well for six-well plates; 1 ml per well for 12-well plates).
6. Transfer ticks into the feeding unit using either forceps or a fine paintbrush (*see* **Note 7**). Work quickly and apply appropriate containment measures [2]. Immediately close the unit with a dampened mite plug to prevent ticks from escaping, ensuring that a gap of 1–2 cm remains between the membrane and the mite plug.
7. Place the feeding unit containing the ticks into the blood meal, ensuring it remains suspended approximately 5 mm above the bottom of the beaker or well. Verify that no air bubbles are trapped beneath the membrane.
8. Place the system on a hot plate set to 42 °C inside an incubator set to 27 °C, 80% relative humidity, and 1% CO_2. These conditions work well for species such as *I. ricinus* but may be adjusted if needed (*see* **Note 8**). Alternatively, the beakers or well plates may be placed in a water bath set at 38 °C, ensuring that the water does not flow into the beakers or wells. The blood meal should be kept at approximately 37–38 °C. To offset heat dissipation, the hot plate surface is set to a slightly higher temperature, typically around 42 °C.

3.6 Maintenance of the Artificial Feeding

1. For the routine blood replacement, ideally performed at 12:12 h intervals, though 14:10 h intervals are generally still effective, begin by preparing and pre-warming a fresh blood meal as described above.
2. Remove the feeding units from the previous blood meal and then rinse both the outer membrane surface and the underside of the unit with autoclaved 0.9% NaCl using a wash bottle, continuing until all clotted blood and visible irregularities are removed. Discard the old blood meal. Temporarily suspend the units in an empty, sterile beaker or well for inspection under a stereomicroscope.
3. Record the number of attached ticks and remove any dead individuals as well as accumulated faeces. Faeces can be removed by holding the feeding unit with the open side facing downward over a paper towel and gently tapping the unit

several times. Engorged and detached ticks can also be collected at this stage. Refill autoclaved beakers or well plates with freshly prepared blood meal, suspend the cleaned feeding units in the blood as described above, and return them to the incubator or water bath.

3.7 Management of Fungal Contamination on the Membrane

If fungal growth is observed on the underside of the feeding unit, typically appearing as small clumps of coagulated or fibrous material, a treatment with nystatin can extend the usable lifespan of the membrane and improve feeding success.

1. Prepare a nystatin working solution by dissolving 10,000 U/mL in 1 × PBS (5 mL per feeding chamber), followed by sterile filtration through a 0.2-µm syringe filter.
2. Rinse the underside of the feeding unit with 0.9% NaCl to remove residual blood.
3. Immerse the underside of the feeding unit in the nystatin solution for 1 min.
4. Rinse again with 0.9% NaCl before placing the unit back into fresh blood.
5. Repeat the antifungal treatment every 2–3 days, depending on the severity of contamination, until the experiment concludes.

4 Notes

1. Goldbeater's skin is recommended for feeding ticks with short mouthparts due to its thinness, but it can also be used for species with longer mouthparts as an alternative to lens cleaning paper.
2. Adding color paste tints the membrane, improving contrast and making it easier to visually estimate membrane thickness.
3. Using a paint roller instead of a scraper to distribute the silicone mixture over the membrane has been reported to create a slightly uneven surface, which may positively influence tick attachment and feeding [21].
4. Doubling the glucose concentration to 4 g per liter of blood has been reported to potentially improve the feeding success of *D. reticulatus* adults [22].
5. Increasing the CO_2 level may stimulate tick activity and improve feeding success [22, 23].
6. Adding broad-spectrum antibiotics, such as gentamicin, to the blood meal may help prevent spoilage and improve feeding success. Bacterial and fungal contamination are among the main

challenges in artificial tick feeding, as stasis and the required blood temperature of approximately 38 °C create ideal conditions for microbial growth. However, antibiotics may also affect the tick microbiome, particularly endosymbionts, which should be taken into consideration.

7. When selecting ticks for in vitro feeding, ensure that only active ticks showing host-seeking behavior (e.g., extending their forelegs or moving toward heat/CO_2) are selected, as these are more likely to attach.
8. Numerous parameters in the artificial feeding process can be optimized to improve feeding success for different tick species. These include, but are not limited to, the membrane base material, membrane thickness and silicone composition; the type of attachment stimuli used (e.g., hair extracts, clipped hair, tick frass); the choice and concentration of phagostimulants (e.g., ATP, glucose, glutathione); the positioning of the blood meal, which is placed above the ticks in some systems (e.g., 14), and environmental conditions such as temperature, light, relative humidity, and CO_2 levels.

Competing Interests

The authors have no conflicts of interest to declare that are relevant to the content of this chapter.

References

1. Levin ML, Schumacher LBM (2016) Manual for maintenance of multi-host ixodid ticks in the laboratory. Exp Appl Acarol 70(3):343–367. https://doi.org/10.1007/s10493-016-0084-8
2. Allan SA (2014) Tick rearing and in vitro feeding. In: Sonenshine D, Roe RM (eds) Biology of ticks. Oxford University Press, Oxford, pp 444–473
3. Eckert J (1997) Alternatives to animal experimentation in parasitology. Vet Parasitol 71(2–3):99–120. https://doi.org/10.1016/s0304-4017(97)00027-7
4. Kröber T, Guerin PM (2007) In vitro feeding assays for hard ticks. Trends Parasitol 23(9):445–449. https://doi.org/10.1016/j.pt.2007.07.010
5. Nijhof AM, Tyson KR (2018) In vitro feeding methods for hematophagous arthropods and their application in drug discovery. In: Meng CQ, Sluder AE (Eds.) Ectoparasites: Drug discovery against moving targets. Wiley-VCH Verlag GmbH & Co., Weinheim, 187–204
6. Hindle E, Merriman G (1912) The sensory perceptions of *Argas persicus* (Oken). Parasitol 5(3):203–216. https://doi.org/10.1017/S0031182000000305
7. Schwan EV, Hutton D, Shields KJ, Townson S (1991) Artificial feeding and successful reproduction in *Ornithodoros moubata moubata* (Murray, 1877) (Acarina: Argasidae). Exp Appl Acarol 13(2):107–115. https://doi.org/10.1007/BF01193661
8. Fourie JJ et al (2019) Transmission of *Anaplasma phagocytophilum* (Foggie, 1949) by *Ixodes ricinus* (Linnaeus, 1758) ticks feeding on dogs and artificial membranes. Parasit Vectors 12(1):136. https://doi.org/10.1186/s13071-019-3396-9
9. Hart T, Yang X, Pal U, Lin YP (2018) Identification of Lyme borreliae proteins promoting vertebrate host blood-specific spirochete survival in *Ixodes scapularis* nymphs using artificial feeding chambers. Ticks Tick Borne Dis 9(5):1057–1063. https://doi.org/10.1016/j.ttbdis.2018.03.033

10. Macaluso KR, Sonenshine DE, Ceraul SM, Azad AF (2001) Infection and transovarial transmission of rickettsiae in *Dermacentor variabilis* ticks acquired by artificial feeding. Vector Borne Zoonotic Dis 1(1):45–53. https://doi.org/10.1089/153036601750137660
11. Tajeri S, Razmi G, Haghparast A (2016) Establishment of an artificial tick feeding system to study *Theileria lestoquardi* Infection. PLoS One 11(12):e0169053. https://doi.org/10.1371/journal.pone.0169053
12. Young AS, Waladde SM, Morzaria SP (1996) Artificial feeding systems for ixodid ticks as a tool for study of pathogen transmission. Ann N Y Acad Sci 791:211–218. https://doi.org/10.1111/j.1749-6632.1996.tb53527.x
13. Cotte V et al (2008) Transmission of *Bartonella henselae* by *Ixodes ricinus*. Emerg Infect Dis 14(7):1074–1080. https://doi.org/10.3201/eid1407.071110
14. Vimonish R et al (2020) Quantitative analysis of *Anaplasma marginale* acquisition and transmission by *Dermacentor andersoni* fed in vitro. Sci Rep 10(1):470. https://doi.org/10.1038/s41598-019-57390-y
15. Thomas K, Guerin PM (2007) An in vitro feeding assay to test acaricides for control of hard ticks. Pest Manag Sci 63(1):17–22. https://doi.org/10.1002/ps.1293
16. Lew-Tabor AE, Bruyeres AG, Zhang B, Rodriguez Valle M (2014) *Rhipicephalus (Boophilus) microplus* tick in vitro feeding methods for functional (dsRNA) and vaccine candidate (antibody) screening. Ticks Tick Borne Dis 5(5):500–510. https://doi.org/10.1016/j.ttbdis.2014.03.005
17. Kuhnert F, Diehl PA, Guerin PM (1995) The life-cycle of the bont tick *Amblyomma hebraeum* in vitro. Int J Parasitol 25(8):887–896. https://doi.org/10.1016/0020-7519(95)00009-q
18. Elati K et al (2024) In vitro feeding of all life stages of two-host *Hyalomma excavatum* and *Hyalomma scupense* and three-host *Hyalomma dromedarii* ticks. Sci Rep 14(1):444. https://doi.org/10.1038/s41598-023-51052-w
19. Militzer N et al (2021) Artificial feeding of all consecutive life stages of *Ixodes ricinus*. Vaccines (Basel) 9(4):385. https://doi.org/10.3390/vaccines9040385
20. Duron O et al (2018) Tick-bacteria mutualism depends on B vitamin synthesis pathways. Curr Biol 28(12):1896–1902e1895. https://doi.org/10.1016/j.cub.2018.04.038
21. Trentelman JJ, Kleuskens JA, van de Crommert J, Schetters TP (2017) A new method for in vitro feeding of *Rhipicephalus australis* (formerly *Rhipicephalus microplus*) larvae: a valuable tool for tick vaccine development. Parasit Vectors 10(1):153. https://doi.org/10.1186/s13071-017-2081-0
22. Krull C, Bohme B, Clausen PH, Nijhof AM (2017) Optimization of an artificial tick feeding assay for *Dermacentor reticulatus*. Parasit Vectors 10(1):60. https://doi.org/10.1186/s13071-017-2000-4
23. Voigt WP et al (1993) In vitro feeding of instars of the ixodid tick *Amblyomma variegatum* on skin membranes and its application to the transmission of *Theileria mutans* and *Cowdria ruminatium*. Parasitol 107(Pt 3):257–263. https://doi.org/10.1017/s0031182000079233

Check for updates

Chapter 4

Tick Saliva Collection

Ard M. Nijhof

Abstract

Tick saliva contains a complex mixture of biologically active compounds that enable prolonged blood feeding by modulating host hemostasis, inflammation, and immune responses. In addition, saliva serves as a key interface for the transmission of tick-borne pathogens, with specific salivary molecules facilitating pathogen survival and host infection. Tick saliva may also contain glycosylated proteins bearing galactose-α-1,3-galactose (α-Gal), which have been implicated in the development of α-Gal allergy in humans, as well as neurotoxic components responsible for tick paralysis. Given the central role of tick saliva in tick feeding, pathogen transmission, and tick-associated allergic or toxic reactions, reliable methods for its collection are essential, and we here present a protocol for the collection of saliva from ticks.

Keywords Allergy, Alpha-Gal, Tick, Pathogen, Pilocarpine, Saliva

1 Introduction

Tick saliva is an extraordinarily complex and dynamic cocktail of biologically active molecules that enables ixodid ticks to remain attached and blood-feed for prolonged periods [1]. Through the concerted action of anti-hemostatic, anti-inflammatory, and immunomodulatory components, saliva effectively suppresses host defense mechanisms at the bite site and creates a permissive microenvironment for feeding. Beyond its role in blood acquisition, tick saliva constitutes a critical interface for the transmission of tick-borne pathogens, as specific salivary factors promote pathogen survival, immune evasion, and successful establishment of infection in the vertebrate host. Moreover, tick saliva can contain glycosylated proteins bearing galactose-α-1,3-galactose (α-Gal), which have been implicated in the induction of α-Gal syndrome in humans, as well as neurotoxic salivary components that underlie tick paralysis.

Over sixty years ago, it was shown that small amounts of tick saliva could be collected by mechanically stimulating the chelicerae of *Dermacentor andersoni* ticks freshly removed from the host through the placement of a tightly fitting capillary tube over the

José de la Fuente and Alejandro Cabezas-Cruz (eds.), *Allergic Disease Models*, Methods in Molecular Biology, vol. 3092,
https://doi.org/10.1007/978-1-0716-5635-8_4,

tick's mouthparts. Compression of the chelicerae elicits a salivation response, and small amounts of saliva could be directly collected from the capillary tube [2]. It was subsequently shown that salivation could also be pharmacologically induced by administering pilocarpine, a muscarinic acetylcholine receptor agonist, in the genital orifice of the soft tick *Ornithodoros savignyi* [3]. Building on these observations, Tatchell showed that saliva yield could be markedly increased in hard ticks as well when this approach was combined with the injection of pilocarpine dissolved in Levy's solution into female *Rhipicephalus (Boophilus) microplus* ticks [4]. Later studies showed that topical application of pilocarpine dissolved in solvents such as acetone and applied on the tick cuticle was also effective in inducing salivation [5], even leading to the development of an automated instrument for tick saliva collection [6].

Other substances, such as dopamine [7] and arecolin hydrobromide [8], have also been reported to induce salivation when injected into ticks, with dopamine reportedly resulting in darker-colored saliva [9]. It is important to note that a proportion of the pilocarpine or dopamine used to induce salivation is co-secreted in the saliva, which may influence downstream bioassays and should therefore be appropriately controlled for [10, 11].

Saliva yield and composition can vary considerably and depend, among other factors, on the tick's life stage and physiological state, the time elapsed between detachment and treatment, and the pharmacological effectors used to induce secretion [7, 12]. The highest volumes are typically obtained from females during the rapid feeding phase, whereas only small amounts can be collected from unfed ticks.

2 Materials

2.1 Preparation and Decontamination of Ticks

1. Fine forceps
2. Commercial bleach, diluted to 1% (optional)
3. DNA-free water (optional)

2.2 Tick Saliva Collection

1. Glass microscope slides
2. 10 µl microcapillary tubes with inner diameter of approx. 0.5 mm (e.g., Drummond 2-000-010) with aspirator tube, or alternative (*see* **Note 1**)
3. Pipette for expelling the fluid from the aspirator tube
4. Double-sided adhesive tape
5. Modeling clay (non-toxic), e.g. plasticine
6. Pilocarpine HCl (CAS no.: 54-71-7)

7. Ethanol (95%) when applying pilocarpine topically, or phosphate-buffered saline (PBS), pH 7.4, when injecting the pilocarpine
8. 1 mL syringe with a 27 G needle (optional, when injecting the pilocarpine)
9. 10 μl Hamilton syringe with 33 G needle, or equivalent, when injecting the pilocarpine
10. Incubator or humid chamber
11. Protease inhibitor cocktail

3 Methods

3.1 Preparation of Ticks

1. Clean ticks by removing any remaining cement, skin, or other debris attached to the mouthparts using fine forceps.
2. To minimize microbial contamination, ticks can be washed in 1% commercial bleach for 30 s, followed by three successive rinses in baths of DNA-free water, and subsequent drying [13].

3.2 Tick Saliva Collection

1. Fix a piece of double-sided adhesive tape to one end of a glass microscope slide.
2. Attach a small piece of modeling clay to the opposite end of the glass slide.
3. Using a fine tool such as forceps creates a small notch in the center of the modeling clay to securely hold the microcapillary or alternative saliva collection tube in place.
4. For topical application of pilocarpine: Place the tick onto the adhesive tape with the ventral side facing upwards. Apply 2–5 μl of 5% pilocarpine (50 mg/mL in 95% ethanol) to the posterior ventral surface of the tick. Once the applied solution has cleared from the cuticular surface, carefully fit a 10-μl microcapillary tube or alternative collection tube (*see* **Note 1**) over the hypostome, ensuring that the palps are outside of the tube and the hypostome is not damaged, and secure the tube in the notch of the modeling clay. Make sure that the tube remains properly engaged with the mouthparts and reposition if necessary. Placing the tube before complete clearance of the pilocarpine solution may result in the unintended influx of the solution into the capillary.
5. If pilocarpine is applied by injection: slowly inject the tick through one of the spiracular plates (*see* **Note 2**) with 0.2%–10% pilocarpine in phosphate-buffered saline (PBS) (*see* **Note 3**)

using a 10 μL Hamilton syringe fitted with a 33 G needle. Prepiercing of the spiracular plate with the tip of a 27 G needle may be required. Proceed with placing the tick on adhesive tape and position the saliva collection tube as described above (*see* **Note 4**).

6. Incubate the slide in a humid chamber or high-humidity incubator at room temperature. The slide may be slightly tilted to allow gravity to facilitate saliva flow into the collection tube.
7. Monitor ticks regularly, particularly at the start of salivation, and ensure that the collection tube remains properly engaged to the mouthparts. Salivation typically ceases after 3–4 h.
8. Transfer the collected saliva from the tube into a microcentrifuge tube. This can be performed at regular intervals after the onset of salivation, e.g., every 30 min, to limit reuptake of the expelled saliva by the tick. After each collection, reposition the collection tube onto the hypostome to ensure continued saliva capture.
9. Store saliva at −20 °C or lower until use. The addition of a protease inhibitor cocktail to the saliva is optional and may be considered depending on downstream applications.

4 Notes

1. Alternative saliva collection tubes include a 10-μl pipette tip with a fine tip or a glass Pasteur pipette flame-drawn to a fine tip. An alternative saliva collection method that does not involve placing a capillary over the tick hypostome has previously been described [14]. In this approach, ticks were placed in a 0.5-ml microcentrifuge tube after the topical application of pilocarpine and incubated at 37 °C for 3 h. Saliva was subsequently recovered by creating a small hole in the bottom of the 0.5 ml tube, inserting it into a 1.5-mL microcentrifuge tube, washing the tick with adherent saliva with PBS, and collecting the eluate by centrifugation. This method leads to a dilution of the saliva sample with PBS and may lead to the inclusion of contaminants such as cuticular components, so thorough decontamination of the tick prior to saliva collection is recommended.
2. Injections can also be administered directly through the cuticle, but injecting through the spiracular plate may reduce reflux of hemolymph or tick tissue due to the greater rigidity of this sclerotized structure. The use of a Hamilton syringe mounted on a micromanipulator may improve positional stability and injection accuracy.

3. The amount and concentration of pilocarpine used for injection or topical application depend on the tick's life stage and size. Reported injection volumes range from 1–30 µl and concentrations from 0.2% to 10%. Optimal conditions should be determined empirically.
4. In addition to topical application and injection, salivation in *Ixodes scapularis* nymphs has also been induced by placing 10 µl capillaries preloaded with 5 µl of a 2% pilocarpine solution in PBS over the tick mouthparts, with saliva being collected directly in the same capillary [15].

Competing Interests

The author has no conflicts of interest to declare that are relevant to the content of this chapter.

References

1. Nuttall PA (2019) Wonders of tick saliva. Ticks Tick Borne Dis 10(2):470–481. https://doi.org/10.1016/j.ttbdis.2018.11.005
2. Gregson JD (1960) Morphology and functioning of the mouthparts of *Dermacentor andersoni* Stiles. Acta Trop 17:48–79
3. Howell CJ (1966) Collection of salivary gland secretion from the argasid *Ornithodoros savignyi* Audouin (1827) by the use of a pharmacological stimulant. J S Afr Vet Med As 37:236–239
4. Tatchell RJ (1967) A modified method for obtaining tick oral secretion. J Parasitol 53(5):1106–1107
5. Clarke RH, Hewetson RW (1971) A modification to the collection of saliva from *Boophilus microplus*. J Parasitol 57(1):194–195
6. Binnington KC, Schotz M (1973) Automatic collection of tick saliva. J Aust Entomol Soc 12:78–79
7. McSwain JL, Essenberg RC, Sauer JR (1992) Oral secretion elicited by effectors of signal transduction pathways in the salivary glands of *Amblyomma americanum* (Acari: Ixodidae). J Med Entomol 29(1):41–48. https://doi.org/10.1093/jmedent/29.1.41
8. Purnell RE, Branagan D, Radley DE (1969) The use of parasympathomimetic drugs to stimulate salivation in the tick *Rhipicephalus appendiculatus*, and the transmission of *Theileria parva* using saliva obtained by this method from infected ticks. Parasitology 59(3):709–718. https://doi.org/10.1017/s003118200003122x
9. Oliveira CJ et al (2013) Proteome of *Rhipicephalus sanguineus* tick saliva induced by the secretagogues pilocarpine and dopamine. Ticks Tick Borne Dis 4(6):469–477. https://doi.org/10.1016/j.ttbdis.2013.05.001
10. Ribeiro JM et al. (2004) How much pilocarpine contaminates pilocarpine-induced tick saliva?. Med Vet Entomol 18(1):20–24. https://doi.org/10.1111/j.0269-283x.2003.0469.x
11. Ferreira BR, Silva JS (1998) Saliva of *Rhipicephalus sanguineus* tick impairs T cell proliferation and IFN-gamma-induced macrophage microbicidal activity. Vet Immunol Immunopathol 64(3):279–293. https://doi.org/10.1016/s0165-2427(98)00135-4
12. Bechara GH, Szabo MP, Machado RZ, Rocha UF (1988) A technique for collecting saliva from the cattle-tick *Boophilus microplus* (Canestrini, 1887) using chemical stimulation. Environmental and temporal influences on secretion yield. Braz J Med Biol Res 21(3):479–484
13. Binetruy F, Dupraz M, Buysse M, Duron O (2019) Surface sterilization methods impact measures of internal microbial diversity in ticks. Parasit Vectors 12(1):268. https://doi.org/10.1186/s13071-019-3517-5
14. Lubinga JC et al. (2013) Detection of lumpy skin disease virus in saliva of ticks fed on lumpy skin disease virus-infected cattle. Exp Appl Acarol 61(1):129–138. https://doi.org/10.1007/s10493-013-9679-5
15. Kim TK et al (2021) *Borrelia burgdorferi* infection modifies protein content in saliva of *Ixodes scapularis* nymphs. BMC Genomics 22(1):152. https://doi.org/10.1186/s12864-021-07429-0

Chapter 5

The Zebrafish Model for Alpha-Gal Syndrome

Rita Vaz-Rodrigues, Marinela Contreras, Almudena González-García, David Fernández-Castellanos and José de la Fuente

Abstract

Zebrafish (*Danio rerio*) has emerged as a robust vertebrate model for studying immune and allergic disorders, including food allergies. The alpha-Gal syndrome (AGS) is a tick-borne allergy characterized by an IgE-mediated delayed hypersensitivity to the oligosaccharide galactose-α-1,3-galactose (α-Gal), which is present in tick saliva and mammalian-derived foods and drugs. This chapter describes the experimental procedures used to establish and analyze the validated zebrafish model for AGS. Adult zebrafish were intramuscularly injected with *Ixodes ricinus* tick saliva and maintained on a mammalian meat–based diet to induce allergic responses. Clinical manifestations were monitored and correlated with serum anti-α-Gal IgM titers measured by ELISA, while the intestinal expression of allergic and immune biomarkers (*ctss2.1*, *prkdc*, *il1b*, *il4*, *tnfα*, *tlr2*, *c3a*, and *ifn1*) was analyzed by RT-qPCR. This animal model successfully reproduces key features of AGS such as delayed allergic reactions (hemorrhagic-type), altered behavior, and increased mortality, accompanied by elevated anti-α-Gal IgM antibody levels and modulation of immune gene expression. Importantly, zebrafish also enables the testing of specific salivary biomolecules from ticks or other arthropods to evaluate their allergenic potential. Overall, the zebrafish model provides a reproducible and cost-effective in vivo platform to elucidate molecular mechanisms underlying AGS and arthropod-induced allergies.

Keywords Zebrafish, *Danio rerio*, Alpha-Gal syndrome, Tick, α-Gal, Food allergy

1 Introduction

Zebrafish (*Danio rerio*) has emerged as a powerful vertebrate model in biomedical research, contributing to a broad range of fields such as toxicology, oncology, neurobiology, pharmacology, and behavioral science [1–3]. In recent years, zebrafish have laid the groundwork for disease modeling, advancing the understanding of not only inflammatory allergic disorders, such as asthma and atopic dermatitis [4–6], but also food allergies [7], including alpha-Gal syndrome (AGS) [8–10]. Owing to their conserved innate and adaptive immune systems, zebrafish offer valuable insights into the molecular and cellular mechanisms underlying immune responses and hypersensitivity reactions

José de la Fuente and Alejandro Cabezas-Cruz (eds.), *Allergic Disease Models*, Methods in Molecular Biology, vol. 3092, https://doi.org/10.1007/978-1-0716-5635-8_5,

[11–14]. Additionally, their use in preclinical in vivo studies offers several advantages, including low maintenance costs, a short life cycle with high reproductive capacity, and notable genetic, anatomical and physiological similarities with humans [15–17].

The AGS is an IgE-mediated type I delayed hypersensitivity to the oligosaccharide galactose-α-1,3-galactose (α-Gal) [18, 19], present in tick salivary glycoproteins and glycolipids, as well as in mammalian meat, organs, dairy products, and gelatin derived from mammalian collagen used in pharmaceuticals and vaccines [20–22]. Symptoms typically appear 2–6 h after consuming α-Gal-containing foods or products and include delayed anaphylaxis and cutaneous manifestations like urticaria, angioedema, and pruritus [23]. Additionally, a gastrointestinal (GI) phenotype characterized by delayed abdominal pain, nausea, diarrhea, reflux, and vomiting is increasingly recognized [24, 25]. The delay between exposure and symptom onset, combined with limited awareness and a lack of standardized tests, complicates diagnosis and often leads to AGS being underdiagnosed or misdiagnosed [19, 26, 27].

Despite advances, key aspects of AGS pathogenesis remain unclear [28]. Recently, studies using the zebrafish model have highlighted the involvement of tick biomolecules without α-Gal modifications in the pathophysiology of the disease [10, 29]. The zebrafish model also offers advantages for studying relevant gut microbiome changes associated with the development of allergic reactions in AGS caused by different tick salivary components [30]. Therefore, this chapter discusses the key phenotypic changes observed in the zebrafish model of AGS during the experimental procedure. It also highlights the use of additional immunological analyses, such as measuring IgM antibody levels against α-Gal in zebrafish serum by ELISA and assessing the expression of allergic and immune biomarkers in midgut tissue by quantitative reverse transcription polymerase chain reaction (RT-qPCR).

2 Materials

2.1 Zebrafish Facility and Husbandry

Wild-type adult AB strain zebrafish (6–8-month-old) should be provided by scientific institutions (e.g., Dr. Juan Galcerán Sáez from the Instituto de Neurociencias IN-CSIC-UMH, Sant Joan d'Alacant, Alicante, Spain) and certified by Biosait Europe S.L. (Barcelona, Spain) as free of major fish pathogens. The zebrafish facility (Fig. 1A–C) includes two modules with a total capacity of 1000 zebrafish, featuring automatic systems for water filling and drainage and a total of 18 aquariums with a capacity of 25–30 L. One module is fully automatic, while the other includes four sealed tanks for larval rearing and quarantine, each equipped with thermo-heaters (900 W), common

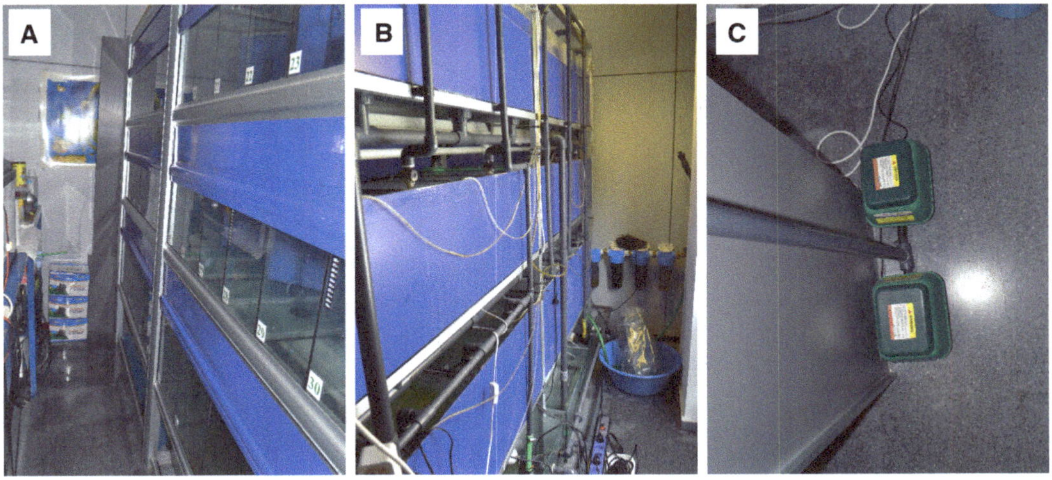

Fig. 1 Zebrafish facility. (**A**), Overview of the zebrafish housing system showing the front side of the stacked aquarium modules. (**B**), Rear view of the facility detailing the water recirculation system. (**C**), Common biological and mechanical filtration systems used for water conditioning

biological and mechanical filters, and an individual biological filter per aquarium, plus a mechanical pumping system. Zebrafish can be housed in groups of 20 fish per 25–30 L tank, maintained in a flow-through water system at 27 °C with a light/dark cycle of 14/10 h (Fig. 2A) and fed twice daily at 9:30 a.m. and 3:30 p.m. with dry fish feed (Premium food tropical fish, DAPC, 50–70 μg/fish). Experiments are conducted in strict accordance with the recommendations of the European Guide for the Care and Use of Laboratory Animals. Zebrafish are housed, and experiments conducted at an experimental facility with the approval and supervision of the Ethics Committee and authorized Organizations (e.g., Animal Experimentation of the University of Castilla-La Mancha PR-2021-09-14 and the Counseling of Agriculture, Environment and Rural Development of Castilla-La Mancha REGA code ES130340000218).

2.2 Experimental Procedures

1. Adult zebrafish and housing systems
2. Premium tropical fish food (DAPC)
3. Portable plastic container for small fish (180 × 110 x 120 mm)
4. Fishing net (10 cm)
5. Disposable polystyrene weighing dishes (135 × 135 × 20 mm)
6. Kern analytical balance
7. Biowater bacteria (100 mL, DAPAC)
8. Sera pH-test analysis kit for aquariums (15 mL)
9. Sera nitrite and ammonium water analysis kits

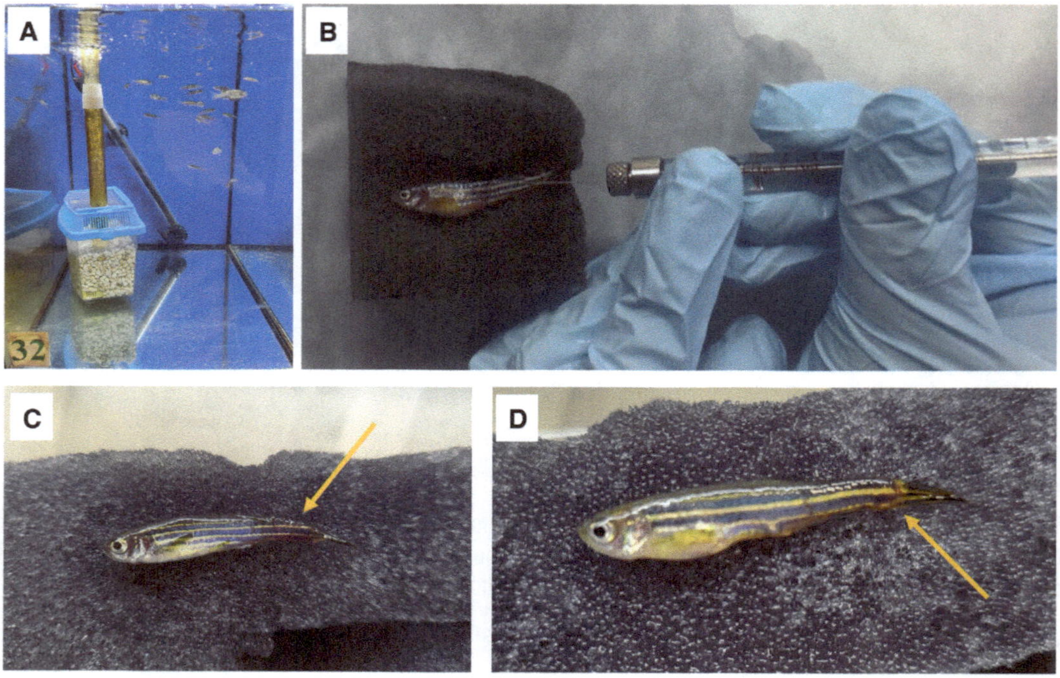

Fig. 2 Zebrafish maintenance and experimental proceedings. (**A**), Housing system: 25-L fish tank with appropriate filtering system. (**B**), Zebrafish intramuscular (IM) treatment injections using a Hamilton microliter syringe. (**C**), (**D**), Allergic reactions (hemorrhagic-type) observed in zebrafish that were administered twice with *Ixodes ricinus* tick saliva and maintained on dog food (red meat–based diet)

10. Sera Aquatan water conditioner (500 mL)
11. Dog food—classic red (Acana, Alberta, Canada)
12. Phosphate-buffered saline (PBS)
13. Pilocarpine hydrochloride
14. Stereo microscope (Zeiss or similar)
15. Double-sided adhesive tape
16. Corning microscope slides
17. Tick saliva (*I. ricinus*), obtained as in Sect. 2.5
18. Capillary glass tubes
19. Pasteur pipette rubber bulbs
20. Petri dishes (100 × 20 mm)
21. Hamilton microliter syringes
22. Hamilton syringe holder
23. Eppendorf microcentrifuge tubes (1.5 mL volume)
24. Centrifuge adapted for Eppendorf microcentrifuge tubes
25. Sponge fish holder (made in-house from scrap foam)

26. Sterile gauze pads (50 × 50 mm)
27. Micropipettes (P10, P20, and P1000)
28. Pipette tips (P10, P20, and P1000)
29. Tricaine methanesulphonate
30. Dissection set
31. Real Statistics extension for Microsoft Excel
32. GraphPad Prism (version 8.0.1)

2.3 ELISA: Anti-α -Gal IgM Titers

1. Micropipettes (P10, P200, and P1000)
2. Pipette tips (P10, P200, and P1000)
3. Multichannel microliter pipette (Eppendorf or similar)
4. Eppendorf microcentrifuge tubes (1.5 mL volume)
5. Absorbent paper
6. Container for wash solution
7. Aluminum foil
8. Nunc MaxiSorp white 96-well plates
9. Orbital shaker
10. Laboratory automated incubator, 37 °C
11. Adjustable 1–25 mL pipettes
12. Alpha-Gal-BSA (NGP0203; Dextra Laboratories, Shinfield, UK)
13. Carbonate-bicarbonate buffer
14. Phosphate-buffered saline (PBS)
15. Tween-20 detergent
16. Skim milk
17. Rabbit anti-zebrafish IgM antibodies (produced *in-house*)
18. Goat anti-rabbit IgG-peroxidase conjugate (A9169; Sigma-Aldrich)
19. 3,3′,5,5′-Tetramethylbenzidine (TMB) One Solution
20. Sulfuric acid (H_2SO_4), 2 N
21. Spectrophotometer with 450-nm filter
22. Real Statistics extension for Microsoft Excel
23. GraphPad Prism (version 8.0.1)

2.4 RNA Extraction and RT-qPCR: Midgut Expression of Allergic and Immune Markers

1. RNA extraction kit
2. Dissection set
3. Petri dishes (100 × 20 mm)
4. Pellet pestle
5. NanoDrop One spectrophotometer

6. iTaq Universal SYBR Green One-Step RT-qPCR Kit
7. Gene-specific primers (Table 1)
8. Micropipettes (P10, P20, and P200)
9. PCR plate seals
10. Eppendorf PCR cooler
11. 96-Well PCR plates
12. Microplate centrifuge
13. Pipette tips P10, P20, and P200
14. Real-time PCR detection system
15. Real Statistics extension for Microsoft Excel
16. GraphPad Prism (version 8.0.1)

2.5 I. ricinus *Tick Saliva Extraction*

1. *I. ricinus* must originate from a laboratory colony, be approved by the Institutional Animal Care and Use Committee (IACUC) and adhere to National and Institutional guidelines and regulations.
2. Depending on the amount of saliva required, select the appropriate number of semi-engorged pathogen-free female ticks that have been fed for 6–7 days on pathogen-free guinea pigs.
3. Position up to ten ticks dorsally on a microscope slide using double-sided adhesive tape (Fig. 3A).
4. Attach a Hamilton microliter syringe into the appropriate holder (Fig. 3A), preloaded with a 2% (wt/vol) solution of pilocarpine hydrochloride in PBS at pH 7.4.
5. Using fine-tip forceps (*see* **Note 1**) and a stereomicroscope, place the syringe tip on the lower left quadrant of the tick's abdomen (Fig. 3B) and administer 5 μL of 2% pilocarpine hydrochloride per tick to stimulate tick saliva secretion.
6. Harvest saliva by carefully inserting the tip of a pulled capillary glass tube inside the tick's hypostome, allowing the maxillary palps to remain alongside the outside of the capillary tube while observing the tick under a stereomicroscope.
7. Place the saliva sample inside a 1.5-mL tube using a Pasteur pipette bulb by expelling it out of the capillary tubes, then reserve it on ice.
8. Saliva can be pooled normally with material collected from 100 to 200 female ticks and then stored at −80 °C until further use (*see* **Note 2**).

Table 1

Oligonucleotide primer sequences and annealing temperatures used for analyzing mRNA expression levels of different gut zebrafish (*D. rerio*) immunity and allergy biomarkers by quantitative reverse transcription polymerase chain reaction (RT-qPCR)

Target (NCBI accession no.)	Amplicon (bp)	Oligonucleotide primers (5′–3′) forward (F) and reverse (R)	Annealing temperature (°C)	Reference
ctss2.1 (NM_001024409.2)	115	F: CGGGTTTATGAGTGACGCCT R: ACGCTGGGATGAACTGTAGC	60	[31]
prkdc (XM_009303401.2)	115	F: GCCTGGCTTGTGACGTTGAC R: ACTGCCACTGTGTCTTGGCT	60	[31]
tnfα (BC165066.1)	170	F: GCTTATGAGCCATGCAGTGA R: TGCCCAGTCTGTCTCCTTCT	57	[32]
il4 (NM_001170740.1)	190	F: GTGAATGGGATCCTGAATGG R: TTCCAGTCCCGGTATATGCT	55	[32]
c3a (NM_131243.1)	264	F: ACGCTCTCTGGATTGAAACA R: TGCCTTCTTGCATGGCAATC	56	[8, 32]
tlr2 (NM_212812.1)	169	F: TGAATGGGTCGAGGAGATTC R: CACAAAGTGCTCCGACAGAA	56	[8, 32]
il1b (NM_212844)	189	F: GCATGTCCACATATGCGTCG R: GCTGGTCGTATCCGTTTGGA	59	[8, 32]
ifn1 (NM_207640.2)	258	F: ATGAGAACTCAAATGTGGAC R: TTACACTCGAGGATTGAC	50	[8]
gapdh (NM_001115114.1)	259	F: CGTGGTGCCAGTCAGAACAT R: AGTCAGTGGACACAACCTGG	59	[8, 31]
actb1 (NM_131031.1)	155	F: AAGCTGTGACCCACCTCACG R: GGCTTTGCACATACCGGAGC	62	[31, 33, 34]

ctss2.1: cathepsin S, ortholog2, tandem duplicate 1; *prkdc*: protein kinase, DNA-activated, catalytic subunit; *tnfα*: tumor necrosis factor alpha, TNF superfamily, member 2; *il4*: interleukin 4; *c3a*: complement C3a, tandem duplicate 2; *tlr2*: toll-like receptor 2; *il1b*: interleukin 1, beta; *ifn1*: interferon phi 1; *gapdh*: glyceraldehyde-3-phosphate dehydrogenase; *actb1*: beta actin 1

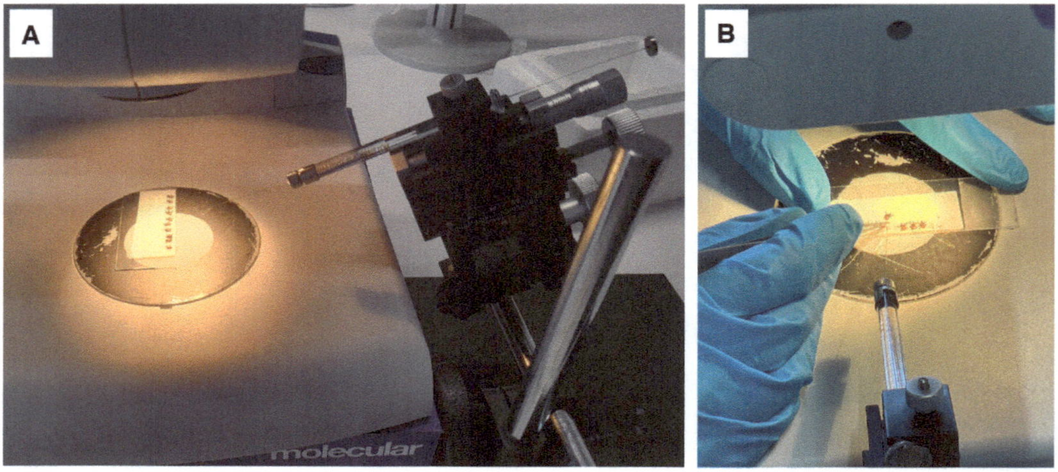

Fig. 3 *I. ricinus* tick saliva extraction. (**A**), Female ticks ($N = 10$) placed dorsally on a microscope slide using white double-sided adhesive tape and a Hamilton microliter syringe coupled to the appropriate syringe holder prepared for treatment administration. (**B**), Example of the administration method by placing the Hamilton syringe tip on the lower left quadrant of the tick's abdomen

3 Methods

3.1 Zebrafish Handling and Maintenance

1. The water tank (Fig. 2A) should be dechlorinated and treated with Biowater (10 mL per 100 L) and Sera Aquatan water conditioner (5 mL per 20 L), then maintained under stable physicochemical conditions for approximately 1 month prior to zebrafish introduction to allow for biological equilibration.
2. Check pH, ammonia, nitrite, and nitrate levels with the appropriate test kits (*see* **Notes 3 and 4**), and monitor temperature, lighting, and oxygenation to ensure optimal and stable environmental conditions.
3. It is recommended to acclimate zebrafish at least 1 week prior to the beginning of the experiment (*see* **Note 5**).
4. A few days before the start of the experiment, carefully separate ten zebrafish per group into individual tanks using a fishing net, maintaining a 1:1 sex ratio and similar body weights (approximately 330 ± 40 mg in our case).
5. At days 0 and 8, record the body weight of the zebrafish, either individually by placing each fish in a pretared weighing dish filled with tank water on an analytic balance, or collectively by placing ten zebrafish per group in a pretared portable fish container filled with tank water.
6. On days 0 and 1, feed the zebrafish with dry fish food twice daily at around 9:30 a.m. and 3:30 p.m. with dry fish feed (50–70 μg/fish).

7. From day 2 to the end of the experiment (day 8), feed the zebrafish twice daily with dog food (50–70 μg/fish), previously mashed with a mortar and pestle (*see* **Note 6**).
8. Perform water-quality checks twice per week (pH, ammonia, nitrite, and nitrate levels) and add weekly approximately 0.25 mL/L of Sera Aquatan to maintain optimal tank conditions.

3.2 Zebrafish Treatment Injections

1. Prepare the appropriate treatment (*see* **Note 7**) depending on the experimental group (Fig. 4A): (1) tick saliva with 1 μL of *I. ricinus* saliva in 9 μL of PBS (1:10) or (2) negative control with 10 μL of PBS and place them on ice.
2. Using a fishing net, carefully collect the ten zebrafish of each group and place them in a portable fish container filled with tank water.
3. Prepare a portable fish container filled with tank water and add tricaine methanesulphonate to a final concentration of 150 mg/L.
4. Introduce each zebrafish individually into the container to be anesthetized by immersion, during which properly anesthetized fish exhibited normal opercular movements but no active swimming, remaining either motionless at the bottom or floating passively in the water.
5. Using a fishing net, carefully collect each zebrafish and place it on a sponge previously moistened with tepid tank water (*see* **Note 8**), handling the fish gently with forceps or by hand and positioning it on its side with the caudal fin to the right (*see* **Note 9**) to allow compound injection (Fig. 2B).
6. Administer the appropriate treatment (*see* **Note 10**) via intramuscular (IM) using a Hamilton microliter syringe on days 0 and 3 of the experiment by inserting the needle approximately 2 mm beneath the skin at a 40°– 45° angle to the horizontal plane and about 5 mm from the zebrafish tail (Fig. 2B).
7. Place the treated zebrafish in a new portable container filled with freshwater tank and monitor them until full recovery before returning them to the housing tank.

3.3 Zebrafish Symptom Monitoring

Zebrafish require twice-daily monitoring for symptoms associated with AGS, which include the development of allergic reactions, increased mortality, altered behavior, and/or changes in feeding patterns (*see* **Note 11**). Symptoms are typically delayed and should be assessed 40–50 min after each feeding with α-Gal-containing dog food. In addition, any changes following treatment administration should be monitored to distinguish those directly related to the allergenic potential of the treatment agent from those associated with AGS that manifest after mammalian meat consumption.

1. Examine the development of hemorrhagic-type allergic reactions (Fig. 2C, D) based on the presence of lesions in the scales and fins of the fish associated with redness, discoloration, hemorrhage, bristling scales, and loss of caudal fin (*see* **Note 12**).
2. All lesions should be classified as qualitative variables and categorized/classified as absent (0), very mild (1), mild (2), moderate (3), intense (4), or very intense (5).
3. Evaluate swimming behavior (*see* **Note 13**) and classify the movement as: slow, normal, rapid, compromised, altered, or zigzag motion. Additionally, note any abnormal positioning within the tank (e.g., remaining excessively near the bottom).
4. Analyze feeding patterns (*see* **Note 14**) to classify the ingestion of food as: absent, slow, normal, fast, or at the bottom of the tank.
5. Annotate cumulative mortality by recording the number of dead fish along with corresponding time and day of death.
6. Incorporate all data into a Microsoft Excel spreadsheet, perform statistical analysis between the two groups using a one-tailed Student's t-test with unequal variance with the Real Statistics extension for Microsoft Excel, and present the results graphically. Representative results are shown in Fig. 4B.

3.4 Zebrafish Sample Collection, Handling, and Processing

1. Using a fishing net, carefully collect the zebrafish of each group and transfer them to a portable fish container filled with tank water.
2. Euthanize the zebrafish individually by prolonged immersion in an overdose of tricaine methanesulphonate (300 mg/L solution) (*see* **Note 15**).
3. With the help of a fishing net, place the euthanized fish on a Petri dish lined with a piece of dry gauze, then cover it with another piece, leaving the tail exposed.
4. Quickly perform tail ablation using a surgical blade and collect the emerging blood with a 20 μL tip attached to a micropipette (*see* **Note 16**).
5. Transfer the aspirated blood into a prelabeled 1.5-mL microcentrifuge tube and reserve it on ice (up to 15 min).
6. Make a sagittal cut along the zebrafish midline using a new surgical blade, carefully collect the intestine to a new 1.5-mL tube and reserve it on ice.
7. Centrifuge the blood at $10\times g$ (around 300 rpm) for 10 min, ensuring this is done within 15 min after collection to avoid hemolysis.
8. Aspirate the serum (top layer) with a 10 μL tip attached to a micropipette and transfer it to a new, prelabeled 1.5-mL microcentrifuge tube.

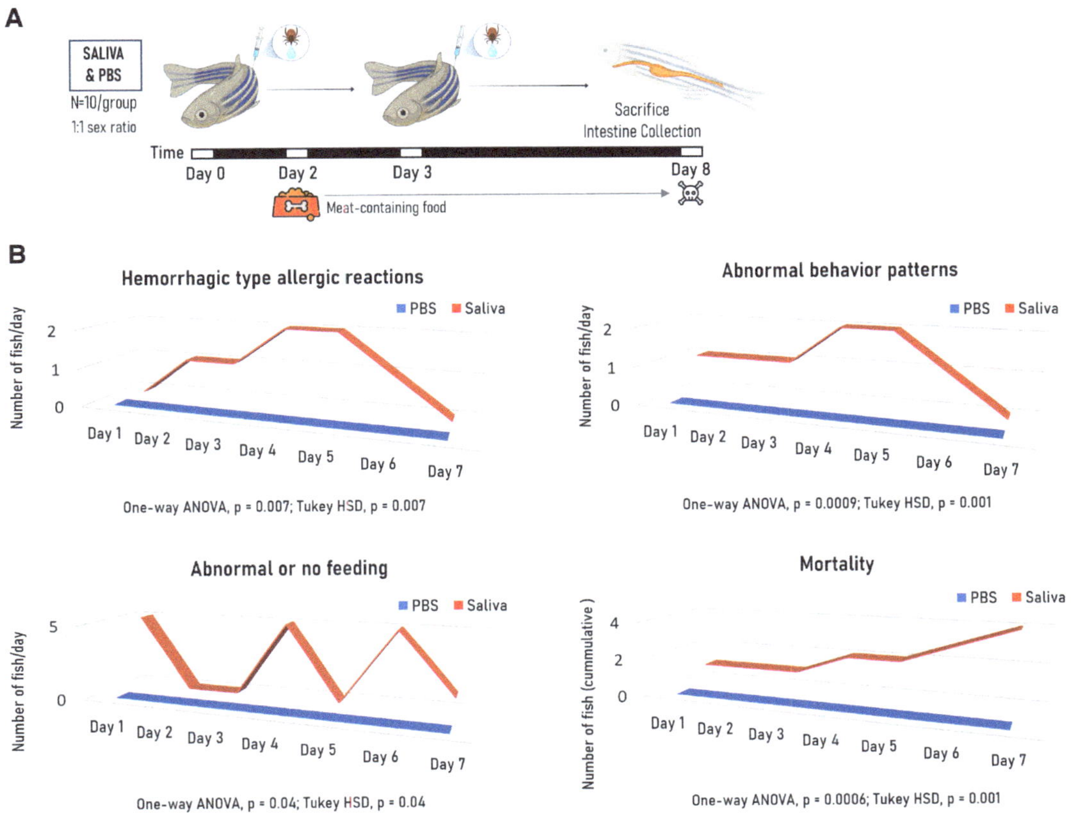

Fig. 4 Experimental design and results. (**A**), Experimental design. (**B**), Representative zebrafish symptom in zebrafish. Zebrafish were injected intramuscularly with tick saliva from *I. ricinus* or PBS (negative control) on days 0 and 3 from the experiment. Clinical manifestations of the alpha-Gal syndrome (AGS) include the development of allergic reactions (hemorrhagic-type), abnormal behavior, changes in feeding patterns, and an increase in cumulative mortality. Representative results were obtained from Vaz-Rodrigues et al. [31]

9. Store the serum samples at −20 or −80 °C until further analysis of anti-α-Gal IgM titers by ELISA.
10. Store the intestinal samples at −80 °C until nucleic acid extraction for subsequent analysis of allergic and immune marker gene expression by RT-qPCR (*see* **Note 17**).

3.5 ELISA: Anti-α-Gal IgM Titers

Measure IgM antibody levels that recognize α-Gal in previously collected zebrafish serum samples (Sect. 3.4) employing an indirect in-house ELISA (*see* **Note 18**).

1. Coat plates with 100 ng of α-Gal-BSA in 100 μL of carbonate/bicarbonate buffer per well and incubate overnight at 4 °C with gentle shaking.

2. Remove the solution from each well and wash the plates four times with 300 μL per well of PBS containing 0.05% Tween 20 (PBST) with a multichannel pipette. After the last wash, invert the plate and blot it against clean absorbent paper to remove any remaining wash buffer (*see* **Note 19**).
3. Block the coated plates with 100 μL/well of 5% skim milk in PBS (blocking solution) for 1 h at room temperature (RT) with gentle shaking.
4. Discard the blocking solution and do not wash.
5. Add 100 μL/well of zebrafish serum samples in duplicates, previously diluted in blocking solution (1:100), cover the plates with aluminum foil and incubate for 2 h at 37 °C. Be sure to include two blank wells (with blocking solution only).
6. Discard the solution and wash the plate four times as directed in Step 2.
7. Add 100 μL/well of rabbit anti-zebrafish IgM antibodies diluted in blocking solution (1:1000) and incubate for 1 h at RT with gentle shaking.
8. Discard the solution and wash the plate four times as directed in Step 2.
9. Add 100 μL/well of goat anti-rabbit IgG-peroxidase conjugate diluted in blocking solution (1:3000) and incubated for 1 h at RT with gentle shaking.
10. Discard the solution and wash the plate five times as directed in Step 2.
11. Incubate with 100 μL/well of 3,3′,5,5′-tetramethylbenzidine (TMB) One Solution for 20–30 min at RT in the dark, with wells covered with aluminum foil.
12. Stop the reaction with 50 μL/well of 2 N H_2SO_4. Gently tap the side of the plate to mix as the solution changes from blue to yellow and immediately read the optical density (OD) at 450 nm in a spectrophotometer.
13. Use antibody levels (OD450) for analysis after background with blank subtraction (coated wells incubated with blocking solution).
14. Analyze the results between the two groups using a one-tailed Student's *t*-test with unequal variance, performed with the Real Statistics extension for Microsoft Excel.
15. Represent the results graphically with GraphPad Prism (version 8.0.1) as shown in Fig. 5A.
16. Perform a Pearson's correlation analysis (Fig. 5B) between the anti-α-Gal IgM antibody titers and the symptoms associated with AGS, as mentioned in Sect. 5.3.4.

3.6 Zebrafish Midgut Total mRNA Isolation

1. Disrupt up to 30 mg of zebrafish intestinal tissue per sample with a scalpel on a Petri dish and homogenize with a tissue grinder until lysates are completely clear.
2. Isolate total mRNA using the RNeasy mini extraction kit or equivalent following manufacturer's instructions.
3. Assess concentration (ng/μL) and purity of the mRNA samples by measuring the optical density at 260 nm (OD260) and the ratio of absorbance at 260/280 nm using the NanoDrop One spectrophotometer or equivalent.
4. Standardize mRNA concentrations based on the obtained results, typically to 20–50 ng/μL.
5. Store all samples at −80 °C until further analysis.

3.7 RT-qPCR: Midgut Expression of Allergic and Immune Markers

1. Prepare gene-specific primers (Table 1) from the 100 μM stock solution to a 10 μM working solution by mixing 10 μL of stock with 90 μL of nuclease-free water.
2. Depending on the number of samples, prepare a PCR mixture (total volume of 18.5 μL per well) containing: 10 μL of SYBR Green Master Mix, 0.25 μL of reverse transcriptase, 1 μL (10 μM) each of forward and reverse primers and 6.25 μL of nuclease-free water. Include two to three technical replicates per sample and two negative controls using nuclease-free water.
3. Add 18.5 μL of PCR mixture to each well of a 96-well PCR plate, avoiding bubble formation. Then add 1.5 μL (30–150 ng) of sample mRNA or control to the corresponding wells. Perform all steps on a PCR cooler to maintain samples

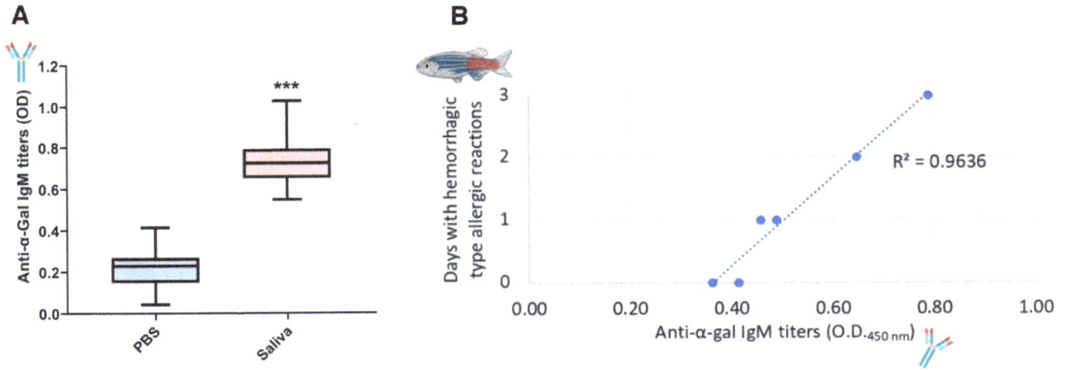

Fig. 5 Expected results for anti-α-Gal IgM titers measured by ELISA. **A**, Characterization of IgM antibody titers against α-Gal in serum samples from zebrafish treated with *I. ricinus* tick saliva compared to the control group (PBS; $p < 0.001$), both maintained on a α-Gal-containing diet. **B**, Correlation analysis between hemorrhagic type allergic reactions and anti-α-Gal IgM antibody titers in zebrafish treated with *I. ricinus* tick saliva ($R^2 = 0.96$, $p < 0.001$; $n = 6$). Representative results were adapted from Contreras et al. [9] and Vaz-Rodrigues et al. [10]

and reagents at low temperature. Seal the plate with adhesive PCR plate seals.

4. Centrifuge the PCR plate at 3000 rpm for 2 min at 18 °C to ensure all the mixture settles at the bottom of the wells and no bubbles remain.
5. Perform RT-qPCR utilizing the real-time PCR detection system and the following PCR conditions: a reverse transcription reaction at 50 °C for 10 min, followed by an initial denaturation step at 95 °C for 1 min, amplification by 40 cycles of 95 °C for 20 s and different annealing temperatures (Table 1) for 1 min, followed by a dissociation curve analysis.
6. Perform the process for each target biomarker (Table 1), including the two housekeeping genes (glyceraldehyde 3-phosphate dehydrogenase and β-actin 1).
7. Apply the generalized qBase model [35] to calibrate, normalize and obtain the relative expression values of the targeted biomarkers utilizing β-actin 1 (actb1) and glyceraldehyde 3-phosphate dehydrogenase (gapdh) as housekeeping genes.
8. Analyze statistical differences between the two groups (saliva vs. control) for each allergic/immune biomarker using a one-tailed Student's *t*-test with unequal variance (*see* **Note 20**), performed with the Real Statistics extension for Microsoft Excel.
9. Represent results with GraphPad Prism (version 8.0.1) as illustrated in Fig. 6.

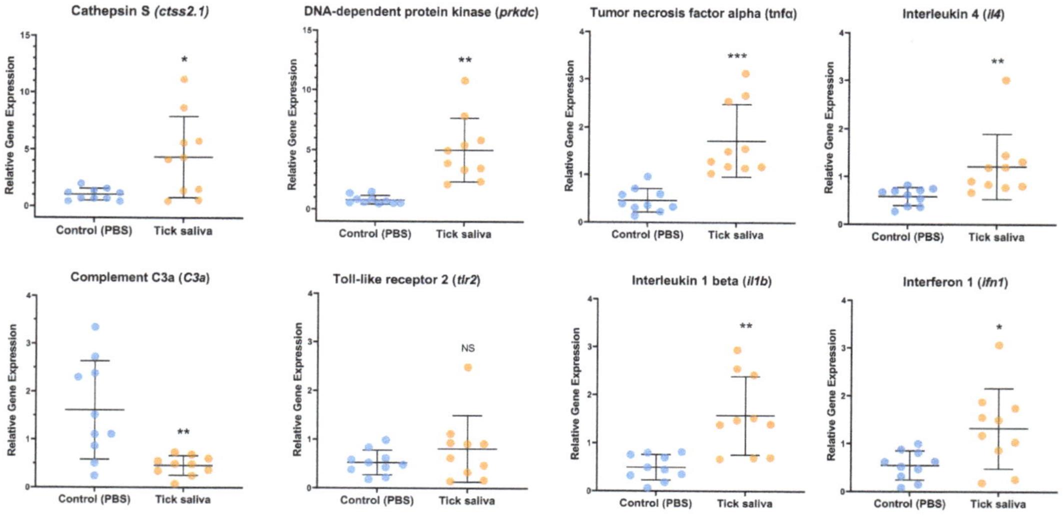

Fig. 6 Expected results for zebrafish midgut gene expression. Gene expression of allergic and immune markers was measured by reverse transcription quantitative polymerase chain reaction (RT-qPCR). Representative results were adapted from Vaz-Rodrigues et al. [10]

4 Notes

1. When handling ticks with forceps, avoid applying excessive pressure when gripping them to prevent physical damage.
2. Each tick can regurgitate up to 0.6 μL of saliva following pilocarpine injection. Generally, between 100 and 200 female ticks are used to collect saliva.
3. Ammonia and nitrite levels in the water tank should be near 0 ppm to maintain optimal water quality and prevent zebrafish mortality. If elevated, replace 50% of the water and add a nitrite and ammonia conditioner like Seachem AmGuard (250 mL) at 5 mL per 200 L daily until levels reach 0 ppm.
4. The pH level in the water tank should be adjusted to the species' requirements and kept as stable as possible, ideally between 6.8 and 7.5 for zebrafish and not exceeding 9. If the pH is too high, use SERA pH/KH Minus (250 mL) at 5 mL per 20 L of water; if too low, use SERA pH/KH Plus (250 mL) at the same dosage. This procedure should be repeated daily until the desired pH value is reached.
5. Only introduce zebrafish into the water tank once pH, ammonia, nitrite, and nitrate levels are stable and within the recommended range. Place the unopened transport bags in the tank for 30 min to equalize temperatures and minimize stress or thermal shock. Then, open the bags while they remain in the water, allowing the fish to swim out freely before removing the bags.
6. As the goal of this experiment is focused on the measurement of allergic markers and serological changes, it is imperative to change the fish feed to dog food containing mammalian meat or alternatively incorporate pig kidney extracts in their diet.
7. If needed, tick saliva protein concentration can be determined by using a Pierce BCA Protein Assay Kit with BSA protein standards following manufacturer's recommendations.
8. When handling anesthetized zebrafish outside the water, the procedure should be performed quickly and carefully to avoid physical damage and to ensure the fish does not become excessively dry.
9. When performing zebrafish treatment injections, always position the fish in the same manner to ensure compound injection in a consistent direction and to facilitate lesion control.
10. Any other desired salivary biomolecules from ticks or other arthropods such as hymenopters (bees, wasps, ants), dipterans (mosquitoes, flies), flees, and trombiculid mites can also

be administered (Sect. 3.2) and evaluated for their allergenic potential using this animal model.

11. It is recommended that zebrafish symptom monitoring be conducted by a professional with experience and knowledge of zebrafish behavior, allowing them to interpret changes based on their background and familiarity with this animal model.
12. The examination of hemorrhagic-type allergic reactions 40–50 min after feed intake should include noting their location and further categorizing them as described in Sect. 3.3. If necessary, zebrafish can be observed under anesthesia using a magnifier or stereomicroscope.
13. Normal swimming behavior of zebrafish is characterized by smooth, coordinated, and active movement throughout the tank, with regular exploration and quick responses to external stimuli. The refined novel tank diving test (NTT) can also be used to determine anxiety-like behaviors in zebrafish [36].
14. Normal zebrafish feeding behavior includes rapid and coordinated swimming toward the fish feed, active pecking at food particles near the water surface and short bursts of competitive activity among individuals, followed by a return to calm exploratory swimming once feeding ceases.
15. Tricaine methanesulphonate effectively induces euthanasia in adult fish by stopping gill ventilation. Zebrafish should be left in the solution for at least 30 min following cessation of opercular movement.
16. The volume of blood collected by tail ablation depends on the zebrafish's size and the adequacy of anesthesia, typically ranging from 5 to 20 μL.
17. Gene expression analysis of relevant allergic and immunological biomarkers is conducted in the intestine due to its central role in zebrafish immunity and further relevance in AGS pathophysiology.
18. Measurement of anti-α-Gal IgM antibody titers and gene expression analysis are the recommended procedures for obtaining immunological data. However, alternative techniques may be employed as needed.
19. Complete removal of liquid at each step of the ELISA is essential to good assay performance.
20. If the experiment includes more than two groups, the statistical analysis should be performed using a one-way ANOVA followed by a Tukey's post hoc test.

Competing Interests

The authors have no conflicts of interest to declare that are relevant to the content of this chapter.

References

1. Eisen JS (2020) Chapter 1—history of zebrafish research. In: Cartner SC, Eisen JS, Farmer SC, Guillemin KJ, Kent ML, Sanders GE (Eds.) Eds. The zebrafish in biomedical research, American College of Laboratory Animal Medicine. Academic Press, 3–14
2. Choi T-Y, Choi T-I, Lee Y-R, Choe S-K, Kim C-H (2021) Zebrafish as an animal model for biomedical research. Exp Mol Med 53:310–317
3. Teame T, Zhang Z, Ran C, Zhang H, Yang Y, Ding Q, Xie M, Gao C, Ye Y, Duan M et al. (2019) The use of zebrafish (*Danio rerio*) as biomedical models. Anim Front 9:68–77
4. Progatzky F, Cook HT, Lamb JR, Bugeon L, Dallman MJ (2016) Mucosal inflammation at the respiratory interface: a zebrafish model. Am J Physiol Lung Cell Mol Physiol 310:L551–L561
5. Harms M, Habib MMW, Nemska S, Nicolò A, Gilg A, Preising N, Sokkar P, Carmignani S, Raasholm M, Weidinger G et al. (2021) An optimized derivative of an endogenous CXCR4 antagonist prevents atopic dermatitis and airway inflammation. Acta Pharm Sin B 11:2694–2708
6. Progatzky F, Jha A, Wane M, Thwaites RS, Makris S, Shattock RJ, Johansson C, Openshaw PJ, Bugeon L, Hansel TT et al. (2019) Induction of innate cytokine responses by respiratory mucosal challenge with R848 in zebrafish, mice, and humans. J Allergy Clin Immunol 144:342–345.e7
7. Bao J, Qiu Y, Xu X, Fu X, Song J, Wang L, Huang L, Zhang W (2024) Towards an optimized model of food allergy in zebrafish. Mol Immunol 173:110–116
8. Contreras M, Pacheco I, Alberdi P, Díaz-Sánchez S, Artigas-Jerónimo S, Mateos-Hernández L, Villar M, Cabezas-Cruz A, de la Fuente J (2020) Allergic reactions and immunity in response to tick salivary biogenic substances and red meat consumption in the zebrafish model. Front Cell Infect Microbiol 10:78
9. Contreras M, Vaz-Rodrigues R, Mazuecos L, Villar M, Artigas-Jerónimo S, González-García A, Shilova VN, Bovin NV, Díaz-Sánchez S, Ferreras-Colino E et al. (2023) Allergic reactions to tick saliva components in zebrafish model. Parasites Vectors 16:242
10. Vaz-Rodrigues R, Mazuecos L, Contreras M, González-García A, Rafael M, Villar M, de la Fuente J (2025) Tick salivary proteins metalloprotease and allergen-like p23 are associated with response to glycan α-Gal and mycobacterium infection. Sci Rep 15: 8849
11. Boehm T (2012) Evolution of vertebrate immunity. Curr Biol 22:R722–R732
12. Sunyer JO (2013) Fishing for mammalian paradigms in the teleost immune system. Nat Immunol 14:320–326
13. Ye J, Kaattari IM, Ma C, Kaattari S (2013) The teleost humoral immune response. Fish Shellfish Immunol 35:1719–1728
14. Flajnik MF, Kasahara M (2010) Origin and evolution of the adaptive immune system: genetic events and selective pressures. Nat Rev Genet 11:47–59
15. Larijani B, Hamidpour SK, Tayanloo-Beik A, Shahbazbadr A, Yavari H, Namazi N, Biglar M, Arjmand B (2022) An overview of zebrafish modeling methods in drug discovery and development. Adv Exp Med Biol 1387:145–169
16. Howe K, Clark MD, Torroja CF, Torrance J, Berthelot C, Muffato M, Collins JE, Humphray S, McLaren K, Matthews L et al. (2013) The zebrafish reference genome sequence and its relationship to the human genome. Nature 496:498–503
17. Hemlata H, Murali N, Roy S, Kumar S, Tandon R, Betal S, Minocha S (2025) Zebrafish as a versatile preclinical model for evaluating nanomedicine in biomedical research: Recent advances and applications. J Drug Deliv Sci Technol 113:107348
18. Peterson CJ, Mohankumar P, Tarbox JA, Nugent K (2025) Alpha-Gal syndrome: a review for the general internist. Am J Med Sci 369:313–320
19. Vaz-Rodrigues R, Mazuecos L, de la Fuente J (2022) Current and future strategies for the diagnosis and treatment of the alpha-Gal syndrome (AGS). J Asthma Allergy 15:957–970
20. Zhan M, Yin J, Xu T, Wen L (2024) Alpha-Gal syndrome: an underrated serious disease and a potential future challenge. Glob Chall 8:2300331
21. Aiuto B, Cirrincione S, Giuffrida MG, Cavallarin L, Portesi C, Rossi AM, Borreani G, Rolla G, Geuna M, Nicola S et al. (2024) Milk

fat globule proteins are relevant bovine milk allergens in patients with α-Gal syndrome. Mol Nutr Food Res 68:2300796

22. Milano F, Masi A, Madaghiele M, Sannino A, Salvatore L, Gallo N (2023) Current trends in gelatin-based drug delivery systems. Pharmaceutics 15:1499.
23. Fischer J, Yazdi AS, Biedermann T (2016) Clinical spectrum of α-Gal syndrome: from immediate-type to delayed immediate-type reactions to mammalian innards and meat. Allergo J Int 25:55–62
24. Propst SBH, Thompson DK (2025) Alpha-gal syndrome and the gastrointestinal reaction: a narrative review. Front Allergy 6:1535103
25. Lee CJ, McGill SK (2023) Food allergies and alpha-gal syndrome for the gastroenterologist. Curr Gastroenterol Rep 25:21–30
26. Nalçacı M (2024) Mysterious allergy caused by tick bite: alpha-Gal syndrome. Turkiye Parazitol Derg 48:195–207
27. de la Fuente J (2024) The alpha-gal syndrome is underdiagnosed. Actas Dermosifiliogr 115:T116–T117
28. Erickson LD, Wilson JM, Cramton K, Rival CM (2025) Pre-clinical strategies and emerging technologies driving advances in the alpha-gal syndrome. Allergol Int 74(3):397–405
29. Vaz-Rodrigues R, Mazuecos L, Villar M, Contreras M, González-García A, Bonini P, Scimeca RC, Mulenga A, de la Fuente J (2024) Tick salivary proteome and lipidome with low glycan content correlate with allergic type reactions in the zebrafish model. Int J Parasitol 54:649–659
30. Díaz-Sánchez S, Vaz-Rodrigues R, Contreras M, Rafael M, Villar M, González-García A, Artigas-Jerónimo S, Gortázar C, de la Fuente J (2024) Zebrafish gut microbiota composition in response to tick saliva biomolecules correlates with allergic reactions to mammalian meat consumption. Microbiol Res 285, 127786.
31. Vaz-Rodrigues R, Mazuecos L, Villar M, Contreras M, Artigas-Jerónimo S, González-García A, Gortázar C, De La Fuente J (2023) Multi-omics analysis of zebrafish response to tick saliva reveals biological processes associated with alpha-Gal syndrome. Biomed Pharmacother 168:115829
32. Pacheco I, Contreras M, Villar M, Risalde MA, Alberdi P, Cabezas-Cruz A, Gortázar C, de la Fuente J (2020) Vaccination with alpha-gal protects against mycobacterial infection in the zebrafish model of tuberculosis. Vaccines (Basel)8:195
33. Thiagarajan SK, Mok SY, Ogawa S, Parhar IS, Tang PY (2022) Receptor-mediated AKT/PI3K signalling and behavioural alterations in zebrafish larvae reveal association between schizophrenia and opioid use disorder. Int J Mol Sci 23:4715
34. Foo YY, Motakis E, Tiang Z, Shen S, Lai JKH, Chan WX, Wiputra H, Chen N, Chen CK, Winkler C et al. (2021) Effects of extended pharmacological disruption of zebrafish embryonic heart biomechanical environment on cardiac function, morphology, and gene expression. Dev Dyn 250:1759–1777
35. Hellemans J, Mortier G, De Paepe A, Speleman F, Vandesompele J (2007) qBase relative quantification framework and software for management and automated analysis of real-time quantitative PCR data. Genome Biol 8: R19
36. Shinkai T, Nakai M, Takeshita U, Morita K, Shimada Y (2025) Refinement of the novel tank diving test: toward standardized and robust analysis of anxiety-like behavior in zebrafish. Front Behav Neurosci:19

Check for updates

Chapter 6

The Humanized Mouse Model of AGS: A Preclinical System for Analyzing Hypersensitivity to α-Gal and Potential Allergenic Epitopes

Lorena Mazuecos

Abstract

Humanized mouse models are indispensable in vivo experimental platforms for investigating human-specific biological processes in a controlled and reproducible environment, serving as a preclinical link between basic research and application in translational research. Alpha-Gal syndrome (AGS) is one of the emerging pathologies that highlight the need for a murine system capable of mimicking the reproduction of key aspects of human immune and allergic responses. Humanized α1,3GalT-knockout (KO) mice have expanded research on immunological basis of the syndrome and advanced in preclinical research. Here we present a detailed experimental protocol of KO validation and studying α1,3GalT-KO in vivo immune responses to α-Gal and/or other epitopes or potential allergens.

Keywords α-Gal, Mice, Immune system, Allergen, Tick, Saliva, Red meat

1 Introduction

Alpha-Gal syndrome (AGS) represents an emerging form of food allergy characterized by a delayed anaphylactic response to galactose-α-1,3-galactose (α-Gal), a carbohydrate epitope abundantly expressed in nonprimate mammalian tissues [1]. Unlike classical food allergies, AGS is triggered by prior sensitization through tick bites of species such as *Amblyomma americanum* and *Ixodes ricinus*, among others [2]. AGS initiates an IgE-mediated immune response against α-Gal after subsequent exposure. The atypical onset, delayed symptomatology, and broad distribution of α-Gal also in food (red meat), pharmaceuticals, and biomedical products present significant challenges for both diagnosis and management [3–6]. As reasonably, humans, along with Old world monkeys and apes, do not express the oligosaccharide α-Gal, due to inactivation of the *Ggta1* gene (or GalT gene) by single-base pair mutations [7], which encodes the enzyme α-1,3-galactosyltransferase. This evolutionary loss

José de la Fuente and Alejandro Cabezas-Cruz (eds.), *Allergic Disease Models*, Methods in Molecular Biology, vol. 3092, https://doi.org/10.1007/978-1-0716-5635-8_6,

(catastrophic-selection mechanism) means that the human immune system recognizes α-Gal as a foreign antigen. As a result, and, interestingly, humans naturally produce natural anti-α-Gal antibodies as part of the normal immune repertoire and conferring a protective activity against other pathogen infections [7]. In contrast to humans and described exceptions, all mammals, including rodents, have retained the expression of *Ggta1*, maintaining the activity of the enzyme α-1,3-galactosyltransferase and synthesizing α-Gal epitope without an immune response against it.

For the study of pathologies such as this one, humanized mouse models have emerged as a powerful preclinical platform. These models recapitulate in vivo key aspects of human-specific biological processes, including immune responses, infectious diseases, cancer, and hypersensitivity reactions, in a widely used animal model for research. The humanized mice model of AGS was first generated by Tearle et al. in 1996 by engineering on mouse with C57BL/6 background and named as α-1,3-galactosyltransferase knockout (KO) mouse (GalT$^{-/-}$) [8]. This homozygous knockout model was generated by targeting the catalytic domain of the enzyme, located in exon 9 of the mouse GalT gene, here in *Ggta1* gene (chromosome 4, GRCm39) (Fig. 1). Gene inactivation was performed by homologous

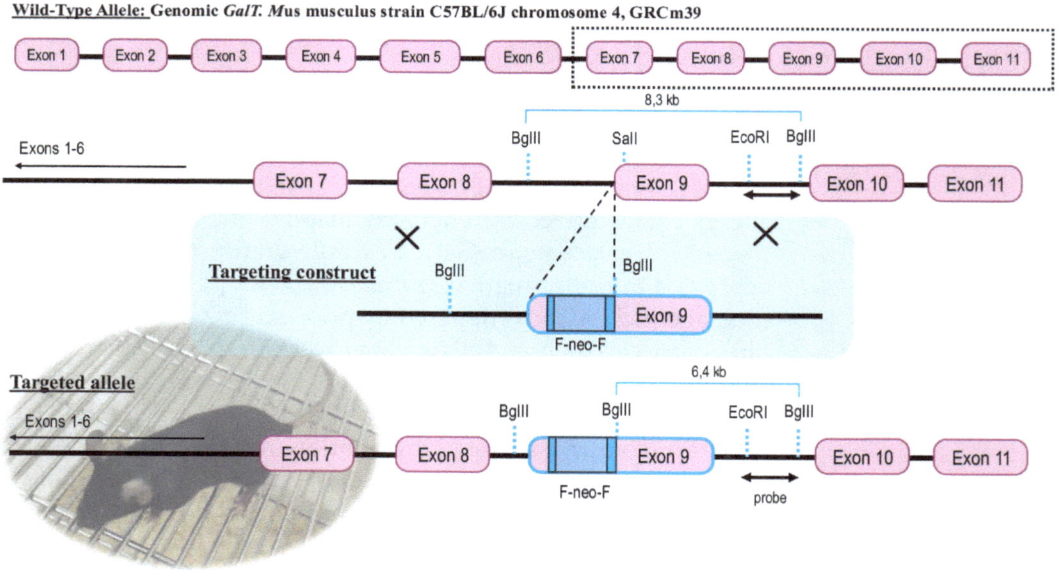

Fig. 1 Disruption of the mouse wild-type *GalT* gene (*Ggta1*) for GalT$^{-/-}$ or α1,3GalT-KO mouse model generation. Diagram shows the wild-type allele for GalT gene region with 11 exons, targeting construct and the targeted allele. The targeting construct was engineered for targeting the catalytic domain of the enzyme, located in exon 9. Gene inactivation was performed by homologous recombination using the neomycin resistance gene (neo^R) cassette as previously reported [8]

recombination using the neomycin resistance gene (neo^R) cassette. The targeting construct was electroporated into murine embryonic stem (ES) cells and resistant colonies were identified by Southern analysis based on the presence of a distinct 6.4 kb *BglII* fragment (Fig. 1). Targeted clones were injected into blastocysts to produce chimeric mice. Chimeric mice derived from two of these ES cell lines were subsequently bred to establish homozygous *Ggta1* knockout (GalT$^{-/-}$ or α1,3GalT-KO) mice [8].

The α1,3GalT-KO mouse model has enabled detailed investigation into the immunological mechanisms involved in sensitization and allergic response. Among multiple research studies, α1,3GalT-KO humanized mice have validated the induction of IgG and IgG anti-α -Gal antibodies in response to tick-feeding exposure or injected tick saliva [9] and after oral administration of mammalian meat in sensitized animals [10, 11]. Indeed, this preclinical model has made it possible to link the immune response to α-Gal sensitization in a Th2-type- and tick-species-dependent manner [12–14] and the impact of gut microbiota in the repertoire of natural antiglycan antibodies production [15].

Here we described a protocol for studying in vivo immune response to tick saliva, and/or other epitopes or potential allergens in both sensitized α1,3GalT-KO female and male mice. This method includes dietary interventions, subcutaneous injections, and blood-tail vein extractions.

2 Materials

2.1 Humanized α 1,3GalT-KO Mouse Model

Recommend using the humanized α1,3GalT-KO mouse model with C57BL/6 background engineered with α-1,3-galactosyltransferase knockout (KO) for inactivation of α-Gal synthesis as naturally occurs in humans [8]. For experimental protocol, information is disclosed in Sect. 3.2.

2.2 qRT-PCR of Ggta1 Gene

1. Midgut from 8 to 10-weeks-old C57BL/6 wild type and α 1,3GalT-KO mouse (female and/or male).
2. RNA isolation method or commercial kit.
3. Nanodrop or RNA and DNA quantification validated method.
4. Reverse transcription validated method.
5. Reagent solutions used for quantitative PCR (qPCR).
6. Specific primers for *Ggta1* and housekeeping gene expression analysis by qRT-PCR (*see* **Note 1**). For amplification of the *Ggta1* gene in the midgut (NM_001420265.1), the following primers were used in this study (5'–3'): *forward:* CCGCCCGGATGTTTTGACAG, *reverse:* GCAAACAC

TGTCAGCCCCAC. The glyceraldehyde 3-phosphate dehydrogenase gene (*Gapdh*) can be used as housekeeping gene to normalize expression levels of the target, using the following primers (5′–3′): *forward*: AGGGTGGAGCCAAAAGGGTC, *reverse*: GGGGCTAAGCAGTTGGTGGT (*see* **Note 2**).

7. Thermocycler for real-time PCR detection.

2.3 Dietary Intervention

1. Regular chow diet.
2. Porcine meat-containing diet.

2.4 Subcutaneous Administrations and Blood Extractions

All subcutaneous administrations should be dissolved in sterile PBS and injected into a final volume of 100 μL (*see* **Note 3**).

1. 30 G syringes.
2. Capillary tubes.
3. Scalpel or surgical scissors.
4. Optional: hemostatic powder.
5. Rodent hair clipper for laboratory use.
6. Optional: rodent restrainer/tubular holder.
7. Sterile phosphate buffer saline (PBS). If using commercial, dissolved PBS powered in distilled water according to manufacturer's instructions, adjust the pH if necessary (7.2–7.4) and sterilize filtering through a 0.22-μm filter into a sterile container (if components are heat-stable, autoclave PBS solution at 121°C for 15–20 min). If commercial PBS is not available, dissolve 8 g of NaCl, 0.2 g of KCl, 1.44 g of Na_2HPO_4, and 0.24 g of KH_2PO_4 in a beaker with 800 mL distilled water. Stir until completely dissolved, adjust pH (7.2–7.4) and the final volume to 1 L. Sterilized by autoclave or using a 0.22-μm filter. Store at room temperature until use.
8. Tick saliva subcutaneous administration: dissolved 25 μg of tick saliva protein extract (up to 10 μL) to a final volume of 100 μL sterile PBS.
9. Potential protein allergens: dissolve 10 μg of antigen to a final volume of 100 μL sterile PBS.
10. Nonprotein fraction (NPF) of tick saliva: dissolve 10 μL of saliva nonprotein fraction to a final volume of 100 μL sterile PBS.
11. Mixture of two or more protein allergens: dissolve equal amounts of each antigen (μg) in sterile PBS to a final volume of 100 μL. For instance, when preparing a mixture of two protein allergens, add 5 μg of each (10 μg) total in a final volume of 100 μL.

12. Mixture of a protein allergen and saliva non-protein fraction (NPF): dissolve 10 μg of antigen and 10 μL of NPF into a final volume of 100 μL of sterile PBS.
13. Control subcutaneous administration (PBS group): 100 μL sterile PBS.

3 Methods

Carry out all procedures at room temperature unless otherwise specified. All experimental procedures involving rodents must receive prior approval from the institution's Ethics Committee for Animal Experimentation.

3.1 Validation of α 1,3GalT-KO Mouse Model by qRT-PCR Analysis of Ggta1 Gene

If this is the first time your laboratory is using the α1,3GalT-KO mouse, we strongly recommend validating it beforehand corroborating the successful inactivation of this gene. In this report, we describe a protocol for analyzing *Ggta1* gene expression levels in the midgut by qRT-PCR; however, alternative methods such as Southern blotting using genomic DNA or flow cytometric analysis (FACS) detecting expression of the Gal epitope have also been reported for the same purpose [8].

1. Sacrifice male and/or female C57BL/6 wild-type (WT) and α 1,3GalT-KO mice under anesthetized conditions.
2. Carefully dissect midgut (or other targeted tissues if needed or desire) (*see* **Note 4**).
3. Isolate mRNA from both WT and α1,3GalT-KO midgut (or targeted tissue) and quantified it.
4. Perform reverse transcription of mRNA into cDNA following the manufacturer's protocol of the selected method (*see* **Note 5**).
5. Conduct quantitative PCR (qPCR) using 300 ng of synthesized cDNA following the standard amplification protocol routinely used in the laboratory. Employ the desired primers, ensuring their specificity.
6. Use the $2^{-\Delta\Delta Ct}$ method [16] to calibrate, normalize, and determine the relative expression levels in wild-type and α 1,3GalT-KO animals (Fig. 2).
7. Apply appropriate statistical tests to evaluate differences between experimental groups. For comparison between two groups, use a Student's *t*-test, considering $p < 0.05$ as statistically significant.

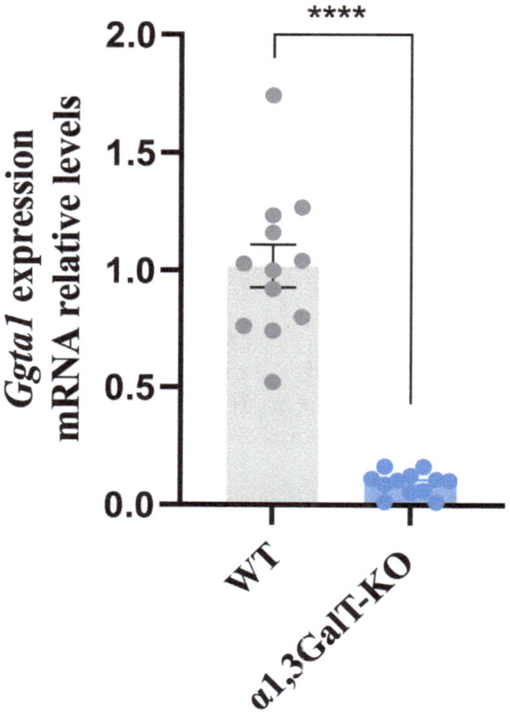

Fig. 2 Gene expression levels of *Ggta1* gene in the intestine of wild-type (WT) and α1,3GalT-KO of a group animals with a 1:1 female/male ratio ($n = 6$ per gender, $n = 12$ per group) "****" means p value < 0.0001 (T-test analysis)

3.2 Experimental Protocol for Studying Immune Response in α1,3GalT-KO Mice

Mice should be maintained under normal housing conditions and kept in ventilated, climate-controlled quarters at a temperature of 20–25 °C, humidity of 50%–55 %, and a 12-h light/dark cycle (8 a.m. to 8 p.m.). Both male and female could be maintained in social groups in the same cage separated by gender (*see* **Note 6** and Fig. 3 for graphical representation of the experiment).

1. Maintain animals under *ad libitum* conditions with a standard chow diet and provide free access to water.
2. Randomly assign animals to the required experimental groups at least 24 h prior to day 0 of the experiment. If sex is considered as a study variable, ensure a balanced 1:1 female-to-male ratio in their respective groups. Individually identify each animal using ear tags or tail markings through a marker to guarantee accurate group assignment and tracking.
3. Day 0, before any subcutaneous injection, proceed to blood extraction from the tail vein (less than 10% of total body weight) (*see* **Note 7**). Obtain serum by centrifugation at 3000 rpm for 10 min and store at −80°C until analysis.

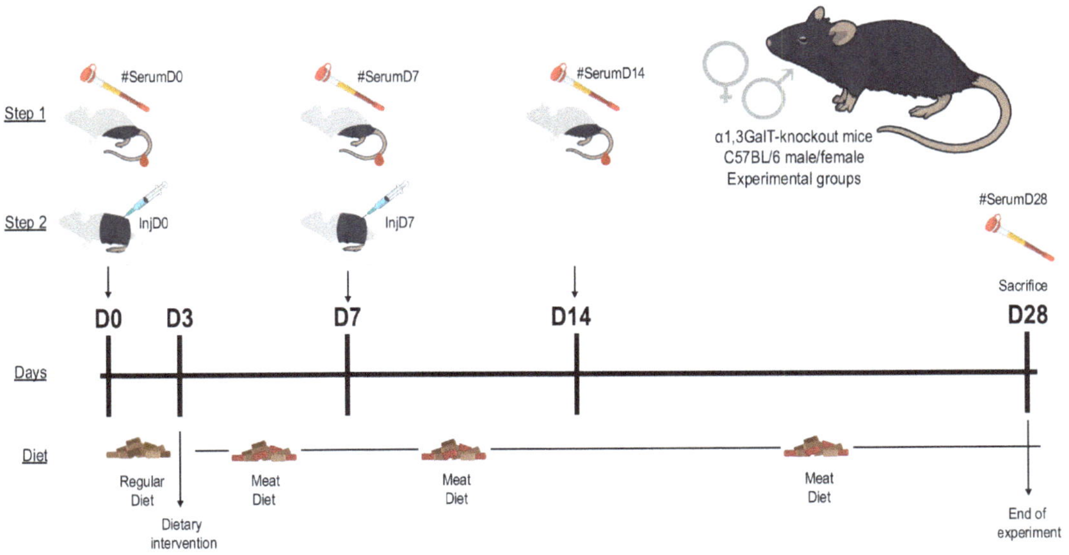

Fig. 3 Experimental protocol design proposed to investigate the immune response to α-Gal and other potential allergens in α1,3GalT-knockout mice. Step 1: blood extraction (serum) from the tail vein at day 0 (#SerumD0), day 7 (#SerumD7), day 14 (#SerumD14), and day 28 (sacrifice, #SerumD28). Step 2: subcutaneous injection of PBS or allergen at day 0 (InjD0) and day 7 (InjD7)

4. Day 0, first subcutaneous injection (InjD0): following blood extraction, sensitized animals by 100 μL of subcutaneous injections with a 30 G syringe of the potential allergens into a previously shaved flank (1 × 1 cm) (*see* **Notes 8** and **9**). One of the experimental groups must be a control, injected subcutaneously with PBS. Monitored animals until fully recovered of the subcutaneous injection.
5. 24 h after the first injection, monitor the flanks of the animals for any signs of allergic reactions or lesions, especially after days of injections. Evident signs of skin abnormalities including edema, erythema, scabbing, or abrasions should be recorded at the injection site systematically. From now on, flanks, body weight, and food intake should be monitored daily to ensure the health and well-being of the animals.
6. Day 3, dietary intervention: replace the regular chow diet with a porcine meat-containing diet.
7. Day 7, before subcutaneous injection, proceed to blood extraction as described in Step 3.
8. Day 7, second subcutaneous injection (following blood extraction, proceed to the second subcutaneous sensitization of experimental groups as described in Step 4.
9. Day 14, proceed to blood extraction as described in Step 3.

10. Day 28, proceed to blood and serum extraction and sacrifice animals under anesthetized conditions and store dissected tissues at −80 °C until use.
11. Measure serum levels of specific immunoglobulins and cytokines using ELISA or an alternative validated method of choice.

4 Notes

1. Specific primers for the targeted genes could be designed with the primer-BLAST online tool [17].
2. qPCR conditions for these specific primers comprise an initial denaturation step at 95 °C for 1 min, amplification by 45 cycles of 95 °C for 20 s and annealing varying between 58 and 63°C for 1 min, followed by a dissociation curve analysis.
3. We recommend preparing injections on the same day of subcutaneous administration to ensure protein stability.
4. For RNA or protein isolation, approximately 300 mg of tissue may be sufficient. However, most commercial RNA/protein isolation kits recommend using around 100 mg of tissue per preparation.
5. Sample could be standardized at a desired concentration and store at −20 °C until use.
6. On several occasions, male α1,3GalT-KO mice may exhibit aggressive behavior. We strongly recommend not housing males from different litters together and monitoring the animals daily for any signs of fighting or injury by an experience technician. If aggression occurs, the dominant mouse or mice should be separated immediately.
7. We highly recommend using capillary tubes for blood collection. If tail vein collection causes excessive stress to the animals, we suggest using a rodent restrainer. Additionally, we strongly advise applying hemostatic powder to the tail wound after blood collection to prevent undesirable bleeding.
8. Before performing the injection, the animal's flank should be shaved using a rodent-specific clipper. We recommend manually restraining the animal by an experienced individual or trained personnel.
9. The mouse's skin can get caught between the clipper blades; therefore, it is important to stretch the skin properly and shave the area carefully to avoid causing wounds that could interfere with the evaluation of allergic reactions.

Competing Interests

The author has no conflicts of interest to declare that are relevant to the content of this chapter.

References

1. Peterson CJ, Mohankumar P, Tarbox JA, Nugent K (2025) Alpha-Gal syndrome: a review for the general internist. Am J Med Sci 369:313–320
2. Young I, Prematunge C, Pussegoda K et al (2021) Tick exposures and alpha-gal syndrome: a systematic review of the evidence. Ticks Tick Borne Dis 12:101674
3. Çelebioğlu E, Akarsu A, Şahiner ÜM (2021) IgE-mediated food allergy throughout life. Turk J Med Sci 51:49–60
4. Román-Carrasco P, Lieder B, Somoza V et al (2019) Only α-Gal bound to lipids, but not to proteins, is transported across enterocytes as an IgE-reactive molecule that can induce effector cell activation. Allergy 74:1956–1968
5. Milano F, Masi A, Madaghiele M et al (2023) Current trends in gelatin-based drug delivery systems. Pharmaceutics 15:1499
6. Mullins RJ, James H, Platts-Mills TAE, Commins S (2012) Relationship between red meat allergy and sensitization to gelatin and galactose-α-1,3-galactose. J Allergy Clin Immunol 129:1334–1342.e1
7. Galili U (2019) Evolution in primates by "Catastrophic-selection" interplay between enveloped virus epidemics, mutated genes of enzymes synthesizing carbohydrate antigens, and natural anti-carbohydrate antibodies. Am J Phys Anthropol 168:352–363
8. Tearle RG, Tange MJ, Zannettino ZL et al (1996) The alpha-1,3-galactosyltransferase knockout mouse. Implications for xenotransplantation. Transplantation 61:13–19
9. Araujo RN, Franco PF, Rodrigues H et al (2016) Amblyomma sculptum tick saliva: α-Gal identification, antibody response and possible association with red meat allergy in Brazil. Int J Parasitol 46:213–220
10. Choudhary SK, Karim S, Iweala OI et al (2021) Tick salivary gland extract induces alpha-gal syndrome in alpha-gal deficient mice. Immun Inflamm Dis 9:984–990
11. Chandrasekhar JL, Cox KM, Loo WM et al (2019) Cutaneous exposure to clinically relevant Lone Star ticks promotes IgE production and hypersensitivity through CD4+ T Cell- and MyD88-dependent pathways in mice. J Immunol 203:813–824
12. Saunders MN, Rival CM, Mandal M et al (2024) Immunotherapy with biodegradable nanoparticles encapsulating the oligosaccharide galactose-alpha-1,3-galactose enhance immune tolerance against alpha-gal sensitization in a murine model of alpha-gal syndrome. Front Allergy 5:1437523
13. Sharma SR, Choudhary SK, Vorobiov J et al (2023) Tick bite-induced alpha-gal syndrome and immunologic responses in an alpha-gal deficient murine model. Front Immunol 14:1336883
14. Hils M, Hoffard N, Iuliano C et al (2024) IgE and anaphylaxis specific to the carbohydrate alpha-gal depend on IL-4. J Allergy Clin Immunol 153:1050–1062.e6
15. Bello-Gil D, Audebert C, Olivera-Ardid S, et al (2019) The formation of glycan-specific natural antibodies repertoire in GalT-KO mice is determined by gut microbiota. Front Immunol10.
16. Rao X, Huang X, Zhou Z, Lin X (2013) An improvement of the 2ˆ(–delta delta CT) method for quantitative real-time polymerase chain reaction data analysis. Biostat Bioinforma Biomath 3:71–85
17. Ye J, Coulouris G, Zaretskaya I et al (2012) Primer-BLAST: a tool to design target-specific primers for polymerase chain reaction. BMC Bioinformatics 13: 134

Check for updates

Chapter 7

Proteomics Analysis for Discovery of Biomarker of Tick Allergy

Margarita Villar, Sara Artigas-Jerónimo and José de la Fuente

Abstract

The alpha-Gal syndrome (AGS) is an allergic disease with IgE-mediated delayed hypersensitivity to the oligosaccharide galactose-α-1,3-galactose (α-Gal) present in tick saliva and mammalian-derived foods and drugs. The incidence of AGS is increasing worldwide and thus requires the identification of serum biomarkers and tick biomolecules for the development of preventive, treatment, and control interventions. In this chapter, we describe step-by-step methods for proteomics analysis of human serum samples for AGS-related biomarker identification and characterization, characterization of molecular pathways involved in tick-induced allergic reactions, and advance in diagnostics, prognostics, and therapeutic strategies in this emerging and clinically relevant field.

Keywords Allergy, Alpha-Gal syndrome, α-Gal, Biomarker, Proteomics, Serum, Tick

1 Introduction

Ticks are obligate hematophagous ectoparasites that transmit pathogens, which cause diseases in humans and animals with growing incidence worldwide [1]. Beyond their role in pathogen transmission, tick saliva contains a complex mixture of bioactive molecules that modulate host hemostasis and immune responses, enabling prolonged blood feeding. These salivary compounds not only facilitate pathogen survival but also directly shape host immunity, creating a unique and beneficial context for tick bites [2].

In the early 2010s, the discovery of how ticks affect human health through allergic reactions associated with IgE antibodies against the glycan galactose-alpha-1-3-galactose (α-Gal) triggered by tick bites and mammalian products (known as α-Gal syndrome; AGS) [3, 4] opened a new pathway of disease incidence by these arthropods.

Despite growing awareness of tick-induced allergies, the molecular mechanisms driving these reactions remain incompletely understood, and standardized methodologies for biomarker discovery

José de la Fuente and Alejandro Cabezas-Cruz (eds.), *Allergic Disease Models*, Methods in Molecular Biology, vol. 3092, https://doi.org/10.1007/978-1-0716-5635-8_7,

are still scarce. To develop and improve diagnosis, prognosis, and treatment of tick-involved diseases in both humans and animals, omics technologies are established as very powerful tools to snap a concrete physiological condition in a holistic manner, and they have transformed biological research in the last three decades. Besides, omics technologies allow us to identify and characterize biomarkers of diseases in response to many physiological situations and treatments.

Although proteomics was one of the last original omics technologies to appear, it has turned out to be one of the most powerful ones because it sheds more information about biological processes and functional relevance than its sister approaches [5] including the result of translational processes, post-translational modifications, and protein folding, all features with a high relevance to the functionality of the proteome [6].

In this chapter, we describe step-by-step methods for proteomic analysis in human samples for biomarker identification and characterization related to tick allergy. By providing a rigorous, detailed framework for proteomic studies, this chapter addresses a critical methodological gap, equipping researchers with the tools to uncover novel biomarkers, clarify molecular pathways involved in tick-induced allergic reactions, and ultimately advance diagnostics, prognostics, and therapeutic strategies in this emerging and clinically relevant field.

2 Materials

2.1 Reagents, Consumables, Kits, Equipment, and Software

1. All reagents used for these protocols should be of HPLC–MS grade. The solutions are prepared with HPLC–MS grade water and stored at 4 °C, except for the solutions containing sodium dodecyl sulfate (SDS) that are stored at room temperature (RT) to avoid precipitation. All solutions must be prepared using glassware, which should never be washed with detergent, as this may leave residues that interfere with the subsequent proper digestion of the proteome.

2.2 Buffers

1. Buffer of extraction: 50 mM Hepes-KOH (pH 7.4), 150 mM NaCl, 2 mM EDTA, 1% Triton X-100 with proteases inhibitors (1 mM PSMF, 2 mM NAF, 10 μg/μL Leupeptin, and 2 mM Na_3VO_4). Add 5.96 g of HEPES free acid, 4.38 g of NaCl, and 0.37 g of disodium EDTA dihydrate to 400 mL of ultrapure water in a clean glass beaker and stir until dissolved. Adjust the pH to 7.4 with KOH. Add 5 mL of Triton X-100 and mix thoroughly. Bring the volume to 500 mL with ultrapure water. Immediately before use, add 5 mL of 100 mM PMSF stock solution, 1 mL of 1 M NaF stock, 500 μL of 10 mg/mL Leupeptin stock, and 5 mL of 200 mM Na_3VO_4 stock. Mix gently and keep the buffer on ice during use.

2. Two percent (2%) SDS: add 2 g of sodium dodecyl sulfate to 100 mL of HPLC-grade H_2O.
3. Resolving buffer: 1.5 M Tris-HCl, pH 8.8. Weigh 181.7 g of Tris, add water to a volume of 900 mL, adjust to pH 8.8 with HCl, and make up to 1 L.
4. Stacking buffer: 1.5 M Tris-HCl, pH 6.5. Weigh 181.7 g of Tris, add water to a volume of 900 mL, adjust to pH 6.5 with HCl, and make up to 1 L.
5. Tris-glycine SDS-PAGE buffer (10×): 0.25 M Tris-HCl, pH 8.3, 1.92 M glycine, and 1% SDS. This buffer is diluted ten times with water before use.
6. Phosphate-buffered saline 1×, pH 7.4: add 4 g of NaCl, 0.1 g of KCl, 0.72 g of Na_2HPO_4, and 0.122 g of KH_2PO_4 to 450 mL of HPLC-grade water, mix, and adjust pH to 7.4. Complete to 500 mL with water.
7. 2% pilocarpine. Add 2 g of pilocarpine to 80 mL of 1× phosphate-buffered saline (PBS) and mix. Complete to 100 mL with PBS.
8. Tris-buffered saline (TBS) buffer, pH 7.5: add 6.05 g of Tris-HCl and 8.76 g of NaCl in 800 mL of milliQ water and mix. Adjust pH to 7.5. Complete to 1000 mL with milliQ water.
9. 1% bovine serum albumin (BSA). Add 0.5 g of BSA in 45 mL of TBS buffer and mix. Complete to 50 mL with TBS buffer.
10. Ammonium bicarbonate, 50 mM, pH 8.8. Add 0.04 g of ammonium bicarbonate to 9 mL of LC–MS grade water, mix, and adjust pH to 8.8 with 5 N ammonium hydroxide. Complete to 10 mL with water to obtain a 50 mM final solution.
11. 10 mM dithiothreitol (DTT): add 1.54 mg of DTT to 1 mL of 50 mM ammonium bicarbonate, pH 8.8.
12. 50 mM iodoacetamide: add 9.3 mg of iodoacetamide to 1 mL of 50 mM ammonium bicarbonate, pH 8.8.

2.3 Samples

1. Samples may be suitable for use, including tissue, cell cultures, serum, and even culture medium. Once the samples are collected, proper preservation must be ensured by freezing and storing them at −80°C until proteome isolation (*see* **Note 1**).

2.4 SDS-PAGE Gel Components

1. Resolving gel (12%): mix 2.5 mL of 1.5 M Tris-HCl, pH 8.8, 4 mL of 30% acrylamide/bis-acrylamide solution (Bio-Rad, Hercules, CA, USA), 100 μL of 10% SDS, and 3.4 mL of water and mix well. Add 100 μL of fresh 10% ammonium persulfate and 5 μL of TEMED and cast gel.
2. Make 1.5-mm-thick gels with wells of 5 mm wide for conventional one-dimensional gels and 1.2 cm wide for proteome band concentration.

3 Methods

3.1 General Considerations

Experimental procedures for the analysis of protein differential production for biomarkers identification from control and allergic patients' sera, and the identification of human disease biomarker candidates using 2D gel proteome analysis in tick salivary glands and saliva are represented in Figs. 1 and 2, respectively.

1. It is crucial to emphasize the importance of maintaining strict cleanliness and maximizing sterility during sample handling. This includes always wearing gloves, a lab coat, and keeping hair tied back to prevent contamination of the samples with keratins from the handler's skin and hair. Potential contamination could result in a high detection of these keratins, potentially masking experimental results.
2. There are several approaches to digest the proteome into peptides for subsequent proteomic analysis, depending on the proteins of interest in each experiment, as well as different types of analyses based on the final goal of the study. The in-gel digestion method allows the digestion and recovery of all proteins present in the proteome, regardless of their molecular weight, thereby enabling a comprehensive analysis of the proteome. Alternative, faster methods can optimize digestion time and effort. However, they involve filtration steps that prevent the recovery of the entire proteome, leading to the loss of

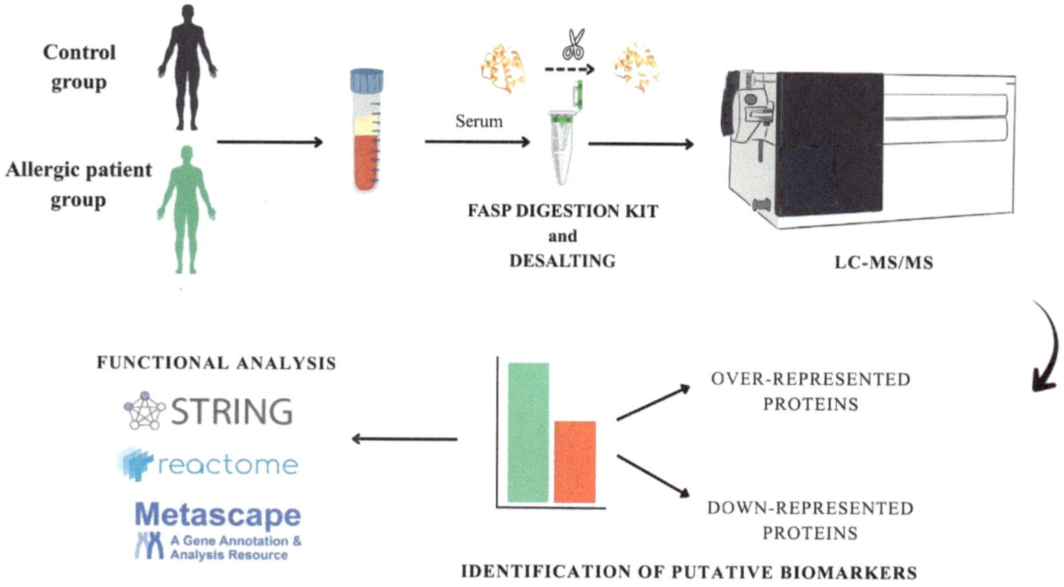

Fig. 1 Schematic representation I. Procedures for the analysis of protein differential production for biomarker identification from control and allergic patients' sera

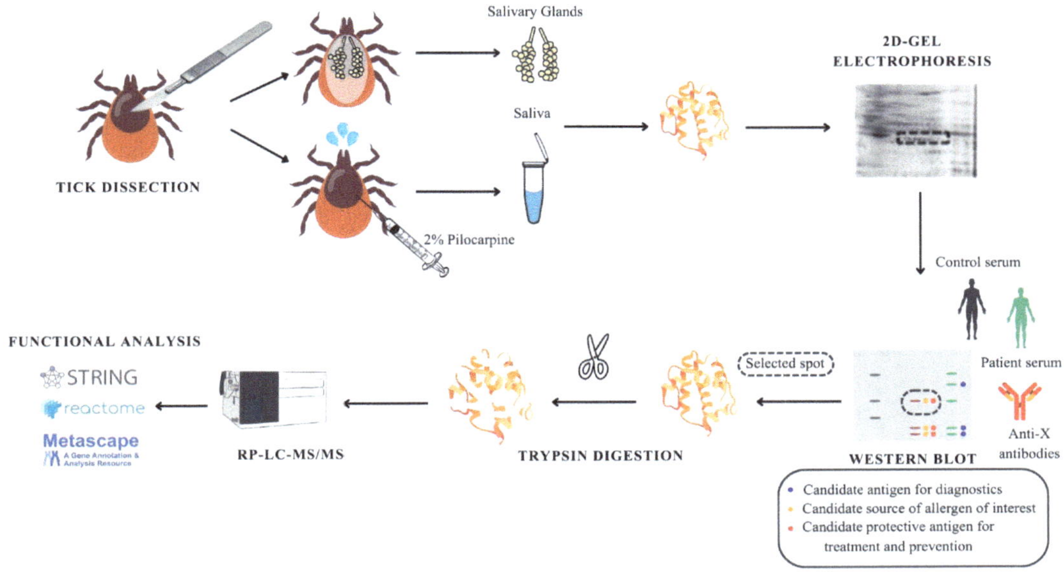

Fig. 2 Schematic representation II. Procedure for the identification of human disease biomarker candidates using 2D gel proteome analysis in tick salivary glands and saliva

low-molecular-weight proteins or peptides, as is the case with the FASP Protein Digestion kit. Both methods have been optimized as described in the following sections.

3.2 Protocols for Tick Allergy's Biomarker Discovery

1. Different methodological approaches could be applied to biomarker discovery through proteomic analysis, but these two methodologies have been optimized depending on the final goal and initial samples. Protein differential expression analysis for biomarker identification from control and allergic patients' serum. The final goal of this protocol is to be able to perform a differential expression analysis of the proteins from the serum of control and tick-allergic individuals, leading to the identification of those differentially expressed proteins between treatments that represent putative disease biomarkers. For this protocol, we present step-by-step the FASP protein digestion method, taking its limitation mentioned above into consideration.

3.3 Human Serum Samples

1. Serum samples should be collected following the corresponding Helsinki Declaration and the approval of the Ethical and Scientific Committees.
2. Blood from healthy and allergic patients is obtained from peripheral blood and maintained in a standing position in a sterile tube without anticoagulant at room temperature (RT) for clotting (20–30 min).
3. Centrifuge at 1500 x*g* for 20 min at RT.

4. Collect the human sera and conserve them at −20°C until used for analysis.
5. Protein quantification should be performed to evaluate the protein concentration in each sample. Several methods can be addressed, such as Bradford method or commercial kits such as Pierce BCA (Thermo Fisher Scientific) and RC-DC (Bio-Rad Laboratories, Inc.) quantification kits, following the manufacturer's recommendations (*see* **Note 1**).

3.4 FASP Protein Digestion and Peptide Desalting

1. Prepare your samples at a maximum concentration of 400 μg of protein in up to 30 μL of 10 mM PBS with 2% SDS (*see* **Note 2**).
2. Add dithiothreitol (DTT) reagent in all samples at a final concentration of 50 mM.
3. Incubate at RT for 10 min with 900 rpm shaking.
4. Boil the samples for 3 min.
5. Continue with the protocol following the manufacturer's recommendations.
6. After protein digestion and dilution from the filter, resultant peptides should be dried using a vacuum.
7. Once peptides are completely dried, peptides should be desalted using OMIX pipette tips C18 following the manufacturer's instructions, vacuum-dried again, and stored at −20°C until mass spectrometry analysis.

3.5 Label-Free Quantitation by Sequential Window Acquisition of All Theoretical Mass Spectra (SWATH-MS)

1. Resuspend the protein-digested peptides in 2% acetonitrile and 5% acetic acid in water (final concentration of 2 μg/μL) and analyze by reverse-phase liquid chromatography coupled online to mass spectrometry (RP-LC–MS/MS) in an nLC 415 system coupled with a 6600 TripleTOF mass spectrometer through Information-Dependent Acquisition (IDA) followed by Sequential Windowed data-independent Acquisition of the Total High-Resolution Mass Spectra (SWATH-MS).
2. Concentrate peptides (online) by reverse-phase chromatography using a 0.1 ×20 mm C18 RP precolumn and then separate peptides using a 0.075×250 mm C18 RP column operating at 300 nL/min.
3. Perform peptide elution using a 60-min gradient from 5% to 30% solvent B followed by a 15-min gradient from 30% to 60% solvent B (solvent A, 0.1% formic acid in water; solvent B, 0.1% formic acid in acetonitrile) and inject directly into the mass spectrometer for analysis.
4. For IDA experiments, generate a representative mixed sample by joining together an equal amount of all samples and use it for the generation of the reference spectral ion library as part of

SWATH-MS analysis. A total amount of 4 μg is recommended to inject in duplicate.

5. Set the mass spectrometer to scanning full spectra from 350 to 1400 m/z (250 ms accumulation time) followed by up to 50 MS/MS scans (100–1500 m/z). Isolate for fragmentation candidate ions with a charge state between +2 and +5 and counts per second above a minimum threshold of 100. One MS/MS spectrum was collected for 100 ms, add those precursor ions to the exclusion list during 15 s. Turn off dynamic background subtraction and acquire data in high-sensitivity mode with rolling collision energy on and a collision energy spread of 5.
6. For SWATH quantitative analysis, subject independent samples (4 μg each) to the cyclic Data Independent Acquisition (DIA) of mass spectra using the SWATH variable windows calculator and a SWATH acquisition method editor.
7. Construct a set of 50 overlapping windows (1 m/z for window overlap), covering the precursor mass range of 400–1250 m/z based on data from the IDA runs previously acquired. For these experiments, acquire a 50-ms survey scan (350–1400 m/z) at the beginning of each cycle, and collect SWATH-MS/MS spectra from 100 to 1500 m/z during 70 ms at high sensitivity mode.

3.6 Proteins Differential Expression Analysis for Biomarkers Identification

1. Create a spectral library of all detectable peptides in the samples, combining the IDA MS raw files and subject them to database search in unison using ProteinPilot software with the Paragon algorithm.
2. Perform spectra identification by searching against the corresponding Uniprot sample type proteome database (www.uniprot.org) with the following parameters: iodoacetamide cysteine alkylation, trypsin digestion, identification focus on biological modification, and thorough ID as search effort and selecting the detected protein threshold at $p = 0.05$.
3. To assess the quality of identifications, use the independent False Discovery Rate (FDR) analysis with the target-decoy approach provided by ProteinPilot considering positive identifications when identified proteins reach a 1% global FDR.
4. For SWATH processing, perform the spectral alignment and targeted data extraction of DIA samples using the SWATH Acquisition MicroApp2.0 in the PeakView v. 2.2 software (ABSciex) with the reference spectral library. The parameters recommended are up to ten peptides per protein, seven transitions per peptide, 15 ppm ion library tolerance, 5-min XIC extraction window, 0.01 Da XIC width, and considering only peptides with at least 99% confidence and excluding those which were shared or contained modifications.

5. To ensure reliable quantitation, only proteins with three or more peptides available for quantitation can be selected for XIC peak area extraction and exported for analysis, for example, MarkerView v. 1.3 software (ABSciex).
6. Additionally, perform global normalization to the total area sums of all detected proteins in the samples.
7. Once proteins have been identified in all samples with a relative intensity-quantification value, differential expression analysis can be performed to identify the differential represented proteins across treatments for biomarker identification.
8. The *z*-score transformation is a commonly used normalization strategy in quantitative proteomics to render protein abundance profiles comparable across samples and conditions. It consists of centering and scaling the intensity values of each protein using the formula $z = (x - \mu)/s$ where x is the observed intensity, μ the mean, and σ the standard deviation of the values (*see* **Note 3**).
9. Differential expression analysis can be performed using the free Perseus Software (Max–Planck Institute) (https://maxquant.net/perseus/) which is a user-friendly tool to analyze omics data (*see* **Note 4**).
10. Once the software is installed, the dataset is imported in the appropriate format with values already transformed into *z*-scores (*see* **Note 5**).
11. The first step is to annotate the rows categorically and assign appropriate names to the columns.
12. Next, the most suitable statistical analysis for the dataset is selected. The software offers a wide range of statistical tests tailored to the characteristics of experimental design. To identify significantly expressed proteins after statistical testing, it is essential to apply filtering based on significance using the "Filter rows based on categorical column" option.
13. Finally, the software provides tools for generating a variety of representative visualizations of the differential analysis, such as hierarchical clustering, volcano plots, among others.
14. These analyses allow us to explore our proteomic results, identifying the most up- and down- represented ($\log_2$ fold change < 0) proteins in response to allergic reactions due to tick bites in humans and, thus, identifying putative biomarkers of allergic diseases and reactions through our raw data.

3.7 Identification of Human Disease Biomarker Candidates Using 2D Gel Proteome Analysis in Tick Salivary Glands and Saliva

The methodology for the identification of human disease biomarker candidates using 2D gel proteome analysis in tick salivary glands and saliva has been validated and published in several articles, mostly leading to the identification and characterization of putative AGS disease biomarkers [e.g., [7, 8]], but it can be performed for biomarker discovery in other tick-associated allergies and in other arthropod-borne allergies.

3.8 Dissection of Tick Salivary Glands and Proteome Extraction

1. Salivary glands should be dissected from ticks using a size 11 scalpel and fine-tipped forceps.
2. For protein extraction, salivary glands are homogenized with a 27 G needle in extraction buffer until complete lysis of the tissue (*see* **Note 6**).
3. Perform two to three sonication cycles if necessary for total sample lysis.
4. Centrifuge the lysates at 10,000 rpm for 10 min at 4 °C.
5. Collect the supernatant containing the extracted proteome and transfer it into a clean Eppendorf tube. Store at −80 °C until protein quantification or proteome digestion.
6. Protein quantification should be performed to evaluate the protein concentration in each sample using methods mentioned above, such as BCA or RC-DC quantification kits.

3.9 Tick Saliva Extraction (see Note 7)

1. Wash female ticks with PBS and place them dorsal side down on double-sided tape on a glass slide.
2. Induce salivation by injecting 1–3 μL of 2% pilocarpine hydrochloride (Sigma-Aldrich, St. Louis, MO, USA) in PBS, pH 7.4, on the ventral side adjacent to the fourth leg coxa using a 34 G angle-beveled needle on a model 701 Hamilton syringe. 2. Collect saliva every 15–30 min using a Hamilton syringe for approximately 4 h at RT and store at −80°C.
3. Evaluate protein concentration as described in previous sections.

3.10 Salivary Glands and Saliva Proteomes Analysis Using 2D Western Blot

1. Precipitate 200 μg of salivary gland protein extract and saliva with the 2D-Clean up kit (GE Healthcare, Little Chalfont, UK) following the manufacturer's instructions.
2. Resuspend the precipitate in 125 μL of DeStreak Rehydration Solution supplemented with 0.5% of IPG buffer 3-11NL pH range.
3. Load resuspended proteins onto IPG strips (pH 3-11 NL Drystrip 7 cm).

4. Perform the isoelectrofocusing at 20 °C in an Ettan IPGphor 3 with the following conditions: 30 V for 3 h, 60 V for 3 h, 60–300 V for 30 min, 300 V for 30 min, 300–1000 V for 1 h, 1000 V for 1 h, 1000–5000 V for 5 h and 5000 V for 5 h.
5. Reduce and alkylate focused proteins by successive incubations in 50 mM Tris-HCl pH 8.8, 6 M urea, 30% v/v glycerol, 2% w/v SDS, 0.2% bromophenol blue, supplemented with 0.5% w/v dithiothreitol (DTT) for the first incubation or supplemented with 4.5% w/v iodoacetamide for the second incubation, during 15 min each with gentle rocking.
6. For the second dimension, place the IPG strips onto homogeneous 12% SDS polyacrylamide gels and conduct electrophoresis for 1 h at 120 V.
7. Transfer the gel/gels to nitrocellulose. One gel should be stained with SYPRO Ruby Protein Gel Stain.
8. Block membranes with 1% BSA (Sigma) in Tris-buffered saline (TBS).
9. Wash three times with TBS buffer for 5 min with shaking.
10. Incubate them overnight at 4 °C with healthy control or allergic patients' serum (dilution 1:200). Also, incubate with specific primary antibody against the allergen of interest.
11. Wash three times with TBS buffer for 5 min with shaking.
12. Next morning, incubate membranes with goal anti-human IgE-conjugated peroxidase (dilution 1:1000) at RT for 1.5 h with shaking.
13. Wash six times with TBS buffer for 5 min with shaking.
14. Visualize immunoreactive proteins with TMB reagent.
15. Protein spots recognized by patients and control sera and by anti-allergen of interest-specific antibodies should be manually excised from the stained gel.
16. Destain gel pieces at RT, alternating 50 mM ammonium bicarbonate/acetonitrile (1:1, v/v) with neat acetonitrile, incubating for 10 min at room temperature with occasional vertexing until the gel pieces become white and shrink.
17. Once gel spots are white, add pure acetonitrile for 5 min, discard, and add 10 mM DTT in 50 mM ammonium bicarbonate for 25 min at 60 °C with 500 rpm shaking. Discard supernatant and add pure acetonitrile for 5 min at RT and after that, discard supernatant and add 50 μL of 50 mM iodoacetamide in 50 mM ammonium bicarbonate for 30 min at RT in the dark with 500 rpm shaking.
18. Discard supernatant and perform a 5-min wash at 500 rpm shaking with pure acetonitrile, 50 mM ammonium bicarbonate, and pure acetonitrile prior to vacuum dry gel pieces.

19. Cover gel spots with sequencing-grade trypsin in 50 mM ammonium bicarbonate, pH 8.8 at a 7.5:1 protein/trypsin (w/w) ratio and leave it in an ice bucket for 1 h to saturate them with trypsin. Digest overnight at 37 °C (*see* **Note 8**).
20. Add trifluoroacetic acid to a final concentration of 1% to stop the digestion.
21. Desalt the samples using OMIX pipette tips C18 following the manufacturer's instructions, vacuum dry, and store at −20°C until the mass spectrometry analysis.

3.11 Mass Spectrometry Analysis

1. Perform Steps 1–5 as previously described in Sect. 3.5.
2. For spectra identification, search the IDA MS raw files against the corresponding Uniprot proteome database using ProteinPilot software with the Paragon algorithm with the following recommended search parameters: iodoacetamide cysteine alkylation, trypsin digestion and gel-based ID as a special factor, and selecting the detected protein threshold at $p = 0.05$.
3. To assess the quality of identifications, use the independent FDR analysis with the target-decoy approach provided by ProteinPilot considering positive identifications when identified proteins reach a 1% global FDR.

3.12 Identification of Putative Disease Biomarkers

1. When analyzing the reactive tick proteins against control and patient sera and specific antibodies for the allergen of interest, it is important to compare treatment-specific western blot between them.
2. Tick proteins labeled with patient but not control sera could be used for diagnostics of a predisposing condition for allergic disease to the allergen of interest.
3. Additionally, tick proteins labeled with specific antibodies for the allergen of interest may be selected as candidate protective antigens for the treatment and prevention of allergic reactions together with other tick-borne diseases.
4. Besides, tick proteins labeled with specific antibodies for the allergen of interest enable us to identify the putative candidates' source of this allergen that may trigger the anaphylactic reactions.

3.13 Biomarkers Functional Characterization

1. Currently, there are many free and user-friendly software for ontology characterization of proteins such as Metascape, Reactome, STRING, among others, and OmicBox that allow us to identify and characterize the biological processes in which proteins of interest are involved and the physical and/or functional protein–protein interactions. This analysis is very useful to provide a functional meaning to a set of proteins identified

through proteomics analysis and to get a very good snap of the biological and functional situation of our experimental design.

2. If you have a sufficiently large and well-curated set of differentially expressed proteins, we recommend using Metascape (https://metascape.org/) due to its ease of use and the broad range of options it offers [9].
3. Enter the identifiers of the proteins of interest (e.g., Gene ID, UniProt ID, symbol) in the designated field. You can also drop your own file in different formats (*see* **Note 9**).
4. Explore different types of analysis such as "Express Analysis."
5. The software will provide you with all the results related to your proteins of interest.

4 Notes

1. If the samples are derived from serum, prior proteome isolation is not required for proteomic analysis. However, accurate quantification of the proteins present in the samples is essential in order to digest an approximately equivalent amount from all samples.
2. When using serum samples, we recommend digesting 200 μg of protein in a 20 μL volume of water.
3. The *z*-score transformation is one of several normalization methods available, and it is the approach selected for this protocol; however, alternative methods can also be applied depending on the experimental design and analytical objectives.
4. There are multiple software tools available for the analysis of omics data, including proteomics, as well as for performing differential analyses. For this protocol, we selected Perseus because it is free of charge and offers a very user-friendly interface.
5. Another option is to import the dataset with raw intensity-relative quantification values and perform the data transformation directly within the software.
6. If only a limited amount of sample is available and the simultaneous isolation of DNA or RNA is required for complementary genomics or transcriptomics analyses, the commercial AllPrep DNA/RNA/Protein Mini Kit (Qiagen, Valencia, CA, USA) may be used for proteome isolation following the manufacturer's instructions.
7. Several female ticks are needed to obtain enough saliva for protein analysis. Furthermore, to achieve better salivation

induction, we recommend collecting saliva from medium-fed female ticks.

8. It is recommended to use 50–75 μL of trypsin solution for small gel pieces and 75–100 μL for larger ones. To preserve the integrity of the digestion, seal the tubes with Parafilm after adding the enzyme solution.
9. To explore the ontology of your proteins of interest, including the biological processes and the metabolic mechanisms, it is important to decide which experimental planning you are going to follow. Our recommendation is that you may first include the overrepresented proteins and then the under-represented proteins in response to the allergic process. In this manner, you will gain insight into the biological processes underlying your experimental condition, as well as the interactions and coordinated actions of your proteins of interest.

Competing Interests

The authors have no conflicts of interest to declare that are relevant to the content of this chapter.

References

1. Artigas-Jerónimo S, De La Fuente J, Villar M (2018) Interactomics and tick vaccine development: new directions for the control of tick-borne diseases. Exp Rev Proteom 15(8):627–635. https://doi.org/10.1080/14789450.2018.1506701
2. Černý J, Arora G (2024) Proteases and protease inhibitors in saliva of hard ticks: Biological role and pharmacological potential. Adv Parasitol 126:229–251. https://doi.org/10.1016/bs.apar.2024.09.001
3. Steinke JW, Platts-Mills TA, Commins SP (2015) The alpha-gal story: lessons learned from connecting the dots. J Allergy Clin Immunol 135(3):589–597. https://doi.org/10.1016/j.jaci.2014.12.1947
4. Vaz-Rodrigues R, Mazuecos L, Villar M, Contreras M, González-García A, Bonini P, Scimeca RC, Mulenga A, de la Fuente J (2024) Tick salivary proteome and lipidome with low glycan content correlate with allergic type reactions in the zebrafish model. Int J Parasitol 54(12):649–659. https://doi.org/10.1016/j.ijpara.2024.07.002
5. Kahn P (1995) From genome to proteome: looking at a cell's proteins. Science (New York, N.Y.) 270(5235):369–370. https://doi.org/10.1126/science.270.5235.369
6. Hasin Y, Seldin M, Lusis A (2017) Multi-omics approaches to disease. Genome Biol 18(1):83. https://doi.org/10.1186/s13059-017-1215-1
7. Cabezas-Cruz A, Mateos-Hernández L, Pérez-Cruz M, Valdés JJ, Mera IGF, Villar M, de la Fuente J (2015) Regulation of the Immune Response to α-Gal and Vector-borne Diseases. Trends Parasitol 31(10):470–476. https://doi.org/10.1016/j.pt.2015.06.016
8. Villar M, Pacheco I, Mateos-Hernández L, Cabezas-Cruz A, Tabor AE, Rodríguez-Valle M, Mulenga A, Kocan KM, Blouin EF, de La Fuente J (2021) Characterization of tick salivary gland and saliva alphagalactome reveals candidate alpha-gal syndrome disease biomarkers. Expert Rev Proteom 18(12):1099–1116. https://doi.org/10.1080/14789450.2021.2018305
9. Zhou Y, Zhou B, Pache L, Chang M, Khodabakhshi AH, Tanaseichuk O, Benner C, Chanda SK (2019) Metascape provides a biologist-oriented resource for the analysis of systems-level datasets. Nat Commun 10(1):1523. https://doi.org/10.1038/s41467-019-09234-6

Check for updates

Chapter 8

Analysis of Metabolomics Response to Tick Saliva in Human Cells

Sara Samino, Rita Vaz-Rodrigues, Almudena González-García, José de la Fuente and Núria Amigó

Abstract

This chapter describes a workflow for the analysis of the metabolomic response of the human basophilic cell line KU812 exposed to *Ixodes ricinus* tick saliva using ^{1}H-nuclear magnetic resonance (NMR) spectroscopy. The protocol includes procedures for cell culture, harvesting, washing, and metabolite extraction, followed by the separation of aqueous and lipid fractions. Detailed instructions for sample preparation and NMR data acquisition are provided for both fractions. Spectral processing and metabolite quantification are performed using targeted profiling approaches based on constrained line-shape fitting, enabling accurate analysis of complex biological samples. Metabolite identification is supported by spectral databases and literature references, and absolute concentrations are obtained using internal standard–based quantification methods. Finally, data normalization based on cell number ensures comparability across samples. Overall, this protocol provides a robust and reproducible strategy for investigating metabolic alterations induced by tick saliva in human cells.

Keywords Tick saliva, Allergy, Metabolomics, ^{1}H-NMR spectroscopy, KU812 cells, Targeted profiling, Spectral deconvolution, Lipidomics, Metabolic response

1 Introduction

Ticks are hematophagous arthropods capable of modulating host physiological and immune responses through the secretion of a complex mixture of bioactive molecules in their saliva [1]. These molecules facilitate blood feeding and contribute to pathogen transmission, while also inducing a wide range of cellular responses in the host [2]. Among these responses are immune sensitization and shifts toward allergic phenotypes, processes that position ticks as modulators of allergy development, including in conditions such as alpha-Gal syndrome (AGS) [3, 4]. Understanding the molecular mechanisms underlying tick–host interactions is therefore essential for elucidating vector biology and revealing how tick saliva modulates host immunity and allergic responses [5, 6].

José de la Fuente and Alejandro Cabezas-Cruz (eds.), *Allergic Disease Models*, Methods in Molecular Biology, vol. 3092,
https://doi.org/10.1007/978-1-0716-5635-8_8,

In recent years, increasing attention has been directed toward the metabolic alterations induced in host cells upon exposure to tick saliva [7], particularly in the context of allergic sensitization and AGS. Metabolomics provides a powerful approach to capture these changes, offering a comprehensive snapshot of cellular biochemical activity and enabling the identification of metabolic pathways involved in host responses. In particular, the study of metabolomic responses in human cells can provide insights into allergy, inflammation, immune modulation, and cellular adaptation mechanisms triggered by tick saliva [1, 5, 8].

Nuclear magnetic resonance (NMR) spectroscopy represents a robust and highly reproducible platform for metabolomics analysis [9, 10]. Compared to other analytical techniques, NMR offers several advantages, including minimal sample preparation, nondestructive analysis, and high quantitative accuracy [11]. Moreover, NMR enables the simultaneous detection of a broad range of metabolites, making it particularly suitable for studying complex biological systems such as host–vector interactions [12, 13].

NMR-based metabolomics can be applied to both aqueous and lipid fractions of biological samples, providing complementary information on cellular metabolism [11]. While aqueous extracts allow the characterization of polar metabolites involved in central metabolic pathways, lipid extracts provide insights into membrane composition, energy storage, and signaling processes [12, 14]. The combined analysis of these fractions enables a more comprehensive understanding of the metabolic response of human cells to external stimuli such as tick saliva [1] or monitoring of health-related interventions [15, 16].

Accurate metabolite quantification in NMR metabolomics relies on the proportionality between signal intensity and metabolite concentration, which allows the direct conversion of spectral data into quantitative information when appropriate internal standards are used. Targeted profiling approaches based on spectral deconvolution further enhance the ability to resolve overlapping signals and quantify metabolites in complex biological samples.

In this chapter, we describe a workflow for analyzing the metabolomic response of the human basophilic cell line KU812 exposed to *Ixodes ricinus* tick saliva using ^{1}H-NMR spectroscopy (Fig. 1) [17, 18]. The protocol includes procedures for cell harvesting and sample preparation, metabolite extraction, acquisition of NMR spectra from both aqueous and lipid fractions, and subsequent data processing and quantification using targeted profiling approaches [19]. Particular attention is given to critical steps that affect data quality and reproducibility, as well as to practical considerations for interpreting metabolomics data.

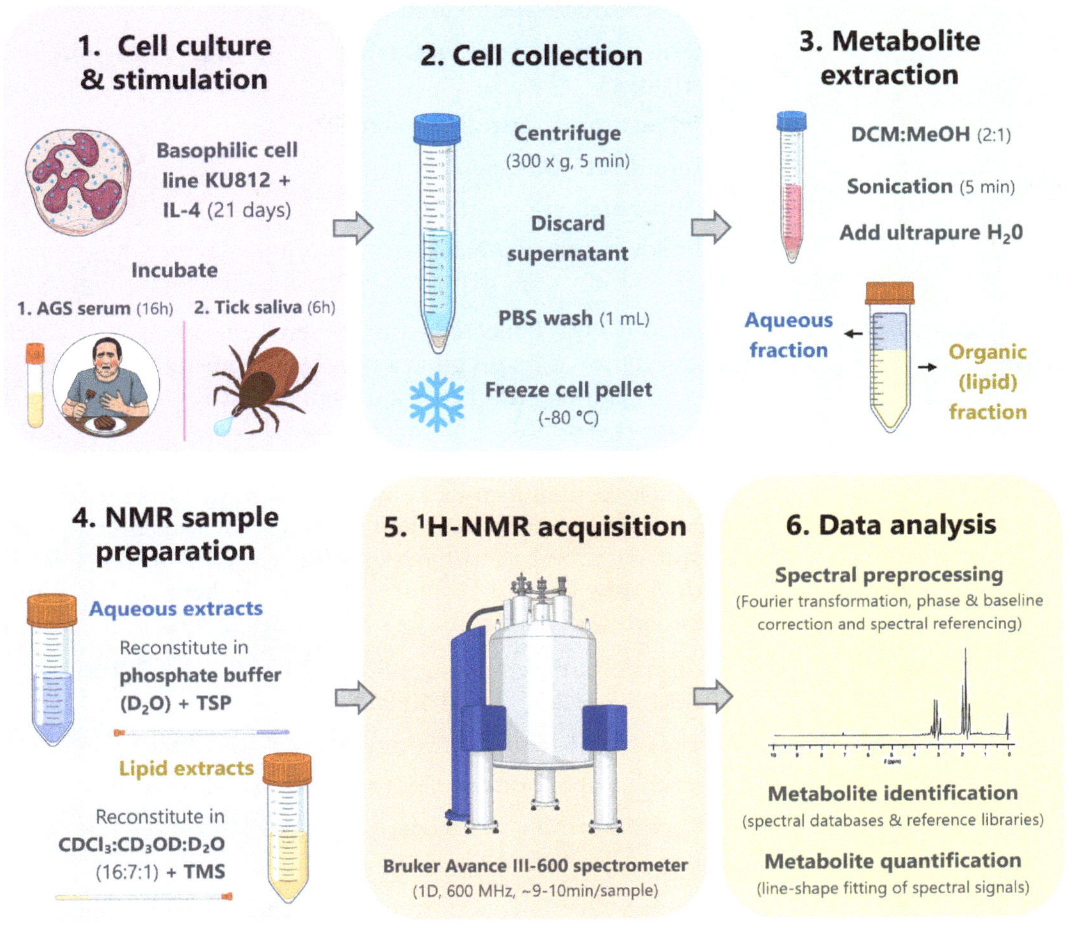

Fig. 1 Experimental workflow. Schematic steps carried out to perform metabolomics analysis in human basophilic cell line KU812 exposed to *I. ricinus* tick saliva using ^{1}H-NMR spectroscopy

2 Materials

2.1 KU812 Cell Culture

1. Roswell Park Memorial Institute (RPMI) 1640 medium.
2. X-VIVO® 15 serum-free hematopoietic cell medium.
3. Fetal bovine serum (FBS), 5% (v/v).
4. Recombinant human IL-4 (1 ng/mL; PeproTech®).
5. Serum from patients diagnosed with alpha-Gal syndrome (AGS).
6. Phosphate-buffered saline (PBS): 137 mM NaCl, 2.7 mM KCl, 10 mM Na_2HPO_4, 2 mM KH_2PO_4, pH 7.4.

7. Sterile, vented 25 cm^2 culture flasks.
8. Sterilization system: peristaltic pump with 0.22-μm filter cartridges.
9. Temperature-controlled water bath.
10. Humidified CO_2 incubator.
11. Giemsa staining reagents.
12. Hemocytometer o Neubauer counting chamber.

2.2 I. ricinus *Tick Saliva*

I. ricinus saliva was collected from 50 semi-engorged female ticks that had fed for 5 days on guinea pigs by administering 5 μL of 2% pilocarpine hydrochloride per tick to stimulate saliva secretion. Ticks were positioned dorsally on a microscope slide using double-sided adhesive tape, and pilocarpine was injected into the lower left quadrant of the abdomen using a Hamilton microliter syringe under a stereomicroscope. Saliva was then harvested by inserting a pulled glass capillary tube into the hypostome, expelled into a 1.5-mL Eppendorf tube using a Pasteur pipette bulb, kept on ice during collection, pooled, and stored at −80°C until further use.

2.3 Cell Harvesting and Washing

1. Phosphate-buffered saline (PBS): 137 mM NaCl, 2.7 mM KCl, 10 mM Na_2HPO_4, 2 mM KH_2PO_4, pH 7.4.
2. Sterile conical 15-mL Falcon tubes.
3. Freezer capable of maintaining samples at −80°C.
4. Dry ice for sample shipment.

2.4 Metabolite Extraction Reagents

Metabolite extraction is performed using the following solvents:

1. Dichloromethane:methanol solution (2:1, v/v), prepared fresh before use.
2. Ultrapure water.

2.5 NMR Sample Preparation Solutions

Different solutions are required depending on whether aqueous or lipid extracts are analyzed.

2.5.1 Aqueous Extracts

For the preparation of aqueous samples, the following components are used:

1. Phosphate buffer (45 mM) prepared in D_2O and adjusted to pH 7.4.
2. Trimethylsilylpropionic acid (TSP) purchased as powder and added to the phosphate buffer at a final concentration of 2.32 mM.
3. Deuterium oxide (D_2O) NMR grade.

2.5.2 Lipid Extracts

Lipid samples are prepared using a deuterated solvent mixture and internal standard:

1. Solvent mixture: $CDCl_3$:CD_3OD:D_2O (16:7:1, v/v/v), prepared immediately before use (*see* **Note 1**).
2. Tetramethylsilane (TMS) added to the solvent mixture at a final concentration of 2 mM.
3. Deuterated chloroform ($CDCl_3$): NMR grade.
4. Deuterated methanol (CD_3OD): NMR grade.

2.6 Equipment and Instrumentation

The following equipment is required for sample processing and NMR data acquisition:

1. Ultrasonic bath.
2. Refrigerated centrifuge capable of reaching 25,100 × *g*.
3. Vacuum concentrator (SpeedVac).
4. 5-mm NMR tubes.
5. Bruker Avance III-600 NMR spectrometer operating at a proton frequency of 600.20 MHz.
6. Temperature control unit for NMR acquisition.

2.7 Software

Data acquisition, processing, and analysis are carried out using the following software tools and databases:

1. Bruker TopSpin software for NMR data acquisition and initial spectral processing.
2. *In-house* developed software tools for spectral processing, metabolite identification, and quantification of ^{1}H-NMR data.
3. The *in-house* software is based on targeted profiling methodologies and spectral deconvolution approaches adapted from previously described tools such as LipSpin [20] and Dolphin [21].
4. Spectral databases for metabolite identification, including BBioref (Bruker), Chenomx, and the Human Metabolome Database (HMDB).
5. The software is not publicly available.

3 Methods

3.1 Human Cell Culture and Exposure to Tick Saliva

Human cells are cultured and exposed to tick saliva under the desired experimental conditions, as described below.

1. Culture KU812 cells in RPMI 1640 medium supplemented with 5% FBS and incubate at 37°C with 5% CO_2 and controlled humidity. Maintain the culture at 3×10^5–3×10^6 cells/mL and passage twice weekly at a 1:2 dilution to preserve viability.
2. Differentiate KU812 cells into a basophil-like phenotype by adding IL-4 (1 ng/mL) to the culture medium. This treatment enhances the number of high-affinity IgE receptors (FcεRI) on the cell surface. Continue differentiation for 21 days to allow full maturation of basophil-like characteristics.
3. Count cells using a hemocytometer or equivalent device. For each experimental well, use 5×10^6 cells, corresponding to a density of 1×10^6 cells/mL in 5 mL medium (*see* **Note 2**). Replicate to ensure consistency.
4. Transfer cells to sterile conical 15-mL Falcon tubes and centrifuge at $200 \times g$ for 5 min to remove culture medium.
5. Carefully remove the supernatant and resuspend the cell pellet in X-VIVO serum-free medium to minimize serum interference during stimulation.
6. Seed 5×10^6 cells in 5 mL medium directly into sterile, vented 25 cm^2 culture flasks and gently swirl to evenly distribute the cells (do not invert vented flasks).
7. Add 1 mL/flask of diluted serum (1:1000) from a patient diagnosed with alpha-Gal syndrome (AGS) and incubate overnight at 37 °C in a 5% CO_2 to sensitize the cells. The volume of serum added should be adjusted according to the patient's specific IgE levels against alpha-Gal. In this case, sIgE = 61.9 kU/L.
8. Transfer cells to sterile conical 15-mL Falcon tubes and centrifuge at $200 \times g$ for 5 min.
9. Carefully remove the supernatant, perform three cell washes with 1 mL phosphate-buffered saline (PBS) and resuspend the cells in X-VIVO serum-free medium to a final density of 1×10^6 cells/mL.
10. Seed 5×10^6 cells in 5 mL medium directly into sterile, vented T-25 flasks and gently swirl to evenly distribute the cells.
11. Add tick saliva to the cells at the desired dose: 3 μL per flask (dose 1) or 15 μL per flask (dose 2). Make sure to include an untreated control group consisting of KU812 cells maintained in standard medium without any added compounds.
12. Incubate the cells with *I. ricinus* tick saliva for 6 h at 37 °C in a humidified incubator with 5% CO_2.

3.2 Cell Collection

Cells are collected following standard procedures, with particular care to preserve cell integrity and metabolic state.

1. Following incubation with tick saliva, transfer cells to sterile conical Falcon 15 mL tubes and centrifuge at 300 × *g* for 5 min to pellet the cells while avoiding excessive mechanical stress (*see* **Note 3**).
2. Carefully aspirate and discard the supernatant without disturbing the pellet (*see* **Note 4**).
3. Add 1 mL phosphate-buffered saline (PBS) to wash the cells and remove residual culture medium.
4. Gently resuspend the pellet and centrifuge again under the same conditions.
5. Carefully remove the PBS. Repeat the washing step if necessary to ensure complete removal of culture medium (*see* **Note 4**).
6. Immediately freeze the cell pellets and store them at −80 °C until metabolite extraction (*see* **Note 5**).
7. Ship frozen samples on dry ice if analysis is performed at a different facility (*see* **Note 6**).

3.3 Metabolite Extraction

Metabolites are extracted using a biphasic solvent system to separate aqueous and lipid fractions.

1. Keep all samples and solvents at low temperature throughout the procedure to minimize metabolic degradation (*see* **Note 7**).
2. Add 1440 μL of dichloromethane:methanol (2:1, v/v) to each frozen cell pellet.
3. Perform sonication for 5 min using an ultrasonic bath to promote cell disruption.
4. Shake the samples briefly to ensure proper mixing.
5. Repeat the sonication step for an additional 5 min.
6. Add 400 μL of ultrapure water to induce phase separation.
7. Mix thoroughly and centrifuge at 25,100 × *g* for 5 min at 4 °C.
8. Carefully separate the aqueous and organic (lipid) phases into new tubes (*see* **Note 8**).
9. Dry both extracts completely using a vacuum concentrator (SpeedVac).
10. Store dried extracts at −80 °C until NMR analysis.

3.4 Sample Preparation for NMR Analysis

Prior to NMR acquisition, dried extracts are reconstituted in appropriate deuterated solvents depending on the fraction analyzed.

3.4.1 Aqueous Extracts

1. Reconstitute dried aqueous extracts in 550 μL of phosphate buffer (45 mM, prepared in D_2O) containing 2.32 mM trimethylsilylpropionic acid (TSP) as an internal standard.
2. Transfer the solution into 5-mm NMR tubes.
3. Ensure samples are free of bubbles and properly homogenized prior to acquisition (*see* **Note 9**).

3.4.2 Lipid Extracts

1. Reconstitute dried lipid extracts in a mixture of $CDCl_3$:CD_3OD:D_2O (16:7:1, v/v/v) containing tetramethylsilane (TMS) as internal reference.
2. Transfer the solution into 5-mm NMR tubes.

3.5 ^{1}H-NMR Data Acquisition

NMR spectra are acquired under standardized conditions to ensure quantitative reproducibility.

3.5.1 Aqueous Extracts

1. Acquire the ^{1}H-NMR spectra using a Bruker Avance III-600 spectrometer operating at a proton frequency of 600.20 MHz.
2. Use a standard one-dimensional NOESY-presaturation pulse sequence (noesypr1d) for water suppression.
3. Perform measurements at 300 K to ensure optimal spectral resolution and stability (*see* **Note 10**).
4. Set the relaxation delay (D1) to 5 s to minimize signal attenuation due to longitudinal relaxation effects (*see* **Note 11**).
5. Set the acquisition time to approximately 3.4 s, resulting in free induction decays (FIDs) of 64k complex data points.
6. Use a spectral width of approximately 16 ppm.
7. Acquire 64 scans per sample.
8. Use four dummy scans prior to acquisition to ensure signal stabilization.
9. The total acquisition time per sample is approximately 10 min.

3.5.2 Lipid Extracts

1. Acquire the ^{1}H-NMR spectra using a Bruker Avance III-600 spectrometer operating at a proton frequency of 600.20 MHz.
2. Use a standard one-dimensional pulse sequence with water presaturation (zgpr).
3. Perform measurements at 286 K in order to shift the residual water signal to a noninformative spectral region (~4.5 ppm) (*see* **Note 10**).

4. Set the relaxation delay (D1) to 5 s to minimize signal attenuation due to longitudinal relaxation effects (*see* **Note 11**).
5. Set the acquisition time to approximately 3 s, resulting in free induction decays (FIDs) of 64k complex data points.
6. Use a spectral width of 18.6 ppm.
7. Acquire 64 scans per sample.
8. Use four dummy scans prior to acquisition to ensure signal stabilization.
9. The total acquisition time per sample is approximately 9 min.

3.6 Spectral Processing, Metabolite Identification and Quantification

Spectral processing and metabolite quantification are generally performed using in-house developed software tools specifically designed for ^{1}H-NMR metabolomics data analysis. These tools are based on previously described targeted profiling methodologies for quantitative ^{1}H-NMR metabolomics, including approaches similar to those implemented in LipSpin [20] and Dolphin [21], but incorporate additional proprietary optimizations for improved robustness and accuracy.

The approach combines targeted profiling with constrained line-shape fitting, enabling accurate quantification of metabolites in complex and highly overlapping spectral regions (*see* **Note 12**). The fitting procedure incorporates constraints on peak position, line width, and signal multiplicity based on prior knowledge of metabolite spectral signatures.

3.6.1 Aqueous Extracts

Metabolite profiling of aqueous extracts (Fig. 2) follows a targeted NMR workflow approach such as Dolphin [21].

1. Import raw free induction decays (FIDs) or processed spectra in Bruker format.
2. Apply standard preprocessing steps including Fourier transformation, phase correction, and baseline correction (*see* **Note 13**).

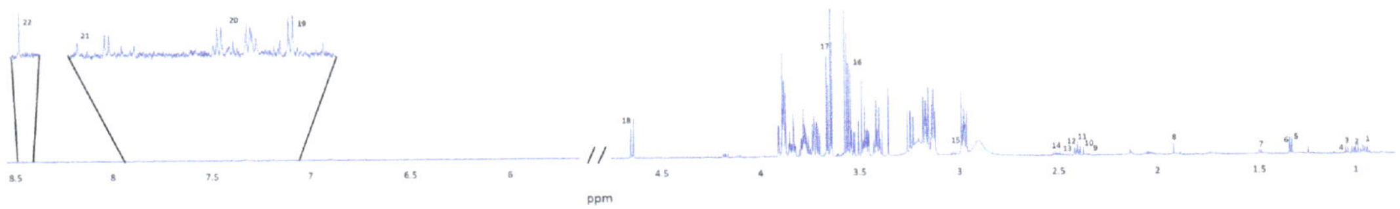

Fig. 2 NMR spectrum. Representative ^{1}H-NMR spectrum of the aqueous extract obtained from human cell pellets following biphasic extraction. Numbered signals correspond to the metabolites identified and quantified in the spectrum: (1) leucine, (2) isoleucine, (3) valine, (4) isobutyrate, (5) lactate, (6) threonine, (7) alanine, (8) acetate, (9) methionine, (10) glutamate, (11) pyruvate, (12) succinate, (13) pyroglutamate, (14) glutamine, (15) lysine, (16) glycine, (17) glycerol, (19) tyrosine, (20) phenylalanine, (21) histidine, and (22) formate

3. Perform spectral referencing using the TSP signal at 0 ppm, ensuring consistent chemical shift calibration across samples (*see* **Note 14**).
4. Identify metabolites using spectral databases and reference libraries. Several database engines, including BBioref (Bruker), Chenomx, and the Human Metabolome Database (HMDB) [22], together with literature data [23], were used for resonance assignment and metabolite identification (*see* **Note 15**).
5. Perform metabolite quantification using a line-shape fitting approach, in which spectral regions are modeled as a combination of signals corresponding to individual metabolites (*see* **Note 12**).
6. Account for overlapping signals by incorporating neighboring resonances into the fitting model, improving quantification accuracy in crowded spectral regions (*see* **Note 16**).
7. Extract signal areas corresponding to individual metabolites for downstream analysis.

3.6.2 Lipid Extracts

Lipid profiling is performed using a deconvolution-based workflow, similar to those implemented in LipSpin [20].

1. Import NMR spectra in Bruker format.
2. Apply phase and baseline correction using consistent criteria across all samples (*see* **Note 13**).
3. Perform spectral referencing using the TMS signal at 0 ppm (*see* **Note 14**).
4. Define spectral regions corresponding to lipid classes and characteristic resonances.
5. Identify lipid signals based on characteristic chemical shifts and previously reported assignments in the literature [23] (*see* **Note 17**).
6. Perform quantification using a deconvolution-based approach based on line-shape fitting of spectral signals (*see* **Note 12**).
7. Model individual signals using predefined spectral patterns and constraints derived from lipid reference spectra.
8. Extract signal areas corresponding to lipid species or functional groups.

3.6.3 Conversion of Signal Areas to Metabolite Concentrations

Metabolite concentrations are calculated using an internal standard-based quantification approach following established quantitative ^{1}H-NMR methodologies. Concentrations are derived from the ratio between metabolite signal integrals and the reference compound, taking into account the number of contributing protons. This enables direct conversion of signal areas into molar concentrations under quantitative acquisition conditions following the approach proposed by Serkova *et al.* [24] (*see* **Note 18**). The validity of this approach relies on fully relaxed acquisition conditions and consistent experimental parameters across all samples.

3.6.4 Data Normalization and Export

Final data processing includes normalization and export for downstream analysis:

1. Normalize metabolite signal intensities based on the number of cells per sample (*see* **Note 19**).
2. Export quantified data for statistical analysis.

4 Notes

1. Preparation of the solvent mixture for lipid extracts should be performed immediately before use. The $CDCl_3$:CD_3OD:D_2O (16:7:1, v/v/v) solvent mixture must be freshly prepared, as chloroform is highly volatile and its evaporation can alter the solvent ratio over time. Changes in solvent composition may affect extraction efficiency and spectral reproducibility. It is therefore recommended to prepare the mixture immediately prior to sample reconstitution.
2. Consistent cell number across samples is crucial for reliable comparison. Variability in cell number directly impacts metabolite concentration. It is strongly recommended to normalize samples based on cell count prior to extraction.
3. Centrifugation conditions are critical for preserving cell integrity. Excessive centrifugation force may cause cell rupture and leakage of intracellular metabolites, leading to altered metabolomic profiles. Use moderate centrifugal forces sufficient to pellet the cells without inducing mechanical stress.
4. Removal of culture medium is essential to avoid contamination of metabolomic profiles. Residual culture medium contains high concentrations of metabolites such as glucose, amino acids, and vitamins that can interfere with intracellular metabolite detection. Perform additional PBS washes if necessary. Importantly, if extracellular metabolites are to be analyzed, the supernatant should be collected after the initial centrifugation step and prior to washing the cells, in order to preserve metabolites secreted into the medium.
5. Rapid freezing is required to preserve the metabolic state of the cells. Delays between harvesting and freezing may lead to ongoing metabolic activity and degradation of labile metabolites. Freeze samples immediately and store them at −80°C.

6. Sample shipment must ensure uninterrupted cold chain conditions. Temperature fluctuations during transport can result in metabolite degradation. Samples should be shipped on dry ice and handled to prevent thawing.
7. All sample-handling steps should be performed under cold conditions. Maintaining low temperatures minimizes enzymatic activity and metabolite degradation, improving reproducibility and data quality.
8. Phase identification is critical during extraction. Following phase separation, the upper phase corresponds to the aqueous fraction, while the lower phase corresponds to the organic (lipid) fraction. Avoid collecting material from the interphase to prevent crosscontamination.
9. Sample homogeneity is essential for NMR analysis. Incomplete dissolution or the presence of bubbles can lead to poor shimming and spectral distortions. Ensure samples are fully dissolved and visually inspected before acquisition.
10. Temperature selection affects spectral quality and metabolite resolution. Aqueous extracts are typically analyzed at 300 K, while lipid extracts may benefit from lower temperatures (e.g., 286 K) to improve spectral resolution and reduce signal overlap.
11. Relaxation delay (D1) influences signal quantification. Short relaxation delays may lead to incomplete relaxation and biased signal intensities. A delay of approximately 5 s is generally suitable for metabolomic studies.
12. Accurate quantification requires appropriate modeling of spectral signals. Targeted profiling based on constrained line-shape fitting enables reliable quantification in complex biological samples, particularly in spectral regions with significant signal overlap. The quality of the fitting model directly impacts quantification accuracy.
13. Spectral preprocessing must be consistent across all samples. Phase and baseline correction should be applied using identical criteria to all spectra. Inconsistent preprocessing may introduce artificial differences and compromise downstream analysis.
14. Accurate chemical shift referencing is essential for metabolite identification. Small deviations in referencing (TSP or TMS at 0 ppm) may lead to incorrect metabolite assignment, particularly in crowded spectral regions.
15. Metabolite identification should be supported by multiple sources. Assignment of NMR resonances should be confirmed using a combination of spectral databases (e.g., BBioref, Chenomx, HMDB) and literature references. Reliance on a

single source may increase the risk of misannotation, particularly in crowded spectral regions.

16. Signal overlap must be explicitly addressed during quantification. Biological samples typically contain highly overlapping resonances. Proper modeling of overlapping signals is essential to avoid systematic errors in metabolite quantification.
17. Lipid signal assignment relies on characteristic spectral patterns described in the literature. Unlike small polar metabolites, lipid resonances are typically assigned based on well-established chemical shift regions corresponding to functional groups (e.g., methyl, methylene, olefinic protons). Reference to literature data is essential to ensure correct interpretation of lipid signals.
18. Quantification relies on the proportionality between signal area and metabolite concentration. This relationship is valid under conditions of full relaxation and stable acquisition parameters. The use of internal standards allows conversion of signal areas into absolute concentrations.
19. Normalization strategy influences biological interpretation. Normalization based on cell number is recommended for intracellular metabolomics, as it reflects true biological differences and reduces variability introduced by sample preparation.

Competing Interests

The authors have no conflicts of interest to declare that are relevant to the content of this chapter.

References

1. Ali A, Zeb I, Alouffi A, Zahid H, Almutairi MM, Ayed Alshammari F et al (2022) Host immune responses to salivary components—a critical facet of tick–host interactions. Front Cell Infect Microbiol 12:809052. https://doi.org/10.3389/fcimb.2022.809052
2. Wikel S (2013) Ticks and tick-borne pathogens at the cutaneous interface: host defenses, tick countermeasures, and a suitable environment for pathogen establishment. Front Microbiol 4:337. https://doi.org/10.3389/fmicb.2013.00337
3. Platts-Mills TAE, Gangwar RS, Workman L, Wilson JM (2025) The immunology of alpha-gal syndrome: history, tick bites, IgE, and delayed anaphylaxis to mammalian meat. Immunol Rev 332:e70035. https://doi.org/10.1111/imr.70035
4. Vaz-Rodrigues R, Feo Brito F, Guzmán Rodríguez R, Mazuecos L, De La Fuente J (2023) Allergic reactions associated with medically relevant arthropods. Ann Med 55:2286531. https://doi.org/10.1080/07853890.2023.2286531
5. Kitsou C, Fikrig E, Pal U (2021) Tick host immunity: vector immunomodulation and acquired tick resistance. Trends Immunol 42:554–574. 10.1016/j.it.2021.05.005
6. Štibrániová I, Bartíková P, Holíková V, Kazimírová M (2019) Deciphering biological processes at the tick–host interface opens new strategies for treatment of human

diseases. Front Physiol 10:830. https://doi.org/10.3389/fphys.2019.00830
7. Rodrigues V, Fernandez B, Vercoutere A, Chamayou L, Andersen A, Vigy O et al (2017) Immunomodulatory effects of *Amblyomma variegatum* saliva on bovine cells: characterization of cellular responses and identification of molecular determinants. Front Cell Infect Microbiol 7:521. https://doi.org/10.3389/fcimb.2017.00521
8. Megha, Preeti, Prasad T (2021) Metabolomics: a promising tool to study disease biomarkers and host–pathogen interactions. In: Hameed S, Fatima Z (eds) Integrated omics approaches to infectious diseases. Springer Singapore, Singapore. pp 403–423. https://doi.org/10.1007/978-981-16-0691-5_22.
9. Crook AA, Powers R (2020) Quantitative NMR-based biomedical metabolomics: current status and applications. Molecules 25:5128. https://doi.org/10.3390/molecules25215128
10. Nagana Gowda GA, Raftery D (2023) Quantitative NMR methods in metabolomics. Handb Exp Pharmacol 277:143–164. https://doi.org/10.1007/164_2022_612
11. Marshall DD, Powers R (2017) Beyond the paradigm: combining mass spectrometry and nuclear magnetic resonance for metabolomics. Prog Nucl Magn Reson Spectrosc 100:1–16. https://doi.org/10.1016/j.pnmrs.2017.01.001
12. Moco S (2022) Studying metabolism by NMR-based metabolomics. Front Mol Biosci 9:882487. https://doi.org/10.3389/fmolb.2022.882487
13. Abad E, Samino S, Grodzicki RL, Pagano G, Trifuoggi M, Graifer D et al (2021) Identification of metabolic changes leading to cancer susceptibility in Fanconi anemia cells. Cancer Lett 503:185–196. https://doi.org/10.1016/j.canlet.2020.12.010
14. Bassols A, Amigó N, Pérez-Rodado M, Saco Y, Peña R, Pato R et al (2025) Fecal metabolomics to understand intestinal dysfunction in male dairy beef calves at arrival to the rearing farm. Sci Rep 15: 6887. https://doi.org/10.1038/s41598-025-90407-3
15. Martínez-Falguera D, Aranyo J, Teis A, Ferrer-Curriu G, Rodriguez-Leor O, Amigó N et al (2025) Effect of early initiation of empagliflozin with or without sacubitril/valsartan in a swine model of myocardial infarction. Br J Surg 112:Supplement_2:znae322.010. https://doi.org/10.1093/bjs/znae322.010
16. Fernández-Cruz E, de la OV, Fernández-Diaz CM, Matía-Martín P, Rubio-Herrera MÁ, Amigó N et al (2025) Urinary hippuric acid as a sex-dependent biomarker for fruit and nut intake raised from the EAT-Lancet Index and nuclear magnetic resonance analysis. Metabolites 15:348. https://doi.org/10.3390/metabo15060348
17. Uribe-Herranz M, Oliver-Caldés A, Martínez-Micaelo N, Español-Rego M, Val-Casals M, Martínez-Soler R et al (2025) Microbiota shape metabolic and immune determinants of CAR-T therapy and correlate with outcomes in myeloma. Blood Cancer Discov 6:484–504. https://doi.org/10.1158/2643-3230.BCD-24-0203
18. Drevet Mulard E, Gilard V, Balayssac S, Rautureau GJP (2025) Quantitative nuclear magnetic resonance for small biological molecules in complex mixtures: practical guidelines and key considerations for non-specialists. Molecules 30:1838. https://doi.org/10.3390/molecules30081838
19. Debik J, Sangermani M, Wang F, Madssen TS, Giskeødegård GF (2022) Multivariate analysis of NMR-based metabolomic data. NMR Biomed 35:e4638. https://doi.org/10.1002/nbm.4638
20. Barrilero R, Gil M, Amigó N, Dias CB, Wood LG, Garg ML et al (2018) LipSpin: a new bioinformatics tool for quantitative[1] H NMR lipid profiling. Anal Chem 90:2031–2040. https://doi.org/10.1021/acs.analchem.7b04148
21. Gómez J, Brezmes J, Mallol R, Rodríguez MA, Vinaixa M, Salek RM et al (2014) Dolphin: a tool for automatic targeted metabolite profiling using 1D and 2D (1)H-NMR data. Anal Bioanal Chem 406:7967–7976. https://doi.org/10.1007/s00216-014-8225-6
22. Wishart DS, Guo A, Oler E, Wang F, Anjum A, Peters H et al (2022) HMDB 5.0: the human metabolome database for 2022. Nucleic Acids Res 50:D622–D631. https://doi.org/10.1093/nar/gkab1062
23. Vinaixa M, Rodríguez MA, Rull A, Beltrán R, Bladé C, Brezmes J et al (2010) Metabolomic assessment of the effect of dietary cholesterol in the progressive development of fatty liver disease. J Proteome Res 9:2527–2538. https://doi.org/10.1021/pr901203w
24. Serkova N, Fuller TF, Klawitter J, Freise CE, Niemann CU (2005) H-NMR-based metabolic signatures of mild and severe ischemia/reperfusion injury in rat kidney transplants. Kidney Int 67:1142–1151. https://doi.org/10.1111/j.1523-1755.2005.00181.x

Chapter 9

Protective Pathogen Infection Against Tick-Induced Allergy in Zebrafish

Alejandro Cabezas-Cruz and José de la Fuente

Abstract

The α-Gal syndrome (AGS) links tick bites with delayed anaphylaxis to mammalian meat consumption. Recent work in zebrafish shows that the *Ixodes ricinus* salivary metalloprotease (MET) and allergen-like p23 trigger AGS-like responses, whereas subsequent infection with the mycobacterial fish pathogen, *Mycobacterium marinum*, skews immunity towards a protective Th1 profile. Here we provide a step-by-step protocol to: (i) reproduce tick-induced allergy in adult zebrafish, (ii) superimpose water-borne mycobacterial infection, and (iii) quantify clinical, serological, and molecular outcomes. The method allows exploration of host–pathogen–allergen interactions and evaluation of prophylactic interventions in a validated vertebrate animal model.

Keywords Zebrafish, α-Gal syndrome, Tick saliva, Metalloprotease, Allergen-like p23, *Mycobacterium marinum*, Intramuscular immunization, ELISA, RT-qPCR

1 Introduction

Tick saliva delivers a complex mixture of proteins and nonprotein fractions (NPF) that modulate host immunity. In humans, the glycan α-Gal modifications on tick salivary glycoproteins and lipids can induce IgE-mediated allergy known as the α-Gal syndrome (AGS), with a growing public concern [1, 2]. Using adult zebrafish, *Danio rerio*, we recently demonstrated that recombinant *Ixodes ricinus* MET and p23 proteins, together with saliva-derived NPF, reproduce key AGS manifestations—hemorrhagic skin lesions, aberrant feeding, and mortality—after mammalian-meat consumption [3, 4]. Remarkably, exposure to low-dose *Mycobacterium marinum* (α-Gal-positive) on Day 7 dampened these reactions and lowered gut bacterial load, suggesting a pathogen-driven rebalancing of Th2 allergy towards Th1 protective response [3]. Accordingly, zebrafish has been validated as a good model for the AGS [4].

José de la Fuente and Alejandro Cabezas-Cruz (eds.), *Allergic Disease Models*, Methods in Molecular Biology, vol. 3092, https://doi.org/10.1007/978-1-0716-5635-8_9,

2 Materials

2.1 Animals

Use healthy wild-type AB strain adult zebrafish (*D. rerio*) (6–8 months old, ~0.3 g body weight) of both sexes (1:1 male:female). Fish should be certified pathogen-free (free of common zebrafish diseases) and maintained in a flow-through aquarium system at ~27 °C on a 14 h light/10 h dark cycle (Fig. 1). Acclimate fish to laboratory conditions and feed a standard fish diet (*see* Sect. 2.2) twice daily. Ensure zebrafish have been pre-sensitized to tick saliva to induce the alpha-Gal allergy phenotype prior to experiments. All animal procedures must follow ethical guidelines and approved protocols.

2.2 Reagents and Biologicals

1. *I. ricinus* tick saliva (pathogen-free). Fresh or frozen saliva collected from semi-engorged female *I. ricinus* ticks (fed 6–7 days on a guinea pig host). Ticks are stimulated with pilocarpine to induce salivation (*see* below), and saliva is harvested into capillaries. Store tick saliva at −80°C until use. For zebrafish treatments, dilute saliva in sterile PBS (e.g., 1:10 v/v) immediately before injection. Safety: handle tick saliva as a potential biohazard (wear gloves, work in a biosafety cabinet) if not certified pathogen-free.
2. Tick saliva nonprotein fraction (NPF). Low-molecular-weight (<3 kDa) fraction of tick saliva prepared by ultrafiltration. To

Fig. 1 Example of a flow-through aquarium with adult zebrafish. Facility at IREC, Spain

prepare NPF, dilute tick saliva in PBS (e.g., 9 µL saliva + 171 µL PBS for a 1:20 dilution) and spin through a 3 kDa cutoff centrifugal filter (Amicon Ultra-0.5, Merck) at 14,000 × *g* for ~30 min. Collect the filtrate (contains NPF) and keep on ice. The retained >3 kDa retentate contains tick salivary proteins. Store NPF at –80°C until use. Tick saliva NPF has very low α-Gal content (~0.04 ng/µg protein) when compared to the protein fraction (*see* **Note 1**).

3. Pilocarpine hydrochloride. Cholinergic agonist used to induce tick salivation. Prepare a 4% (w/v) sterile aqueous solution of pilocarpine HCl and apply ~5 µL to the dorsal surface of each tick to stimulate saliva release. Store pilocarpine powder desiccated at 4 °C and handle with gloves.

4. Recombinant tick salivary proteins (MET, p23, SUB). Produce tick salivary metalloprotease MET (UniProt ID A0A0K8RCY8) and allergen-like p23 (A0A0K8RKR7) as recombinant proteins using codon-optimized synthetic genes, PCR-amplified and cloned into the expression vector (e.g., pET101/D-TOPO: ThermoFisher). Transform expression plasmids into *Escherichia coli* BL21 (DE3) cells for protein production induced with 0.5–1 mM IPTG when culture OD_{600} ~0.6. Proteins (MET and p23) carry a 6× Histidine tag and are purified under denaturing conditions by Ni^{2+}-affinity chromatography in 7 M urea lysis buffer. After elution, refold each protein by extensive dialysis against PBS, pH 7.4 (at 4 °C). Subolesin (SUB) (UniProt Q4VRW2) is another tick protein used as a negative control and produced in *E. coli* and purified using the same His-tag/urea protocol. Adjust each purified protein to a working concentration convenient for dosing (e.g., 0.5 mg/mL in PBS) and store aliquots at –80°C. Use: Thaw on ice and mix as needed for injection (e.g., 5 µg of MET or p23 + 1 µL NPF in a 10 µL dose). For combined treatments, MET and p23 can be mixed (2.5 µg each) plus NPF in 10 µL PBS.

5. Tick salivary gland extract. Prepare antigen extract from dissected salivary glands of *I. ricinus* ticks, used for ELISA coating. To prepare, aseptically dissect salivary glands from ~8 female ticks (fed ~5 days) and homogenize in PBS containing protease inhibitors. Clarify by centrifugation and determine protein concentration (e.g., BCA assay). Store at –80°C. This extract is used to coat ELISA plates (see Sect. 2.3) at 50 ng/well for IgM titer assays.

6. Mammalian meat diet (dog food). A meat-containing diet is required to trigger allergic responses after tick sensitization. A commercial dog kibble with high mammalian content crafted with 50% premium animal ingredients and balanced with

50% wholesome grains, vegetables, fruit, nutrients, vitamin C, Omega-3 fatty acids, prebiotics, and fiber is used, containing approx. 16% raw beef and 15% lamb meal. Obtain a dry dog food with similar composition and grind or crumble it for feeding zebrafish. Store dog food at room temperature in a dry, sealed container. Feed fish as described in Sect. 3.3 (Days 2–8).

7. Standard zebrafish diet. Use a nutritionally complete fish feed without any mammalian ingredients prior to the meat challenge. Use a commercial tropical fish feed crafted with fish meal, squid meal, shrimp meal, earthworms, spirulina, and vitamins and minerals consisting of fish meal and cereals. Use any high-quality pelleted fish diet that contains no mammalian meat/byproducts. Store feed as per manufacturer (usually room temp, dry). Feed fish this diet during acclimation and up to the start of meat feeding. The fish feed and dog-food diet serve as experimental controls to distinguish baseline nutrition from mammalian meat exposure.

8. *Mycobacterium marinum*. Use virulent *M. marinum* (a fish tuberculosis pathogen) for infection challenge in Experiment 2. Use a characterized strain of *M. marinum* (e.g., ATCC strain M or a GFP-tagged lab strain) grown to mid-log phase in Middlebrook 7H9 broth (with OADC supplement and 0.05% Tween-80). Titer the culture and dilute in system water to ~50 CFU/mL for immersion challenge. Caution: *M. marinum* is a BSL-2 pathogen; handle cultures in a biosafety cabinet with appropriate protective equipment. Maintain stock cultures at –80 °C and thaw/grow fresh cultures for each experiment.

9. DNA/RNA/Protein Mini Kit. Use a commercial kit for simultaneous extraction of genomic DNA, total RNA, and protein from zebrafish tissues. Use this kit to isolate DNA and RNA from each intestine sample following the manufacturer's protocol. Store kit components per instructions (most buffers at room temp; protein additive at 2–8 °C). The DNA is used for *M. marinum* qPCR and RNA for gene expression analysis from the same sample.

10. Primers for *M. marinum* detection. Custom oligonucleotide primers targeting the *M. marinum* 16S rRNA gene. The primer sequences are 16S-F: 5′-ACT GAG ATA CGG CCC AGA CT-3′ and 16S-R: 5′-TCA CGA ACA ACG CGA CAA AC-3′. Order primers from a commercial supplier, resuspend in nuclease-free water to 100 μM stock, and dilute to 10 μM working solutions. Store primers at –20 °C. These primers yield a specific amplicon for quantifying bacterial load by qPCR.

11. Primers for zebrafish gene expression (RT-qPCR). Gene-specific primers are required for zebrafish immune response markers (e.g., *tlr2*, *il1b*, *tnfα*, *il4*, *prkdc*) and housekeeping genes (e.g., *gapdh*, *β-actin*). Obtain or design primers by using published sequences and prepare them as 10 μM working stocks (store at −20 °C). Ensure one pair of primers for each target gene and for at least two reference genes (for normalization).
12. SYBR® Green PCR Master Mix. A ready-to-use SYBR Green qPCR mix for DNA amplification. For example, use Power SYBR® Green PCR Master Mix (Applied Biosystems, cat. no. 4367659) for quantitative PCR assays. This 2× mix contains DNA polymerase, dNTPs, SYBR Green dye, and buffer. Store at 4 °C (protected from light). Combine with primers and template DNA for detecting *M. marinum* DNA (*see* Procedure). For RNA templates gene expression, a one-step RT-qPCR system is used.
13. One-step RT-qPCR reagents. To quantify mRNA expression, use a one-step reverse transcription qPCR kit by adding reverse transcriptase directly into the SYBR Green qPCR mix. Store the RT enzyme at −20 °C and the reaction mix at 4 °C. Mix RNA template with SYBR mix, gene-specific primers, and RT enzyme, and then run the combined reverse transcription at 50 °C and PCR amplification in a real-time cycler.
14. BCA protein assay kit. Use a bicinchoninic acid protein quantification kit for measuring protein concentrations. The BCA assay is used to quantify total protein in tick saliva fractions and recombinant protein preparations. Prepare BCA working reagent according to kit instructions and use bovine serum albumin (BSA) standards for calibration. Store kit reagents as directed (many at room temp).
15. ELISA plate coating antigen reagents. For zebrafish IgM titer ELISAs, prepare the following coating antigens:
    1. α-Gal–BSA conjugate. The Galα1-3Gal–BSA glycoprotein is obtained from Dextra Laboratories (product NGP0203) to represent the α-Gal epitope. Use 50–100 ng per well for coating. Store at −20 °C (powder or solution).
    2. Use recombinant tick proteins MET and p23 for coating at ~100 ng/well. Also prepare SUB for coating control wells.
    3. Use tick salivary gland extract from *I. ricinus* at 50 ng/well.
    4. For the *Mycobacterium bovis* P22 protein complex, use an immunopurified protein sub-complex from bovine tuberculin (PPD) as antigen to assess antimycobacterial IgM.

Obtain P22 antigen and use ~100 ng/well for coating. Store protein antigens at −20 °C.

16. Blocking reagent for ELISA: Skim milk powder (nonfat dry milk). For blocking solution, 5% (w/v) skim milk in PBS is used. We recommend a microbiological grade skim milk (e.g., Condalab, Spain). Store powder at room temperature, sealed; prepare fresh blocking solution on the day of the experiment (Alternatively, 5% BSA or casein can be used as blocking agent).

17. Antibodies for ELISA.
    1. Primary antibody. Rabbit anti-zebrafish IgM polyclonal antibody to detect zebrafish IgM in serum. A suitable antibody can be obtained commercially or produced via rabbit immunization. Use at 1:1000 dilution in blocking buffer. Store antibody aliquots at −20 °C.
    2. Secondary antibody. Goat anti-rabbit IgG–HRP conjugate. Use at 1:3000 dilution. Store at −20 °C and light-protected. Thaw an aliquot just before use and avoid repeated freeze–thaw cycles.

18. Use for TMB substrate solution a 3,3′,5,5′-tetramethylbenzidine one-component HRP substrate. This ready-to-use solution develops a blue color in the presence of peroxidase. Store TMB solution at 4 °C in the dark. Add 100 μL per well after secondary antibody incubation. TMB is light-sensitive; therefore, perform incubations in the dark by covering the plate with aluminum foil. Typical development time is ~15 min at room temperature until a blue color is evident.

19. For a stop solution for ELISA, use 2 N sulfuric acid (H_2SO_4). Prepare by diluting concentrated H_2SO_4 into water in a fume hood and wear eye/hand protection. Use 50 μL per well to stop the TMB reaction, which turns the solution yellow. Prepare fresh on the day of assay and dispose the acid waste properly.

20. General reagents and media to use: Luria–Bertani (LB) broth and agar (for *E. coli* culture); agar plates supplemented with appropriate antibiotic (100 μg/mL ampicillin for pET101) to select transformants. Ampicillin sodium salt (Sigma-Aldrich)—prepare a 100 mg/mL stock in water, filter-sterilize, and store at −20 °C. Isopropyl β-D-1-thiogalactopyranoside (IPTG) for induction—prepare 1 M stock in water, store at −20 °C. Tricaine methane sulfonate (MS-222, Sigma) for fish anesthesia/euthanasia—prepare a 0.2% (w/v) stock in system water. Use for anesthesia and euthanasia as described in Sect. 3.3. Ethanol (70%) and antiseptic (povidone–iodine or

chlorhexidine) for disinfecting instruments and injection sites (optional). Nuclease-free water for PCR and buffer preparation. Micropipette tips, microcentrifuge tubes (1.5 mL), and other standard laboratory consumables should be sterile where applicable.

2.3 Buffers and Solutions

1. The phosphate-buffered saline (PBS), pH 7.4, 1× PBS, is used for dilutions, injections, washes, and dialysis. PBS composition: 137 mM NaCl, 2.7 mM KCl, 10 mM Na_2HPO_4, and 1.8 mM KH_2PO_4. Sterilize by autoclaving. Store at room temperature. For injections and tissue work, use sterile filtered PBS. PBST is PBS plus 0.05% Tween-20.
2. Carbonate–bicarbonate coating buffer, pH 9.6, 50 mM carbonate/bicarbonate buffer for ELISA plate coating. Prepare with 1.59 g Na_2CO_3 and 2.93 g $NaHCO_3$ per liter, pH ~9.6. Store at 4 °C. Use to dilute antigens for coating ELISA plates (50–100 ng antigen per well).
3. PBST wash buffer, PBS with 0.05% Tween-20 (v/v). Prepare by adding 500 μL Tween-20 per liter of PBS. Mix well. Use PBST for washing ELISA plates (3×200 μL washes per well between steps). Store at room temperature.
4. Blocking solution with 5% (w/v) skim milk in PBS. Prepare fresh by dissolving nonfat dry milk powder in PBS with 5 g in 100 mL. Stir until fully dissolved. Use to block ELISA plates for 1 h at RT. This solution is also used to dilute serum samples and antibodies for ELISA to reduce nonspecific binding. Do not store it and make it fresh daily as the milk solution may spoil or precipitate.
5. Urea lysis buffer for denaturing is a buffer containing 7 M urea for solubilizing inclusion bodies and purifying His-tag proteins. Use 7 M urea, 300 mM NaCl, 50 mM NaH_2PO_4, pH 8.0, plus 10 mM imidazole (for binding), or 250 mM imidazole for elution. Prepare fresh before use, and urea solutions should be filtered and used at room temperature, or urea can precipitate at 4°C. After lysis and binding in urea buffer, proteins are refolded by dialysis in PBS.
6. Cell lysis buffer is composed of 50 mM Tris-HCl, pH 8.0, 300 mM NaCl, and 10 mM imidazole supplemented with 1× protease inhibitor cocktail.
7. For Ni-NTA affinity purification, use wash (50 mM Tris-HCl pH 8.0, 300 mM NaCl, 20 mM imidazole) and elution (50 mM Tris-HCl pH 8.0, 300 mM NaCl, 250 mM imidazole) buffers.

8. Tricaine solution (MS-222) in system water for zebrafish anesthesia and euthanasia; prepare and use as described in Sect. 3.3. Adjust pH of the tricaine solution to 7.0–7.5 with sodium bicarbonate before use. Prepare fresh tricaine on the day of use and keep it cold.

2.4 Equipment

1. Aquaria or tanks to house experimental fish at 27 ± 1°C with water filtration and aeration. A flow-through system or static tanks with regular water changes can be used. Ensure a thermometer and heater/chiller to maintain 27 °C, and a timer for controlled 14/10 h light/dark cycles. Include tank covers to prevent fish from jumping out.
2. For fish injection, use a method for intramuscular (IM) injections in zebrafish (see Chapt. 5 for procedure). Options include: a 10-μL Hamilton microsyringe with a 26–30-G needle, or a pulled glass capillary microinjector system (e.g., pneumatic picopump). The injector should deliver ~10 μL volume accurately into the fish muscle. Also have a clean sponge or cradle to gently restrain anesthetized fish during injection.
3. For dissection, use fine scissors, forceps, and a scalpel for fish necropsy and tissue collection. Sterilize tools between samples (e.g., dip in ethanol and flame or use a glass bead sterilizer) to prevent cross-contamination. Also prepare a dissecting tray or plate lined with paraffin or silicone for pinning fish during dissection.
4. A bench-top microcentrifuge capable of ≥14,000 × *g* is required for filtering tick saliva with ultrafilters and for pelleting samples during DNA/RNA extraction. Use 1.5 or 2 mL sealed tubes for high-speed spins. A refrigerated centrifuge is preferable for sensitive samples used at 4 °C for nucleic acid work.
5. Use a micro-volume UV-vis spectrophotometer to quantify DNA/RNA concentrations and purity. Ensure it is calibrated and clean before use to measure OD_{600} for bacterial culture density and $OD_{260/280}$ for nucleic acids.
6. Thermal cyclers.
   a. A quantitative PCR machine with fluorescence detection is required for both *M. marinum* DNA qPCR and RT-qPCR for gene expression. Make sure it supports SYBR Green detection and has appropriate software for Ct analysis.
   b. If performing any endpoint PCR or preparing templates, a regular thermal cycler can be used. In one-step RT-qPCR, the real-time instrument also performs the

reverse transcription step, so a separate machine for cDNA synthesis is not strictly necessary.

7. A microplate spectrophotometer capable of reading absorbance at OD_{450} after the TMB stop reaction is used for ELISA. Any ELISA plate reader with a 450-nm filter will suffice. Ensure it has software to blank and export data.
8. Use a Class II laminar flow biosafety cabinet when handling *M. marinum* cultures, tick saliva, or any infectious material. This provides a sterile workspace and protects the user from aerosols. Required for BSL-2 containment of *M. marinum.*
9. A 28 °C incubator or water bath is used for zebrafish immersion infection to maintain water temperature during *M. marinum* exposure, if room temperature is much lower. A 37 °C shaking incubator is used for growing *E. coli* cultures at 250 rpm orbital shaking recommended for aeration. Also, a 37 °C static incubator is used for overnight ELISA plate coating and antibody incubation steps, or alternatively, you may simply use a refrigerator at 4 °C for overnight incubations.
10. A tissue grinder or electric homogenizer is used for intestine samples. For example, disposable pellet pestles or a motorized tissue grinder can homogenize ~20–30 mg zebrafish intestine in lysis buffer. Alternatively, a bead-beater with sterile beads can be used to disrupt tissues in AllPrep lysis buffer. This is crucial for efficient DNA/RNA extraction.
11. A precision balance (±0.01 g) is required to weigh fish for growth measurement and to weigh reagents like urea and powders for buffers. Ensure calibration weights are available.
12. Required general laboratory equipment includes pipettes (P10, P100, and P1000) with sterile filter tips for precise liquid handling, especially for qPCR setup and ELISA. A vortex mixer to mix reagents and samples. A water bath or dry block heater (50 and 95 °C settings) if needed for certain incubations (though the real-time cycler covers these for RT and PCR). A pH meter to adjust buffer pH (for PBS or tricaine solutions). Timer or stopwatch for infection exposure and ELISA timing. Personal protective equipment (lab coat, gloves, and safety goggles) for all experimental procedures.

2.5 Data Analysis

Data are analyzed with GraphPad Prism v8.0 and the Real Statistics Excel add-in.

3 Methods

Carry out all procedures at room temperature unless otherwise specified. All animal protocols need approval by the appropriate institutional ethics committee and should be in accordance with EU Directive 2010/63. Adult wild-type AB strain zebrafish (6–8-months old, ~0.3 g) are acclimated for 2 weeks before use. Zebrafish are maintained at 27 °C on a 14:10 h light:dark cycle and fed a standard flake diet until experimental start. Zebrafish sensitization with tick saliva in the AGS model is performed when fish are exposed to tick salivary components and mammalian meat to induce α-Gal syndrome.

3.1 Tick Saliva Collection

1. For tick source and feeding, obtain pathogen-free adult female ticks, *I. ricinus*. Feed ticks on a vertebrate host such as a guinea pig for 6–7 days. Monitor tick engorgement daily. Remove semi-engorged ticks and keep them on moist filter paper at 4 °C for immediate saliva collection.
2. For pilocarpine stimulation, prepare and apply pilocarpine hydrochloride as described in Sect. 2.2.3 to induce salivation. Handle ticks individually under a stereomicroscope.
3. For saliva collection, immediately attach a glass capillary tube prechilled on ice to the tick's hypostome/mouthparts. Collect saliva by capillary action for 30–60 min. Keep the tick and capillary assembly on ice during collection. After saliva appears in the capillary, carefully remove the tube.
4. Pool saliva from 10 to 20 ticks into a prechilled microcentrifuge tube. Centrifuge pooled saliva at 4 °C for 5 min at 12,000 × *g* to remove debris. Transfer the supernatant with tick saliva to a new tube.
5. To obtain the saliva nonprotein fraction (NPF), process tick saliva as described in Sect. 2.2.2 and store the collected NPF at −80 °C. Determine the protein concentration of saliva by BCA assay for use in treatment.

3.2 Recombinant Protein Expression and Purification

1. For plasmid construction, obtain synthetic, codon-optimized genes for tick metalloprotease (MET, UniProt A0A0K8RCY8), allergen-like p23 (UniProt A0A0K8RKR7), and subolesin (SUB, UniProt Q4VRW2). Clone each coding sequence without signal peptides into an expression vector to add an N-terminal His_6-tag. Verify constructs by restriction digest and Sanger sequencing.
2. For protein expression, transform each construct into *E. coli* strain such as BL21 (DE3) (Invitrogen). Inoculate a single colony into 5 mL LB containing the appropriate selection antibiotic (e.g., ampicillin for pET101/D-TOPO) and grow overnight at 37 °C with shaking (200 rpm). The next day, dilute

this culture 1:100 into 500 mL fresh LB containing the same selection antibiotic and grow at 37 °C until $OD_{600} = 0.6–0.8$. Induce protein expression by adding IPTG to 1 mM and continue shaking at 30 °C for 4 h.

3. For cell lysis, harvest the cells by centrifugation at 5000 × *g* for 10 min at 4 °C. Discard supernatant and resuspend each pellet in 20 mL cell lysis buffer. Lyse cells by sonication on ice with six cycles of 30 s on/30 s off at 40% amplitude. Keep lysate on ice.
4. For clarification, centrifuge lysates at 15,000 × *g* for 30 min at 4 °C. Transfer the supernatant to a new tube and discard the pellet. Filter the supernatant through a 0.45-μm syringe filter.
5. For Ni-NTA affinity purification, equilibrate Ni-NTA agarose with five column volumes of lysis buffer. Incubate the cleared lysate with 1–2 mL Ni-NTA resin per 20 mL lysate by gentle rotation at 4 °C for 1 h. Load the resin with bound protein into a gravity column. Wash the column with ten column volumes of wash buffer to remove nonspecifically bound proteins. Elute bound protein with elution buffer in 1 mL fractions.
6. Pool fractions containing target protein as assessed by SDS-PAGE are dialyzed overnight at 4 °C against PBS to remove imidazole. Alternatively, use a desalting column or centrifugal filter units.
7. For quality control, determine protein concentration by BCA assay. Assess purity by SDS-PAGE (run 5–10 μg protein on a 12% gel and stain with Coomassie). Expect bands at the predicted sizes. Aliquot purified proteins and store at −80 °C.

3.3 Zebrafish Treatment and Monitoring

1. Assign zebrafish (20 fish/group for Experiment 1; 30 fish/group for Experiment 2) to treatments. Maintain a 1:1 female:-male ratio. Weigh fish before treatment to record baseline weight (~0.3 g).
2. Prepare treatment solutions in sterile PBS: (a) PBS alone (10 μL); (b) tick saliva diluted 1:10 in PBS (10 μL); (c) NPF diluted 1:10 (10 μL); (d) MET recombinant protein (Exp1: 5 μg in 10 μL; Exp2: 2.5 μg in 10 μL) with 1 μL NPF; (e) p23 recombinant protein (Exp1: 5 μg; Exp2: 2.5 μg) with 1 μL NPF; (f) SUB protein (5 μg) with 1 μL NPF (Exp1 only). For Experiment 2, combine MET 2.5 μg + p23 2.5 μg + NPF 1 μL in 10 μL (the "Mix"). Aliquot each solution on ice.
3. For intramuscular injection on Days 0 and 3, anesthetize fish in 0.02% tricaine methane sulfonate (MS-222) buffered with NaHCO_3. Using a 30 G insulin syringe, inject 10 μL of the assigned treatment into the dorsal lateral muscle of each fish. Return fish to recovery tanks with fresh system water. All

injections must be performed quickly (<30 s fish out of water) to minimize animal stress.

4. Feed zebrafish with a standard flake diet without mammalian meat components until Day 2. From Day 2 onward, switch all groups to a diet containing mammalian red meat, such as commercial dog food to provoke an allergic response. Provide the meat diet ad libitum each day (e.g., 2%–4% of body weight). Continue feeding the meat diet until the end of each experiment (Day 8 for Experiment 1 and Day 25 for Experiment 2).
5. For Experiment 1, continue daily monitoring for 8 days. Record any allergic symptoms (e.g., hemorrhagic lesions, lethargy, or abnormal swimming/feeding) and mortality each day. Euthanize all fish on Day 8 by overdose of tricaine (≥300 mg/L) and proceed to sample collection.
6. For Experiment 2, on Day 7 euthanize ten fish/group (tricaine overdose) to collect baseline samples. Challenge the remaining 20 fish/group with *M. marinum* strain ATCC 927 grown in Middlebrook 7H9 + OADC to mid-log phase by exposing fish by immersion in system water containing mycobacteria at ~50 CFU/mL for 1 h. After exposure, return fish to clean tanks and continue feeding a meat diet. Observe daily for 18 additional days through Day 25 by recording hemorrhagic allergic reactions and cumulative mortality. On Day 25, euthanize remaining fish.
7. During all experiments, observe fish at least once daily for clinical signs. Weigh fish individually at sampling times (Day 7 and final day) to assess weight gain/loss. Record mortality daily. Remove and record any dead fish promptly.
8. At euthanasia, collect samples as follows: Draw blood from each fish by severing the caudal peduncle and collecting blood in microhematocrit tubes. Centrifuge tubes at 2000 × *g* for 10 min to separate serum; transfer serum to clean tubes and store at –20 °C for ELISA. Dissect the entire intestine: divide it into two portions. Place one portion in RNAlater for gene expression analysis and place the other in DNA lysis buffer or ethanol for qPCR analysis of *M. marinum*. Store tissues at –80 °C until nucleic acid extraction.

3.4 Enzyme-Linked Immunosorbent Assay (ELISA)

1. For antigen coating, use high-binding 96-well plates. Prepare antigens from tick salivary gland (SG) protein extract from 5-day fed ticks (*see* **Note 1**), α-Gal–BSA conjugate (Dextra Labs NGP0203), recombinant p23, MET, SUB, and *M. bovis*

P22 complex (immunopurified subcomplex of bovine PPD). Dilute tick SG extract and α-Gal conjugate to 0.5 μg/mL (50 ng/100 μL), and recombinant proteins or P22 to 1 μg/mL (100 ng/100 μL) in carbonate–bicarbonate coating buffer pH 9.6 (see Sect. 2.3.2). Add 100 μL per well and incubate overnight at 4 °C with gentle shaking.

2. Plate blocking by removing coating solution and washing wells three times with 200 μL PBST. Block wells with 200 μL/well of 5% skim milk (w/v) in PBS for 1 h at room temperature. After blocking, wash wells 3× with PBST.
3. Dilute zebrafish serum at 1:100 in blocking solution (5% skim milk/PBS). Add 100 μL of each diluted serum per well in duplicate or triplicate. Incubate the plate for 1.5 h at 37 °C. Wash wells 3× with PBST.
4. Add 100 μL of rabbit anti-zebrafish IgM (μ-chain-specific) primary antibody diluted 1:1000 in blocking buffer. Incubate for 1 h at room temperature. Wash plates 3× with PBST.
5. Add 100 μL of goat anti-rabbit IgG–HRP conjugate secondary antibody diluted 1:3000 in blocking buffer. Incubate for 1 h at room temperature. Wash plates 3× with PBST.
6. For detection, add 100 μL/well of 3,3′,5,5′-tetramethylbenzidine TMB substrate. Incubate in the dark at room temperature for 15–20 min until color develops. Stop the reaction by adding 50 μL of 2 N H_2SO_4 to each well. Read absorbance at 450 nm using a microplate reader and subtract background in blank wells incubated without serum.
7. Record OD450 values. The ELISA OD signal is proportional to anti-antigen IgM in the sample. Analyze data after blank subtraction using one-way ANOVA and post hoc tests (*see* **Note 2**).

3.5 Real-Time RT-PCR for Gene Expression

1. For RNA extraction, thaw RNAlater-preserved intestine samples on ice. Remove up to 30 mg of tissue and place in a 2-mL tube with 600 μL RLT buffer (Qiagen AllPrep Kit) containing 1% β-mercaptoethanol. Disrupt tissue with a stainless-steel bead or mortar and pestle until completely homogenized. Perform total RNA extraction using the AllPrep DNA/RNA/Protein Kit according to the manufacturer's protocol. Elute RNA in 30–50 μL RNase-free water.
2. Quality check is conducted by measuring RNA concentration and purity with a NanoDrop spectrophotometer. Ensure A260/A280 ≥1.8. Adjust all RNA samples to 50 ng/μL with RNase-free water and store at −80 °C.
3. Prepare real-time RT-PCR reactions using a one-step kit. For each 20 μL reaction using a SYBR Green One-Step Kit, mix

10 μL of 2× SYBR Green Master Mix with 0.25 μL of iScript reverse transcriptase enzyme mix, 1 μL of 10 μM forward primer, 1 μL of 10 μM reverse primer, and 6.25 μL RNase-free water. Add 1.5 μL RNA sample (75 ng) to each reaction. Include controls without a template and without RT. Prepare all samples for analysis in duplicate.

4. PCR conditions. Run reactions on a real-time thermocycler. Thermal protocol at 50 °C for 10 min for reverse transcription, 95 °C for 1 min for initial denaturation and 40 cycles of 95 °C for 20 s and 60 °C for 1 min annealing/extension. Perform a melt-curve analysis at the end to verify single amplicons. Use melt peak and agarose gel to confirm specific bands.
5. For normalization, analyze raw Cq data using the comparative Cq method (qBase model). Normalize each target gene to the average of two housekeeping genes (β-actin1 and gapdh) measured in the same sample. Calculate relative expression fold-changes versus the PBS control group.

3.6 Quantitative PCR for M. marinum Load

1. Thaw DNA-preserved intestine samples. Use the AllPrep or DNeasy Blood & Tissue Kit to isolate total DNA according to the manufacturer's instructions. Elute DNA in 50 μL buffer.
2. Set up a real-time PCR targeting *M. marinum* 16S rRNA gene. Use 2× SYBR Green PCR Master Mix and primers specific for *Mycobacterium* 16S (e.g., MM16S-F and MM16S-R) [5]. In each 25 μL reaction, include 12.5 μL master mix, 0.5 μL of each primer at 10 μM, 1 μL template DNA (≈50 ng), and water. Run qPCR with denaturation at 95 °C for 3 min, 40 cycles at 95 °C for 15 s each, and 60 °C for 30 s. Include a standard curve using serial dilutions of known CFU equivalents of *M. marinum*.
3. Calculate *M. marinum* load as genome equivalents per mg tissue based on the standard curve. Report data as mean ± SEM of the group.

4 Notes

1. Tick saliva NPF has very low α-Gal content (ca. 0.04 ng/μg protein) compared to the protein fraction.
2. Detailed reagent compositions (buffers, media) and suggested suppliers are listed in Sect. 2. Use nuclease-free reagents and sterile technique for all molecular assays. Data can be analyzed with GraphPad Prism v8.0 and the Real Statistics Excel add-in.

Competing Interests

The authors have no conflicts of interest to declare that are relevant to the content of this chapter.

References

1. Vaz-Rodrigues R, Feo Brito F, Guzmán Rodríguez R, Mazuecos L, de la Fuente J (2023) Allergic reactions associated with medically relevant arthropods. Ann Med 55(2):2286531. https://doi.org/10.1080/07853890.2023.2286531
2. Vaz-Rodrigues R, Rafael M, Carniato D, de la Fuente J (2025) Public perspectives on tick bite exposure, healthcare visits and associated allergies in Iberia. Ann Med 57(1):2499028. https://doi.org/10.1080/07853890.2025.2499028
3. Vaz-Rodrigues R, Mazuecos L, Contreras M, González-García A, Rafael M, Villar M, de la Fuente J (2025) Tick salivary proteins metalloprotease and allergen-like p23 are associated with response to glycan α-Gal and mycobacterium infection. Sci Rep 15(1), 19635. https://doi.org/10.1038/s41598-025-02716-210.1038/s41598-025-93031-3
4. Vaz-Rodrigues R, de la Fuente J (2025) Is zebrafish a good model for the alpha-gal syndrome?. FASEB J 39(9):e70602. https://doi.org/10.1096/fj.202500687R
5. Smith SE, Showers-Corneli P, Dardenne CN, Harpending HH, Martin DP, Beiko RG (2012) Comparative genomic and phylogenetic approaches to characterize the role of genetic recombination in mycobacterial evolution. PLoS One 7(11):e50070. https://doi.org/10.1371/journal.pone.0050070

Check for updates

Chapter 10

Flow Cytometry in Human Cells Exposed to Tick Salivary Proteins

Almudena González-García, Lorena Mazuecos, Rita Vaz-Rodrigues and José de la Fuente

Abstract

The Indirect basophil activation test (iBAT) is a flow cytometry–based assay in which basophils are first passively sensitized with patient-derived IgE and subsequently stimulated with the allergen of interest to assess activation markers on the cell surface. This method was implemented by using the human KU812 cell line to investigate basophil activation mechanisms triggered by tick-derived biomolecules. Unlike the traditional BAT, the iBAT approach avoids variability from non-responder patients and donor-dependent differences, providing a stable and reproducible in vitro model. KU812 cells were differentiated with IL-4, primed with serum from a human patient diagnosed with α-Gal syndrome (AGS), and stimulated with increasing concentrations of tick salivary proteins, including metalloprotease (A0A0K8RCY8) and antigen p23 (A0A0K8RKR7). Activation was assessed by flow cytometry through CD63 and CD203c expression. Dose–response curves were analyzed to determine CDmax, EC_{50}, and AUC, enabling comparison of allergenic potency and overall activation. This iBAT model offers a controlled and reproducible method to explore IgE-mediated mechanisms relevant to tick exposure and the pathophysiology of AGS.

Keywords Basophil, iBAT, KU812 cell-line, Flow cytometry, Tick proteins, α-Gal syndrome

1 Introduction

The basophil activation test (BAT), evaluated by flow cytometry, has emerged as a highly sensitive and informative method for detecting clinically relevant allergies, estimating the severity of reactions, and monitoring therapeutic interventions, with greater clinical value compared to other in vitro diagnostic tests [1, 2]. Importantly, interpretation of BAT requires analysis of allergen dose–response curves, which in basophils are broad, often bell-shaped, and variable among individuals. This variability is influenced by antigen complexity, IgE affinity and distribution, receptor density, and intrinsic properties of basophils themselves [3, 4]. For this reason, testing a range of allergen concentrations rather than a single dose is recommended, and the use of parameters such as maximal reactivity (CDmax), sensitivity

José de la Fuente and Alejandro Cabezas-Cruz (eds.), *Allergic Disease Models*, Methods in Molecular Biology, vol. 3092,
https://doi.org/10.1007/978-1-0716-5635-8_10,

(CDsens, or ED50), or even area under the curve (AUC) has been proposed for more robust comparisons [2, 3]. Basophil reactivity correlates not only with allergen exposure but also with symptom severity in food allergy, highlighting its diagnostic and potential prognostic [5].

Despite its advantages, direct BAT presents important limitations. A proportion of individuals (5–10%) are non-responders, which can be due to intrinsic basophil characteristics or to recent allergen exposure that transiently reduces sensitivity [6]. Additionally, BAT requires fresh blood samples, ideally processed within four hours [5]. These restrictions have led to the development of the indirect BAT (iBAT), in which basophils from healthy donors are passively sensitized with IgE from patient serum. This strategy not only avoids the issue of non-responders but also allows greater flexibility in timing sample collection and processing [7]. The success of this approach depends on factors such as the density of high-affinity receptor for IgE (FcεRI) and the proportion of allergen-specific IgE within the total IgE pool. Low levels of specific IgE or low receptor availability can hinder effective sensitization. Optimization may require high concentrations of purified IgE, longer sensitization times, or enhancing agents to ensure robust activation [4].

In this protocol, the iBAT approach will be applied to test different proteins present in tick saliva that may influence the pathogenesis of α-Gal syndrome (AGS), a tick-borne IgE-mediated delayed hypersensitivity to red meat and derived products [8–10]. Since not all salivary molecules contribute equally to basophil activation, this strategy will allow us to compare their relative capacities to trigger degranulation and upregulation of activation markers. In particular, we will focus on the allergen-like salivary protein p23 (p23, UniProt ID: A0A0K8RKR7) and metalloprotease (UniProt ID: A0A0K8RCY8), which have been identified as key candidates in triggering AGS. [11, 12]. Both proteins have been implicated in allergic reactions to mammalian meat in experimental models, including zebrafish (*Danio rerio*), through the upregulation of proinflammatory genes such as *DNA-dependent protein kinase* (*prkdc*), Toll-like receptor 2 (*tlr2*), tumor necrosis factor α (*tnfα*), interleukin 1b (*il1b*), and cathepsin S (*ctss2.1*) [12]. Further quantitative proteomics analysis of specific tick salivary fractions supports the relevance of p23 and metalloprotease as major components potentially involved in sensitization [1]. Moreover, these proteins, without α-Gal modifications, may also elicit immune responses that modulate both tick infestations and pathogen transmission, suggesting a dual role in host–parasite interactions and allergy induction [13]. By testing p23 and metalloprotease in the iBAT, the aim is to clarify their capacity to activate basophils and thereby contribute to AGS development, with potential implications for diagnosis and targeted therapeutic strategies.

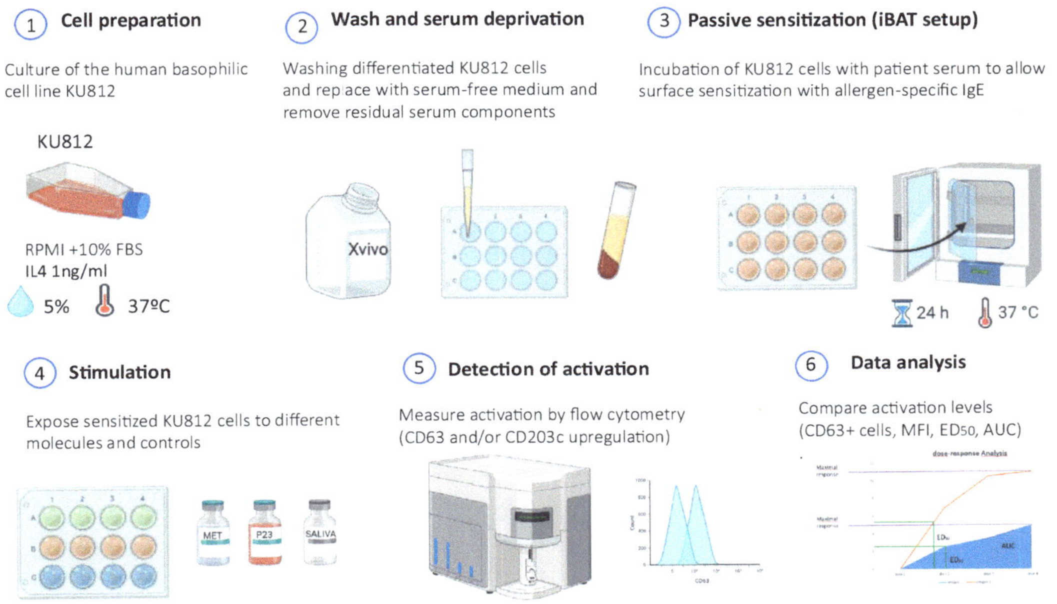

Fig. 1 Experimental design. Schematic steps carried out to detect indirect basophil activation using tick salivary proteins, antigen p23 and metalloprotease. Tick saliva is used as a positive control for activation, and a serum-free medium serves as a negative control

To carry out this study, the human basophilic cell line KU812 will be used [14], which represents a valuable model for immunology and immune system disorder research used to investigate IgE/Fcε RI interactions, given its specific binding to the high-affinity FcεRI receptor despite low receptor density [15–17]. In this study, interleukin-4 (IL-4) was selected, as it markedly increases histamine content and upregulates FcεRI expression, with IL-4–treated cells showing elevated levels of FcεRIα, FcεRIβ, and FcεRIγ mRNA compared to untreated cells [18].

The experimental design is shown in Fig. 1.

2 Materials

2.1 KU812 Cells Culture

1. Roswell Park Memorial Institute (RPMI) 1640 Medium contains the reducing agent glutathione as well as biotin, vitamin B12, and PABA; it also includes high concentrations of the vitamins inositol and choline.
2. 5% Fetal Bovine Serum (FBS) solution.
3. Human IL-4 Recombinant Protein, PeproTech®: add 1 ng/ml. Lyophilized recombinant protein should be reconstituted at room temperature using the appropriate solvent and concen-

tration specified in the lot-specific Certificate of Analysis. The solution should be mixed carefully by gently pipetting up and down, or by securely closing the vial and inverting it slowly several times to ensure complete dissolution.
For short-term storage, the reconstituted protein solution can be kept at 2–8 °C for up to one week. For long-term storage, it should be distributed into aliquots and stored at –20 to –80 °C for 3–12 months (*see* **Note 1**).

4. The standard phosphate-buffered saline (PBS) contains NaCl, 137 mM; KCl, 2.7 mM; Na_2HPO_4, 10 mM; KH_2PO_4, 1.8 mM; adjust pH to 7.2–7.4.
5. X-VIVO® 15 Serum-free Hematopoietic Cell Medium with L-glutamine, gentamicin, and phenol red, xenofree.
6. Filtration equipment for culture media sterilization: Peristaltic pump and filter cartridges with 0.22 µm pore-size membrane (*see* **Note 2**).
7. Temperature-controlled water bath.
8. CO_2 incubator with temperature and humidity control.
9. Giemsa staining.

2.2 Processing of Peripheral Blood for Serum Collection and Storage

1. Syringes and/or other materials required for blood collection.
2. Blood collection tubes with vacuum system, safety cap, and gel separator
3. Sterile cryovials (0.5–2 mL).
4. Sterile pipette tips, with or without filter, compatible with the pipettes used
5. Pipettes capable of handling 0.2–1 mL volumes.
6. Ultra-low-temperature freezer (−80 °C).
7. Human IgE ELISA kit.
8. The ImmunoCAP Phadia 250 automated platform (Thermo Fisher Scientific, Uppsala, Sweden).

2.3 Flow Cytometry Procedures

1. The staining PBS: PBS 1×, 1% FBS, 0.5% Bovine serum albumin (BSA), 0.065% AZIDA.
2. Human antibodies: PE monoclonal human CD203c (E-NPP3) antibody and FITC Monoclonal human CD63 (MEM-259) antibody (*see* **Note 3**).
3. Microcentrifuge.
4. Flow cytometry tubes: polystyrene or polypropylene tubes compatible with the cytometer.
5. Flow cytometer.
6. Neubauer Chamber (*see* **Note 4**).

2.4 Potential Allergy-Inducing Biomolecules

1. Metalloprotease (UniProt ID: A0A0K8RCY8; Genscript Corporation, USA) and p23 (UniProt ID: A0A0K8RKR7, in-house production) should be synthesized as recombinant proteins.
2. Tick saliva (described in Sec. 3.4.1; *see* **Note 5**).

3 Methods

3.1 Cell Preparation

1. Cell line maintenance: Culture KU812 cells in RPMI medium supplemented with 5% FBS and incubated at 37 °C with 5% CO_2 and controlled humidity. The optimal cell density is 3×10^5 viable cells/ml, and the cell concentration should not exceed 3×10^6 cells/ml. Cells should be passaged twice a week at a 1:3 dilution to maintain optimal growth and viability.
2. Differentiation into basophil-like cells: To induce differentiation of KU812 cells toward a basophil-like phenotype, add IL-4 to the culture medium. This treatment promotes an increase in the number of high-affinity IgE receptors (FcεRI) on the cell surface, enhancing the cells' responsiveness to IgE-mediated stimulation. Differentiation is carried out for 21 days to allow full maturation of basophil-like characteristics according to established protocols [18] (*see* **Note 6**).

3.2 Wash and Serum Deprivation

1. Cell counting and calculation for experiments: Cells should be counted using a hemocytometer (Neubauer chamber) or an equivalent device to determine cell density. Based on these counts, calculate the number of cells required for each experiment, considering that 1×10^6 cells per treatment and replicate should be used. This ensures consistent cell numbers across all experimental conditions.
2. Cell washing: Cells should be transferred to sterile conical Falcon tubes and centrifuged at $200 \times g$ for 5 min. This step allows the removal of culture medium and prepares the cells for resuspension in serum-free medium prior to sensitization or stimulation.
3. Resuspension in serum-free medium: After centrifugation, the supernatant should be carefully removed, and the cell pellet will be resuspended in X-VIVO serum-free medium. This prepares the cells for subsequent sensitization and stimulation while minimizing the influence of serum components on basophil activation.
4. Plating of cells: Once resuspended, 1×10^6 cells should be added to each well. Sterile 12-well culture plates will be used, with 1 mL of medium per well to ensure adequate cell density

and nutrient availability. This setup allows proper exposure of the cells to subsequent sensitization or stimulation treatments.

3.3 Passive Sensitization (iBAT Setup)

1. Serum collection and storage: Peripheral blood samples from an AGS patient were collected using sterile vacuum blood collection tubes with a safety cap and clot activator/gel separator by nursing personnel at the General University Hospital of Ciudad Real (HGUCR, Spain). After blood collection, the tubes must be left at RT for 30–45 min to allow clot formation. The samples should be centrifuged at 1500 × *g* for 10 min at 4 °C to separate the serum from the clot. The resulting serum must be carefully transferred into sterile cryovials (0.5–2 mL aliquots) using sterile pipette tips to avoid contamination. Serum aliquots can be stored at −80 °C until use (*see* **Notes 7 and 8**).
2. Serum quantification of total and specific IgE: Before sensitization, it is essential to quantify both total IgE and α-Gal-specific IgE (sIgE) in the patient serum. This information allows accurate calculation of the volume of serum needed to occupy the majority of free FcεRI receptors on KU812 cells [4]. Total IgE levels can be measured by using a human IgE ELISA kit, following manufacturer's recommendations. Anti-α-Gal sIgE levels can be determined using the ImmunoCAP Phadia 250 automated platform (Thermo Fisher Scientific, Uppsala, Sweden) with the commercial ImmunoCAP α-Gal bovine Thyroglobulin kit, according to the manufacturer's instructions. Positive results are considered for anti-α-Gal IgE levels ≥0.1 kU/L [19].
3. Sensitization of KU812 cells with patient serum: In our study, serum sIgE concentration against α-Gal was 671.9 kU/L. To sensitize the KU812 cells, first prepare a 1:1000 dilution of patient serum in serum-free culture medium. Then, add 200 μL of this diluted serum per well containing 1×10^6 KU812 cells in 1 mL of medium. Finally, incubate for at least 24 h at 37 °C in a humidified atmosphere with 5% CO_2 to allow binding of IgE to the FcεRI receptors (*see* **Note 9**).
4. Post-incubation washes: After the sensitization period, the KU812 cells should be transferred to sterile conical tubes and centrifuged at 200 × *g* for 5 min. The supernatant must be carefully removed, and the cells washed twice with cold X-VIVO medium to eliminate any unbound serum. After the washes, the cells should be resuspended in X-VIVO medium at a concentration of 1×10^6 cells/mL and returned to culture plates in preparation for the subsequent stimulation step.

3.4 Stimulation

1. Addition of candidate allergenic proteins or tick saliva: Add candidate allergenic tick salivary proteins to the sensitized

KU812 cells. A dose–response curve should be established using a maximum concentration of 20 µg per one million cells and a minimum concentration of 1 µg per one million cells. Test at least five different concentrations for each protein to generate this dose-response curve. Include a minimum of two replicates per concentration. In this study, the concentrations used for each protein were: 0.5, 1, 2, 5, and 10 µg per 1×10^6 cells. As a positive control, use tick saliva at concentrations ranging from 0.1–10 µg of total protein per 1×10^6 cells. Prior quantification of protein content is necessary to calculate the appropriate dose for experiments. Protein concentration can be determined using a BCA assay kit, following the manufacturer's instructions. This step is essential due to the significant variation in protein concentration among tick species. In this study, tick saliva was obtained from *Amblyomma americanum*, and the maximum dose used was 0.20 µg of protein per one million cells.

2. Incubation: Incubate the cells with the proteins for 3–4 h at 37 °C in a humidified atmosphere with 5% CO_2 (*see* **Note 10**).

3.5 Detection of Activation by Flow Cytometry

1. Sample preparation for flow cytometry: After the stimulation period, transfer the cells from each well to sterile conical tubes. Centrifuge at 200 × *g* for 5 min and remove the supernatant. Resuspend the cell pellet in cold staining buffer. Add fluorophore-conjugated antibodies specific for CD63 and/or CD203c following the manufacturer's recommended concentrations, as well as a viability marker such as propidium iodide (PI) to exclude dead cells from the analysis. Incubate samples for 1 h at 4 °C in the dark to ensure proper staining. After incubation, wash cells once with cold buffer, centrifuge at 200 × *g* for 5 min, discard the supernatant, and finally resuspend the pellet in 300 µL of PBS staining buffer for flow cytometry acquisition.
2. Flow Cytometry Controls: To ensure accurate data acquisition and proper compensation during flow cytometry analysis, include the following controls:
   (a) *Single-stained controls*: prepare separate cell samples stained with each individual fluorophore-conjugated antibody (e.g., CD63, CD203c) and the viability marker (propidium iodide) to allow proper compensation settings on the cytometer.
   (b) *Unstained control*: include a cell sample with no antibody or viability marker staining to assess background fluorescence.

(c) *Fully stained control*: prepare a cell sample stained with all antibodies and the viability marker to verify multicolor panel performance and detect potential spectral overlap.

For these controls, use a positive sample known to express the target markers. In this experiment, tick saliva-stimulated cells were used as the positive sample to ensure sufficient marker expression for proper control setup (*see* **Note 11**).

3. Data Acquisition: During data acquisition on the flow cytometer, follow these steps to ensure accurate gating and analysis of the KU812 basophil population (Fig. 2):

   (a) *Initial population gating* (FSC vs SSC): begin by plotting forward scatter (FSC) on a linear scale versus side scatter (SSC) on a logarithmic scale to visualize cell size and complexity. Use this plot to gate the main cell population, excluding debris and non-cellular events.

   (b) *Doublet exclusion*: to eliminate cell aggregates and doublets generated during sample handling or cytometer error, create a plot of FSC-A versus FSC-H (or FSC-W) and gate only the singlet population.

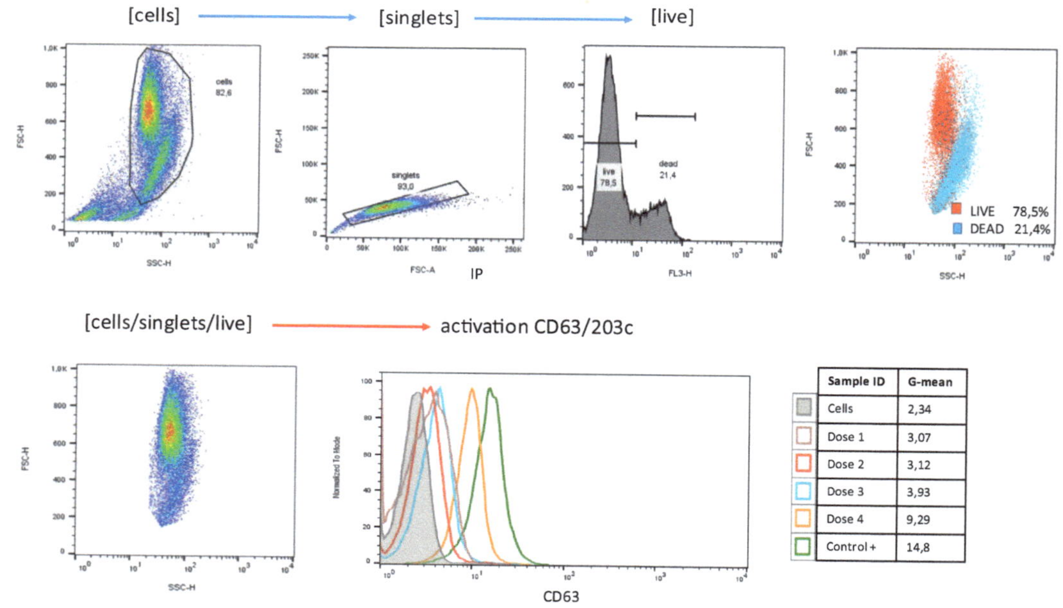

Fig. 2 Representative gating strategy used for flow cytometry analysis of KU812 cells. The first gate selects the cell population based on size and complexity (FSC vs SSC). Singlets are then discriminated using FSC-A vs FSC-H plots to exclude aggregates. Viable cells are identified by excluding PI-positive events. Finally, within the viable singlet population, basophil activation is assessed by measuring the upregulation of CD63 and/or CD203c markers. The plots shown are for illustrative purposes only and do not represent actual experimental data

(c) *Viability gating*: use the viability marker (propidium iodide, PI) to exclude dead cells from the analysis. Select the PI-negative population as the viable cells for downstream activation marker analysis.

(d) *Activation marker analysis*: within the viable singlet population, evaluate the expression of activation markers CD63 and/or CD203c to measure basophil activation in response to allergen stimulation. All fluorescence channels should be acquired on a logarithmic scale.

(e) *Event collection*: acquire a minimum of 10,000 viable events per sample to ensure robust and statistically meaningful analysis (*see* **Note 12**).

3.6 Data Analysis of the KU812 iBAT Assay

1. Generation of Dose–Response Curves: For each allergenic protein selected, a dose–response curve should be constructed by relating the concentration of the stimulant to basophil activation measured through the selected marker (CD63 or CD203c). Activation can be expressed either as the percentage of activated basophils or as the geometric mean fluorescence intensity (Geo MFI) of the marker, since flow cytometry data are acquired on a logarithmic scale. These quantitative parameters allow an accurate assessment of the extent of activation across the tested concentration range.
2. CDmax (Basophil Reactivity): The CDmax represents the maximal activation level achieved in response to the allergen and reflects the overall reactivity of the basophils. It is obtained from the upper plateau of the fitted curve and allows comparison of the maximal response elicited by different stimuli.
3. EC_{50} (Basophil Sensitivity): A sigmoidal four-parameter nonlinear logistic model is used to estimate the EC_{50}, defined as the concentration of allergen that induces 50% of the maximal activation (CDmax). This parameter indicates the sensitivity of the basophil population to the tested stimulus and enables comparison between different allergens or serum conditions (Fig. 3).
4. AUC (Area Under the Curve): The AUC is calculated by integrating the activation response across all tested concentrations. This parameter provides an overall measure of the magnitude and breadth of basophil activation, regardless of the concentration at which maximum activation occurs, and is particularly useful for comparing global activation profiles among treatments (Fig. 3).

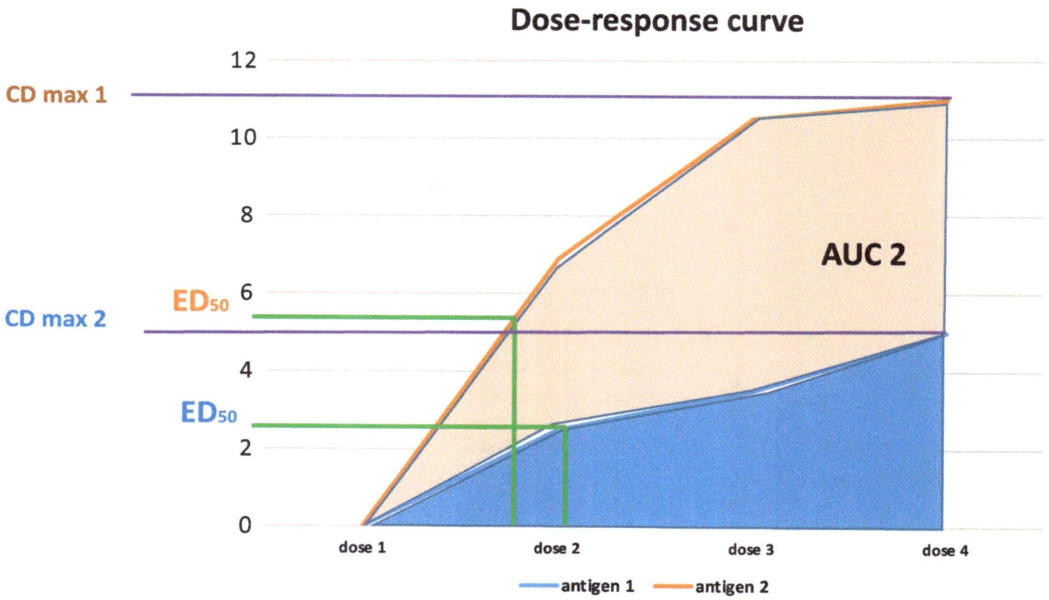

Fig. 3 Representative analysis of basophil activation parameters. Representative dose–response curves illustrating the calculation of key activation parameters: CDmax (maximum basophil reactivity), EC_{50} (concentration producing 50% of maximal activation), and AUC (area under the curve, reflecting total activation) for two different antigens. The data shown are illustrative and not derived from real experimental results

4 Notes

1. When the cell culture environment is considered susceptible to microbial contamination, the culture medium should be supplemented with antibiotics and/or antimycotics to prevent bacterial or fungal overgrowth and thereby ensure the viability and integrity of KU812 cells throughout the experimental procedures.
2. Filtration through 0.22 μm pore-size membranes is widely used for sterilization of culture media, as this pore size effectively removes bacteria and other microbial contaminants while preserving the physicochemical properties of heat-sensitive components that could be degraded by autoclaving.
3. Antibody titration, ideally performed on one million cells, is strongly recommended to define the optimal working concentration. This process enables identification of the antibody dilution that provides the best separation between positive and background populations, thereby enhancing data resolution and clarity. In addition, proper titration minimizes

nonspecific binding and fluorescence spillover into adjacent channels, optimizes the signal-to-noise ratio, and contributes to improved reproducibility while reducing unnecessary antibody consumption.

4. In this flow cytometry panel, only two fluorescence channels should be used. Therefore, any basic flow cytometer equipped with a blue laser is sufficient for performing this technique.
5. It will be necessary to test at least five different concentrations of potential biomolecules to generate a dose–response curve. This approach enables a more reliable assessment of basophil activation and helps capture both suboptimal and maximal responses for accurate interpretation.
6. Morphological changes should be monitored every 2–3 days by Giemsa staining. These changes include increased cytoplasmic granulation, which indicates progression toward a basophil-like phenotype.
7. Repeated freeze–thaw cycles should be avoided to preserve the integrity of immunoglobulins, particularly IgE.
8. All blood collection procedures must be conducted under approved ethical protocols and comply with institutional and regulatory guidelines.
9. This approach ensures that all free FcεRI receptors on the KU812 cells are saturated, allowing subsequent activation assays with tick salivary proteins or other α-Gal-containing antigens to reliably measure basophil degranulation.
10. To determine the maximum protein concentration, it is recommended to perform a cytotoxicity assay, such as the MTT assay (e.g., Abcam kit), to ensure that the protein concentrations used are not within a toxic range for the KU812 cells.
11. If using automated compensation with compensation beads, these controls may not be strictly necessary; however, they remain highly recommended to ensure accuracy and reliability.
12. Ensure proper compensation and voltage settings are applied prior to acquisition, using the controls prepared in Sect. 6.2, to avoid spectral overlap and to obtain reliable measurements.

Competing Interests

The authors have no conflicts of interest to declare that are relevant to the content of this chapter.

References

1. Eberlein B (2020) Basophil activation as marker of clinically relevant allergy and therapy outcome. Front Immunol 11, 1815. https://doi.org/10.3389/fimmu.2020.01815
2. Hemmings O, Kwok M, McKendry R, Santos AF (2018) Basophil activation test: old and new applications in allergy. Curr Allergy Asthma Rep 18:77. https://doi.org/10.1007/s11882-018-0831-5
3. Patil SU, Shreffler WG (2012) Immunology in the clinic review series; focus on allergies: basophils as biomarkers for assessing immune modulation. Clin Exp Immunol 167:59–66. https://doi.org/10.1111/j.1365-2249.2011.04503.x
4. MacGlashan DW (2013) Basophil activation testing. J Allergy Clin Immunol 132:777–787. https://doi.org/10.1016/j.jaci.2013.06.038
5. Hoffmann HJ, Knol EF, Ferrer M, et al (2016) Pros and cons of clinical basophil testing (BAT). Curr Allergy Asthma Rep 16:56. https://doi.org/10.1007/s11882-016-0633-6
6. Ebo DG, Bridts CH, Hagendorens MM et al (2008) Basophil activation test by flow cytometry: present and future applications in allergology. Cytometry B Clin Cytom 74B:201–210. https://doi.org/10.1002/cyto.b.20419
7. Ruinemans-Koerts J, Brouwer ML, Schmidt-Hieltjes Y et al (2022) The indirect basophil activation test is a safe, reliable, and accessible tool to diagnose a peanut allergy in children. J Allergy Clin Immunol Pract 10:1305–1311.e3. https://doi.org/10.1016/j.jaip.2021.12.040
8. Galili U (2019) Evolution in primates by "Catastrophic-selection" interplay between enveloped virus epidemics, mutated genes of enzymes synthesizing carbohydrate antigens, and natural anti-carbohydrate antibodies. Am J Phys Anthropol 168:352–363. https://doi.org/10.1002/ajpa.23745
9. Van Nunen SA, O'Connor KS, Clarke LR et al (2009) An association between tick bite reactions and red meat allergy in humans. Med J Aust 190:510–511. https://doi.org/10.5694/j.1326-5377.2009.tb02533.x
10. Mullins RJ, James H, Platts-Mills TAE, Commins S (2012) Relationship between red meat allergy and sensitization to gelatin and galactose-α-1,3-galactose. J Allergy Clin Immunol 129:1334–1342.e1. https://doi.org/10.1016/j.jaci.2012.02.038
11. Contreras M, Vaz-Rodrigues R, Mazuecos L et al (2023) Allergic reactions to tick saliva components in zebrafish model. Parasit Vectors 16:242. https://doi.org/10.1186/s13071-023-05874-2
12. Vaz-Rodrigues R, Mazuecos L, Contreras M, et al (2025) Tick salivary proteins metalloprotease and allergen-like p23 are associated with response to glycan α-Gal and mycobacterium infection. Sci Rep 15:8849. https://doi.org/10.1038/s41598-025-93031-3
13. Villar M, Pacheco I, Mateos-Hernández L et al (2021) Characterization of tick salivary gland and saliva alphagalactome reveals candidate alpha-gal syndrome disease biomarkers. Expert Rev Proteomics 18:1099–1116. https://doi.org/10.1080/14789450.2021.2018305
14. KU812 - CRL-2099 | ATCC. https://www.atcc.org/products/crl-2099. Accessed 14 Jan 2025
15. Magnusson CGM, Håård J, Matsson P et al (1995) Demonstration of specific high-affinity Fcε-receptors on the human basophil-like leukemia cell line KU812 by flow cytometry. Allergy 50:72–77. https://doi.org/10.1111/j.1398-9995.1995.tb02485.x
16. Passante E (2020) Mast Cell and Basophil Cell Lines: a compendium. Methods Mol Biol Clifton NJ 2163:127–144. https://doi.org/10.1007/978-1-0716-0696-4_10
17. Almlöf I, Nilsson K, Johansson V et al (1988) Induction of basophilic differentiation in the human basophilic cell line KU812. Scand J Immunol 28:293–300. https://doi.org/10.1111/j.1365-3083.1988.tb01451.x
18. Hara T, Yamada K, Tachibana H (1998) Basophilic differentiation of the human Leukemia cell line KU812 upon treatment with interleukin-4. Biochem Biophys Res Commun 247:542–548. https://doi.org/10.1006/bbrc.1998.8816
19. Bircher AJ (2005) Systemic immediate allergic reactions to arthropod stings and bites. Dermatology 210:119–127. https://doi.org/10.1159/000082567

Check for updates

Chapter 11

Evaluation of Alpha-Gal Content and Expression of Galactosyltransferases in Arthropods

José de la Fuente and Rita Vaz-Rodrigues

Abstract

Allergic reactions associated with arthropod bites are a growing concern worldwide. In particular, Alpha-Gal syndrome (AGS) is a red meat allergy associated with IgE antibodies against galactose-alpha-1,3-galactose (alpha-Gal) that is typically linked to tick bites, but other arthropods with alpha-Gal content may also induce allergic reactions. The AGS is underdiagnosed and thus important to evaluate the risks that may be associated with exposure to bites of arthropods with alpha-Gal. To approach this objective, herein are described the materials and methods for the detection of alpha-Gal content and expression of galactosyltransferases in arthropods.

Keywords Arthropod, Alpha-Gal, Alpha-Gal syndrome, ELISA, Galactosyltransferase, RT-PCR, Tick

1 Introduction

Allergens produced by arthropods induce allergic reactions in humans [1–4]. The oligosaccharide galactose-alpha-1,3-galactose (alpha-Gal) is a protein and lipid modification found in mammalian-derived products such as meat and pharmaceuticals such as cetuximab and bioprosthetic valves [2, 5]. Alpha-Gal syndrome (AGS) is underdiagnosed and associated with IgE antibodies to alpha-Gal and allergic reactions sensitized mainly by tick bites but also by other arthropods [1–6].

Alpha-Gal is produced by galactosyltransferases, and hominids evolved with an inactivated capacity to synthesize this biomolecule through a process known as "catastrophic selection" [7]. The inactivation of alpha-Gal synthesis is associated with the capacity to develop anti-alpha-Gal IgM/IgG antibodies that are protective against pathogen infection but carry the risk of developing AGS [7]. To advance in AGS diagnosis, prevention, and control, it is important to evaluate the alpha-Gal content in arthropods associated with

José de la Fuente and Alejandro Cabezas-Cruz (eds.), *Allergic Disease Models*, Methods in Molecular Biology, vol. 3092,
https://doi.org/10.1007/978-1-0716-5635-8_11,

allergic reactions and thus representing a risk for individuals exposed to these organisms [8].

To approach this objective, herein are described the materials and methods for an ELISA to determine alpha-Gal content and a reverse transcriptase (RT) quantitative polymerase chain reaction (qPCR), RT-qPCR, for the characterization of expression of galactosyltransferases in arthropods.

2 Materials

2.1 Arthropods

1. Arthropods. Arthropods are collected in the target study region and morphologically classified by sex and species levels following reference indicators (*see* **Note 1**). If possible, it is recommended to collect unfed arthropods to prevent the effect of host blood-derived biomolecules.
2. Arthropod organs and compounds. Organs from dissected individual phosphate-buffered saline (PBS)-washed arthropods are disrupted with a scalpel and homogenized with a tissue grinder until lysates are completely clear. Fresh saliva is collected from semi-engorged female ticks. Ticks are stimulated with pilocarpine hydrochloride to induce salivation. Saliva is harvested into capillaries. Store organs and tick saliva at −80 °C until use. *Safety:* Handle tick saliva as a potential biohazard (wear gloves, work in a biosafety cabinet) if not certified as pathogen-free. PBS, 137 mM NaCl, 2.7 mM KCl, 10 mM Na_2 HPO_4, 1.8 mM KH_2PO_4, pH 7.4 is used for dilutions, injections, washes, and dialysis. Sterilize by autoclaving and store at room temperature. For injections and tissue work, use sterile filtered PBS.
3. Pilocarpine hydrochloride. Cholinergic agonist used to induce tick salivation. Prepare a sterile aqueous solution (e.g., 4% w/v) of pilocarpine HCl and apply ~5 μL to the dorsal surface of each tick to stimulate saliva release. Store pilocarpine powder desiccated at 4 °C and handle with gloves.
4. Organ extract. Antigen extracts prepared from dissected organs are used for ELISA coating. Dissected organs are homogenized in PBS containing protease inhibitors with a tissue grinder until lysates are completely clear by centrifugation and to determine protein concentration by BCA assay. Store at −80 °C.

2.2 Analysis of DNA/RNA/Proteins

1. DNA/RNA/Protein extraction. For DNA/RNA/Protein extraction, a Mini Kit (e.g., AllPrep, Qiagen, Hilden, Germany) is used for simultaneous extraction of genomic DNA, total RNA, and proteins following the manufacturer's protocol.

Store kit components per instructions (most buffers at room temperature; protein additives at 2–8 °C).

2. RNA and protein concentration. Concentration (ng/μL) and purity of RNA samples are assessed by quantification of the nucleic acids at an optical density of 260 nm (O.D. 260) and the ratio of absorbance at 260/280 nm, using a spectrophotometer (e.g., NanoDrop One, Thermo Scientific, Waltham, USA). Concentrations are then standardized at 50 ng/μL and samples are stored at −80 °C. Total protein concentration is determined with the BCA Protein Assay using bovine serum albumin (BSA) as a standard, and then samples are stored at −80 °C until further analysis.

3. Primers for analysis of galactosyltransferase expression (RT-qPCR). Gene-specific primers for the different galactosyltransferases are designed based on published sequences for each arthropod (Table 1) and prepared as 10 μM working stocks and stored at –20 °C. To quantify mRNA expression, use a one-step reverse transcription SYBR Green qPCR mix (e.g., iScript One-Step RT-PCR Kit with SYBR® Green, Bio-Rad). Store the RT enzyme at –20 °C and the reaction mix at 4 °C. Mix the RNA template with the SYBR mix, gene-specific primers, and RT enzyme, then run the combined reverse transcription (50 °C) and PCR amplification in a real-time cycler.

4. BCA protein assay. Bicinchoninic acid protein quantification kit for measuring protein concentrations (e.g., Pierce BCA Kit, Thermo Scientific). The BCA assay is used to quantify total protein in tick saliva fractions and recombinant protein preparations. Prepare BCA working reagent according to kit instructions and use BSA standards for calibration. Store kit reagents as directed, with most at room temperature.

5. Reagents and solutions for ELISA analysis of alpha-Gal content. Use carbonate–bicarbonate buffer and microbiological-grade skim milk in PBS. The 50 mM carbonate/bicarbonate buffer is used for ELISA plate coating with 1.59 g Na_2CO_3 and 2.93 g $NaHCO_3$ per liter, pH 9.6. Or commercially obtained as a ready buffer and stored at 4 °C. Use to dilute antigens for coating ELISA plates (50–100 ng antigen per well). For blocking solution, freshly prepared 5% (w/v) skim milk in PBS is used. PBS with 0.05% Tween 20 (PBST). 3,3′,5′50-Tetramethylbenzidine (TMB) (e.g., One Solution, Promega, Madison, USA) is a ready-to-use solution that develops a blue color in the presence of peroxidase. Store TMB solution at 4 °C in the dark. Add 100 μL per well after secondary antibody incubation. TMB is light-sensitive and thus perform incubations in the dark (cover the plate with aluminum foil). Typical development time is around 15 min at room

Table 1

Reference arthropod galactosyltransferase primers and RT-qPCR conditions obtained from [4]

Name	Sequence IDs	Organism	Primer sequences (5'–3') and RT-qPCR annealing temperature
beta-1,4-galactosyltransferase 1 (*b4galt1*)	XM_038041878.1 LOC119642653 GQX74_014248	*Glossina fuscipes*	F: AGCTGTGCTCAAAAGGTGGT R: GCCTTGCAGGTCTCATGGAT Annealing: 59 °C
beta-1,4-galactosyltransferase 7 (*b4galt7*)	XM_038036542.1 LOC119639257 GFUI046427	*Glossina fuscipes*	F: CCATCCGATGCTGGACCTTT R: ATCCTCTAGACCCCAACCCC Annealing: 60 °C
alpha-galactosidase (*gla*)	GMOY004750	*Glossina morsitans*	F: TAACCTGGGCGTCCTTCATTT R: GTCTGGTGCGCTAGTCACAATG Annealing: 58 °C
alpha1,4-galactosyltransferase 1 (*a4galt1*)	NM_134893.4 33512 FBgn0031491	*Drosophila melanogaster*	F: CTGGCCCCACCTATCGAATC R: TTTTGACCGCGACAAACGAC Annealing: 58 °C
alpha1,4-galactosyltransferase 2 (*a4galt2*)	NM_143177.4 43124 FBgn0039378	*Drosophila melanogaster*	F: GATTGCCACCATGTGTCTGC R: CTTGAAGCCATTGCAGCGTT Annealing: 59 °C

Sequence IDs are obtained from the National Center for Biotechnology Information (NCBI, https://www.ncbi.nlm.nih.gov for nucleotide and gene sequences) and VectorBase (https://vectorbase.org/vectorbase/app). Abbreviations: F, forward primer; R, reverse primer temperature until a blue color is evident. For the stop solution in ELISA, 2 N sulfuric acid (H_2SO_4) is prepared fresh on the day of the assay by diluting concentrated H_2SO_4 into water (acid into water, in a fume hood; wear eye/hand protection). Use 50 µL per well to stop the TMB reaction, which turns the solution yellow. Dispose of acid waste properly.

6. Antibodies for ELISA. Primary antibody is anti-alpha-Gal epitope monoclonal antibody (Clone M86; Enzo Life Sciences Inc., Farmingdale, NY, USA). Secondary antibodies are goat anti-mouse IgM (µ-chain-specific)-peroxidase antibodies.

2.3 Equipment

1. Centrifuge. Bench-top microcentrifuge capable of ≥14,000 × *g*. This is required for filtering tick saliva (Amicon ultrafilters) and for pelleting samples (e.g., during DNA/RNA extraction). Use 1.5 mL or 2 mL sealed tubes for high-speed spins. A refrigerated centrifuge is preferable for sensitive samples (use at 4 °C for nucleic acid work).
2. Spectrophotometer (NanoDrop). A micro-volume UV–Vis spectrophotometer (e.g., Thermo Fisher Scientific NanoDrop One) to quantify DNA/RNA concentrations and purity. Ensure it's calibrated and clean before use. Use to measure OD_{600} for bacterial culture density (if needed) and $OD_{260/280}$ for nucleic acids.
3. Thermal cyclers for RT-qPCR. A quantitative PCR machine with fluorescence detection (e.g., Bio-Rad CFX96 Touch real-time PCR system).
4. Plate reader. A microplate spectrophotometer capable of reading absorbance at 450 nm after the TMB stop reaction for ELISA results.
5. Biosafety cabinet (Class II). Use a laminar flow biosafety cabinet when handling *M. marinum* cultures, tick saliva, or any infectious material. This provides a sterile workspace and protects the user from aerosols. Required for BSL-2 containment of *M. marinum*.
6. Basic laboratory equipment. Pipettes (P10, P100, P1000) with sterile filter tips for precise liquid handling, especially for qPCR setup and ELISA. A vortex mixer to mix reagents and samples. A water bath or dry block heater (50 °C and 95 °C settings) if needed for certain incubations (though the real-time cycler covers these for RT and PCR). A pH meter to adjust buffer pH (for PBS or tricaine solutions). Timer or stopwatch for infection exposure and ELISA timing. Personal protective equipment (lab coat, gloves, safety goggles) for all experimental procedures.

3 Methods

3.1 Enzyme-Linked Immunosorbent Assay

1. Enzyme-Linked Immunosorbent Assay (ELISA) for alpha-Gal content (*see* **Note 2**). For antigen coating, coat 96-well plates with 500 ng/well of protein from the different samples in 50 μL of carbonate-bicarbonate buffer and incubate overnight at 4 °C with gentle shaking.
2. After three washes with 150 μL/well PBS 1×, block coated plates with 100 μL/well of 5% skim milk in PBS blocking solution for 1 h at room temperature (RT) with gentle shaking. Add anti-alpha-Gal epitope monoclonal antibodies (Clone M86) at 1:50 dilution in 2% skim milk in PBS (assay diluent) and incubate for 1.5 h at 37 °C. Then, wash plates four times with PBST.
3. Add goat anti-mouse IgM (μ-chain-specific)-peroxidase antibodies at a 1:2000 dilution in assay diluent for 1 h at RT. Wash plates 3× with PBST. For detection, add 100 μL/well of TMB solution for 10 min at RT in the dark. Stop reactions with 50 μL/well of 2 N H_2SO_4 and measure the optical density (OD) at 450 nm in a spectrophotometer.
4. For data analysis, subtract the average values of the blanks (coated wells incubated with assay diluent; $n = 4$) from all reads and then calculate alpha-Gal concentration (μg/mL) by using a standard calibration curve ($R^2 > 0.9$) with six concentrations from 0.0 to 1.0 mg/mL of alpha-Gal (Galα1-3Gal-BSA, e.g., Dextra, Shinfield, UK). The limit of detection (LOD) for the assay is approximately 33.28 ng/mL and the limit of quantification (LOQ) is 102.43 ng/mL. Analyze data (after blank subtraction) using a one-way ANOVA and post hoc test. Results are graphically represented using GraphPad Prism (version 8.0.1).

3.2 Real-Time-qPCR

1. Real-Time-qPCR for Galactosyltransferase Gene Expression. Perform total RNA extraction using a DNA/RNA/Protein Mini Kit according to the manufacturer's protocol. Elute RNA in 30–50 μL RNase-free water. For quality control, measure RNA concentration and purity by quantification of the nucleic acids at an optical density of 260 nm (OD260) and the ratio of absorbance at 260/280 nm using a spectrophotometer.
2. Adjust all RNA samples to 50 ng/μL with RNase-free water and store at −80 °C. The analysis of expression of several galactosyltransferase mRNA levels is conducted by RT-qPCR employing a real-time PCR detection system with gene-specific primers and RT-qPCR conditions (Table 1) (*see* **Note 3**). For a total volume of 20 μL, the PCR mixture contains 1.5 μL (75 ng) of

sample mRNA, 10 µL of SYBR Green Master Mix, 0.25 µL of iScript reverse transcriptase, 1 µL (10 µM) each of forward and reverse primers, and 6.25 µL of nuclease-free water. Each PCR reaction has two technical replicates/sample and two negative controls.

3. Complete RT-qPCR conditions comprise a reverse transcription reaction at 50 °C for 10 min, followed by an initial denaturation step at 95 °C for 1 min, amplification by 40 cycles of 95 °C for 20 s and a variable annealing temperature for 1 min, followed by a dissociation curve analysis. For data analysis, employ the generalized qBase model to calibrate, normalize, and obtain the relative expression values of the different samples against the reference sample using tubulin beta-1 chain (β -tubulin) and tubulin alpha-1 chain (α-tubulin) as housekeeping genes (*see* **Note 4**).
4. Perform statistical analysis using a one-tailed Student´s *t*-test with unequal variance via the Real Statistics add-in for Microsoft Excel (version 2310), comparing analyzed samples to tick reference mRNA levels (*see* **Note 3**). Data are expressed as fold change and depicted using GraphPad Prism (version 8.0.1).

4 Notes

1. For arthropod classification, follow guidelines for different species (e.g., [10] and https://www.mayoclinic.org/diseases-conditions/alpha-gal-syndrome/in-depth/tick-species/art-20546861#:~:text=Soft%20ticks%20are%20a%20different,and%20where%20to%20find%20them). The implication of experts in arthropod morphological analysis is recommended.
2. For example, see references [4, 9].
3. Samples from unfed alpha-Gal-positive ticks (e.g., *Ixodes ricinus* or *Amblyomma americanum*) whole-body samples ($n > 4$) are measured as positive controls for comparison using the appropriate primers.
4. See reference to methods in [11, 12].

Competing Interests

The authors have no conflicts of interest to declare that are relevant to the content of this

References

1. Arlian LG (2002) Arthropod allergens and human health. Annu Rev Entomol 47:395–433
2. Vaz-Rodrigues R, Feo Brito F, Guzmán Rodríguez R, Mazuecos L, de la Fuente J (2023) Allergic reactions associated with medically relevant arthropods. Ann Med 55:2286531
3. Bircher AJ (2005) Systemic immediate allergic reactions to arthropod stings and bites. Dermatology 210:119–127
4. Vaz-Rodrigues R,P, Odyek Abila P, Kasaija PD, Kibuuka M, de la Fuente J (2025) Allergic reactions associated with tsetse fly bites. Ann Med 57(1):2514096
5. Kuravi KV, Sorrells LT, Nellis JR et al (2022) Allergic response to medical products in patients with alpha-gal syndrome. J Thorac Cardiovasc Surg 164:e411–e424
6. Shishido AA, Wormser GP (2025) A Review of Alpha-Gal Syndrome for the infectious diseases practitioner. Open Forum Infect Dis 12(8):ofaf430
7. Galili U (2019) Evolution in primates by "Catastrophic-selection" interplay between enveloped virus epidemics, mutated genes of enzymes synthesizing carbohydrate antigens, and natural anti-carbohydrate antibodies. Am J Phys Anthropol 168:352–363
8. Vaz-Rodrigues R, Mazuecos L, de la Fuente J (2022) Current and future strategies for the diagnosis and treatment of the Alpha-Gal Syndrome (AGS). J Asthma Allergy 15:957–970
9. Villar M, Pacheco I, Mateos-Hernández L et al (2021) Characterization of tick salivary gland and saliva alphagalactome reveals candidate alpha-gal syndrome disease biomarkers. Expert Rev Proteom 18:1099–1116
10. Rohde R (2023) Evolution and classification of arthropods. Entomol Ornithol Herpetol 12:322
11. Hellemans J, Mortier G, De Paepe A, Speleman F, Vandesompele J (2007) qBase relative quantification framework and software for management and automated analysis of real-time quantitative PCR data. Genome Biol 8:R19
12. Vandesompele J, De Preter K, Pattyn F et al (2002) Accurate normalization of real-time quantitative RT-PCR data by geometric averaging of multiple internal control genes. Genome Biol 3:research0034.1–11

Chapter 12

Untargeted UHPLC–HRMS Lipidomic Profiling of Ticks for Alpha-Gal Syndrome Research

Hanna Nikolaichuk, Anna Kozub-Pędrak, Joanna Kulisz, Zbigniew Zając, Aneta Woźniak, Alejandro Cabezas-Cruz and Emilia Fornal

Abstract

Lipids are a fundamental group of biomolecules that play diverse roles in biological systems, including energy storage, membrane architecture, and intra- and extracellular signaling. They are classified into eight major categories, and more than 50 000 distinct lipid structures are currently reported in the Lipid Maps Structure Database (LMSD). Given their importance in cellular metabolism and physiology, accurate identification and quantification of lipids are essential. In ticks, lipid metabolism is closely associated with physiological processes related to survival, pathogen transmission, and host interaction. Tick blood-feeding has also been linked to the development of alpha-gal syndrome (AGS), a delayed allergic reaction to mammalian-derived products, suggesting that characterization of tick lipid and glycolipid composition may provide insights into mechanisms of host sensitization. This chapter describes a protocol for untargeted lipidomic analysis of tick samples using ultra-high-performance liquid chromatography coupled with high-resolution mass spectrometry (UHPLC–HRMS). The workflow includes rapid sample freezing, tissue homogenization, solvent-based lipid extraction, protein precipitation, and sample reconstitution prior to analysis. Lipid separation is achieved using reverse-phase chromatography on a C18 column, followed by detection with a quadrupole time-of-flight (Q-TOF) mass spectrometer operating under optimized ionization conditions. The use of blanks and pooled quality control (QC) samples ensures analytical stability and reproducibility. The presented method enables comprehensive profiling of lipids in tick tissues and provides a reproducible platform for studies of tick biology, host–tick interactions, and tick-induced sensitization processes, supporting the mechanistic studies and the identification of potential lipid-based biomarkers.

Keywords Tick lipidomics, Ultra-high pressure lipid chromatography, High-resolution mass spectrometry, Arthropod metabolomics

1 Introduction

Lipids are a crucial group of biomolecules that participate in numerous essential cellular functions, including membrane organization and intra- and extracellular signaling. They are key components of biological membranes and serve as the physical foundation of all

José de la Fuente and Alejandro Cabezas-Cruz (eds.), *Allergic Disease Models*, Methods in Molecular Biology, vol. 3092,
https://doi.org/10.1007/978-1-0716-5635-8_12,

living organisms. Based on their chemical structure, lipids are categorized into eight primary categories: fatty acyls (FA), glycerolipids (GL), glycerophospholipids (GP), sphingolipids (SP), sterols (ST), prenol lipids (PR), saccharolipids (SL), and polyketides (PK). Continuous development of lipidomics resources has enabled systematic classification and annotation of lipids, and the Lipid Maps Structure Database (LMSD) currently contains more than 50,000 characterized lipid structures [1–3].

In ticks, lipid metabolism is closely associated with pathogen acquisition, survival, and transmission. Moreover, tick blood-feeding can lead to the development of alpha-gal syndrome (AGS), a delayed allergic reaction to mammalian-derived products. Recent studies indicate that glycolipids carrying the galactose-α-1,3-galactose (α-Gal) epitope can be recognized by patient IgE antibodies and are capable of activating effector immune cells, suggesting that lipid-bound glycans may contribute to sensitization and delayed immune responses in patients [4, 5]. These findings highlight the importance of comprehensive lipid and glycolipid characterization in ticks.

Given the critical role lipids play in various biological processes, identifying and quantifying them is of utmost importance. The preferred method for this is a combination of ultra-high-pressure lipid chromatography (UHPLC) and high-resolution mass spectrometry (HRMS) due to its high sensitivity and specificity.

This chapter provides a protocol for untargeted lipidomic analysis of ticks using reverse-phase separation and a quadrupole time-of-flight (Q-TOF) mass spectrometer (Fig. 1), which can be applied to characterize lipid and glycolipid fractions and to support mechanistic studies related to tick-induced sensitization and allergy research.

2 Materials

2.1 Solutions

1. Extraction solvent: hexane, LC–MS grade.
2. Reconstitution solvent: methanol–2-propanol (1:1, v/v). Prepare fresh before use.
3. Mobile phase/eluent A: methanol-water (4:1, v/v) containing ammonium formate (5 mM). Dissolve 315.28 mg ammonium formate in 500 mL solvent mixture.
4. Mobile phase/eluent B: 2-propanol-water (499:1, v/v) containing ammonium formate (5 mM). Dissolve 315.28 mg ammonium formate in 500 mL solvent mixture.
5. Needle wash solution: 2-propanol-methanol (4:1, v/v).
6. Blank solution: methanol (100%).

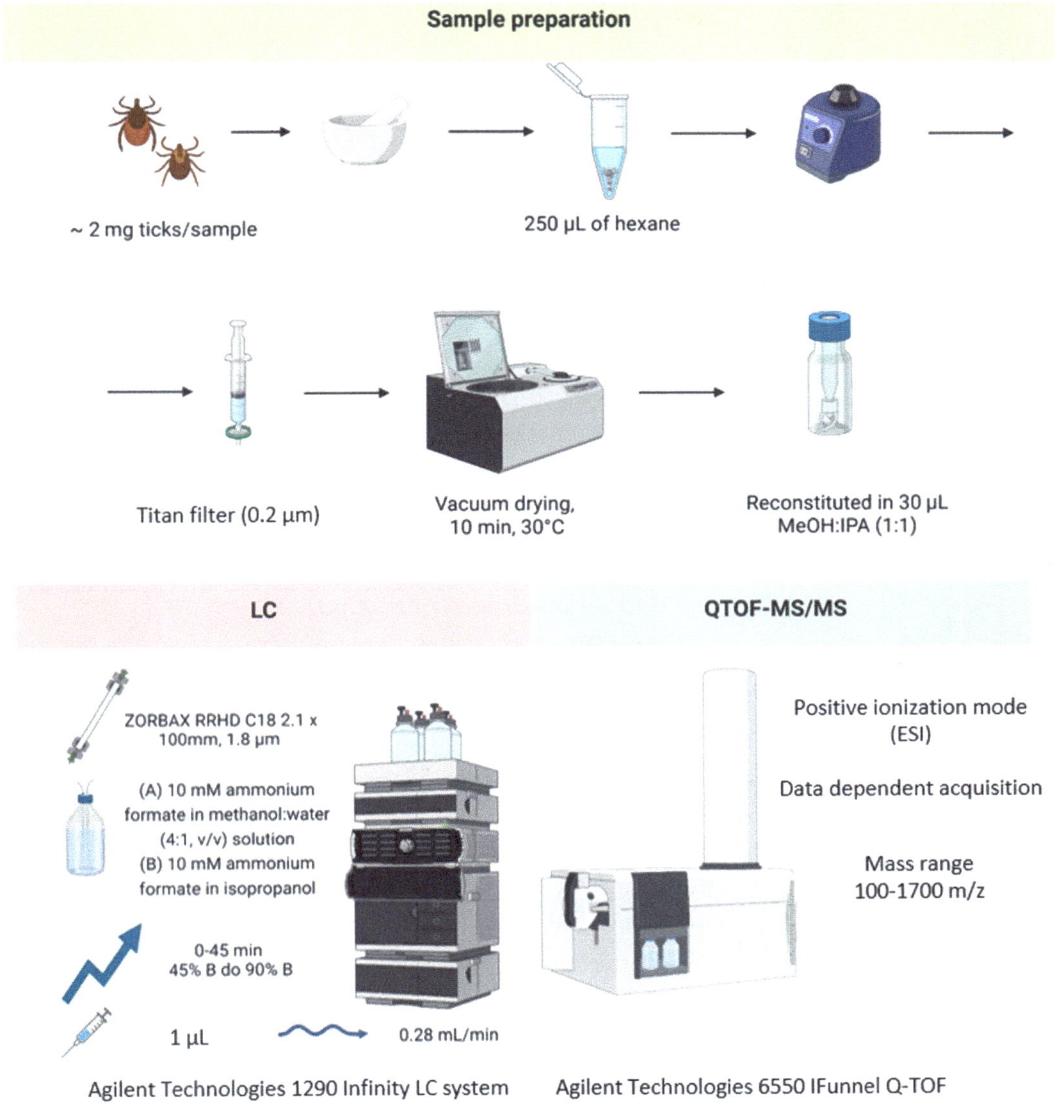

Fig. 1 Workflow for untargeted lipidomic analysis of tick samples. The figure shows three main stages of lipidomic analysis in ticks, e.g., sample preparation, liquid chromatography (LC), and quadrupole time-of-flight (Q-TOF) mass spectrometry. (Created in Biorender.com)

2.2 Equipment

1. Ultra-high-pressure liquid chromatography (UHPLC) system.
2. Quadrupole time-of-flight (Q-TOF) mass spectrometer equipped with an electrospray ionization (ESI) source.
3. Ultra-low temperature freezer (−80 °C).

4. Liquid nitrogen container.
5. Ultra-microbalance.
6. Mortar and pestle for tissue homogenization.
7. Vortex mixer.
8. Ultrasonic bath or probe sonicator.
9. Refrigerated centrifuge.
10. Vacuum concentrator or vacuum dryer.
11. Laboratory shaker.
12. Refrigerator (4 °C).

2.3 Chromatographic Column

1. Reverse-phase C18 analytical column (2.1 mm × 100 mm, 1.8 μm particle size).

2.4 Consumables

1. Aluminum foil boats.
2. Safe-lock microcentrifuge tubes (1.5 mL).
3. Glass autosampler vials.
4. Regenerated cellulose syringe filters.
5. Syringes with needles (small volume).
6. Pipettes and pipette tips.

3 Methods

3.1 Sample Preparation and Lipid Extraction

1. Remove a frozen tick from −80°C freezer and place it on aluminum foil boats (*see* **Notes 1 and 2**).
2. Weigh the sample using an ultra-microbalance scale. Ensure a minimum mass of 2 mg to reach the limits of detection and quantification for LC-MS analysis.
3. Immediately homogenize the frozen sample using a clean mortar and pestle (*see* **Notes 3 and 4**).
4. Transfer the homogenized tissue together with the aluminum foil into a 1.5-mL safe-lock tube.
5. Add 250 μL of hexane to the safe-lock tube (*see* **Note 5**).
6. Vortex the tube for 10 s.
7. Shake the tube for 3 min.
8. Sonicate the tube for 15 min at room temperature using continuous pulsation with a pulse repetition frequency of 97 per 1 s.
9. Place samples in the fridge at 4 °C for 15 min for residual protein precipitation from solution.

10. Centrifuge the samples for 5 min at 12,100×*g* at room temperature.
11. Carefully collect the supernatant using a syringe equipped with a needle; avoid disturbing the pellet at the bottom of the tube.
12. Filter the supernatant through a regenerated cellulose syringe filter and transfer the supernatant to a new 1.5 mL safe-lock tube (*see* **Note 6**).
13. Dry the filtrate under vacuum for 10 min at 37 °C.
14. Reconstitute samples in 30 μL of freshly prepared solution of methanol-2-propanol (1:1, v/v) (*see* **Note 7**).
15. Sonicate samples for 15 min at room temperature.
16. Place samples in the fridge (4 °C) for 15 min incubation.
17. Centrifuge for 5 min at 12,100×*g*, room temperature, using continuous pulsation with a pulse repetition frequency of 97 per 1 s.
18. Collect 25 μL of supernatant using pipette, avoid touching the bottom of a safe-lock tube, and transfer the sample to a glass autosampler vial.
19. Create the quality control (QC) pooled sample by adding aliquots of all samples to a glass autosampler vial and vortex thoroughly (*see* **Note 8**).
20. Store samples at −20 °C until LC-MS analysis (*see* **Note 9**).

3.2 LC-MS Analysis

3.2.1 Liquid Chromatography

1. Degas mobile phases and needle wash solution for 15 min using sonication.
2. Install eluents A and B and needle wash onto the LC system.
3. Prime and purge eluents for 5 min.
4. Prime needle wash ten times.
5. Install the C18 analytical column.
6. Set the column temperature to 45 °C and autosampler temperature to 10 °C.
7. Flush the column with 90% of mobile phase B at a flow rate 0.1 mL/min for 15 min.
8. Gradually increase the flow rate to 0.28 mL/min and continue flushing for 30 min.
9. Equilibrate column with 45% eluent B for 15 min (*see* **Note 10**).
10. Perform gradient elution from 45% to 90% mobile phase B over 45 min.
11. Set equilibration time to 5 min and injection volume to 1 μL.

3.2.2 Mass Spectrometry

1. Calibrate and tune the mass spectrometer before analysis following the manufacturer's instructions (*see* **Note 11**).
2. Choose QTOF-MS/MS data-dependent acquisition (DDA) in positive mode and set acquisition parameters.
3. Set the mass range to m/z 100–1700, and the scan rate to 5 spectra/s for MS and 3 spectra/s for MS/MS.
4. Use a narrow isolation width (~1.3 amu).
5. Apply collision energy with slope 3.6 and offset −4.8.
6. Select a maximum of three precursors per cycle based on abundance.
7. Set drying gas temperature to 200 °C and flow to 14 L/min.
8. Set sheath gas temperature to 300 °C and flow to 12 L/min.
9. Set nebulizer pressure to 241.32 kPa (35 psi).
10. Set the electrospray ionization source (ESI) parameters as capillary voltage to 3000 V, nozzle voltage to 500 V, fragmentor voltage to 200 V, and octopole RF peak voltage to 750 V (*see* **Note 12**).

3.2.3 Acquisition Sequence

1. Randomize sample order before analysis (*see* **Note 13**).
2. Begin and end the sequence with three blank injections.
3. Inject one QC sample after every ten analyzed samples (*see* **Note 8**) .
4. Start acquisition after confirming stable system performance.

4 Notes

1. Immediately freeze ticks after collection to stop metabolic activity and prevent lipid remodeling or degradation.
2. After collection, samples should be flash-frozen in liquid nitrogen and stored at least at −80 °C. The samples should not be refrozen due to ice buildup on the tissue, which can possibly distort the results since they are normalized to the frozen weight.
3. Procedure of sample preparation (see Sect. 3.1) should be carried out at room temperature and should be as fast as possible because lipids are prone to degradation and oxidation.
4. Homogenization of frozen tissue is critical for reproducible extraction efficiency. Incomplete homogenization may lead to biased lipid recovery.
5. Hexane extraction preferentially enriches neutral lipids, including glycerides, and improves signal-to-noise ratio during LC–MS analysis.

6. Ensure complete removal of particulate material before LC–MS injection to prevent column blockage and ion source contamination.
7. Prepare the reconstitution solvent fresh to ensure consistent chromatographic performance.
8. Include pooled QC samples regularly throughout the acquisition sequence to monitor instrument stability and allow post-acquisition signal correction.
9. Lipid extracts should be analyzed as soon as possible after preparation. Long-term storage, even at −20 °C, may result in gradual lipid oxidation.
10. Perform routine mass spectrometer calibration before acquisition to maintain mass accuracy required for lipid annotation.
11. Randomization of sample injection order reduces systematic bias related to instrumental drift.
12. If signal suppression is observed, consider dilution of samples or additional cleanup to reduce matrix effects.
13. Randomization of sample injection order reduces systematic bias related to instrumental drift

Funding

This work was supported by the National Science Center, Poland, Grant No. 2023/07/X/NZ8/00009.

Competing interests The authors have no conflicts of interest to declare that are relevant to the content of this chapter.

References

1. Conroy MJ, Andrews RM, Andrews S, Cockayne L, Dennis EA, Fahy E, Gaud C, Griffiths WJ, Jukes G, Kolchin M, Mendivelso K, Lopez-Clavijo AF, Ready C, Subramaniam S, O'Donnell VB (2023) LIPID MAPS: update to databases and tools for the lipidomics community. Nucleic Acids Res 52:D1677–D1682. https://doi.org/10.1093/nar/gkad896
2. Züllig T, Trötzmüller M, Köfeler HC (2020) Lipidomics from sample preparation to data analysis: a primer. Anal Bioanal Chem 412(10):2191–2209. https://doi.org/10.1007/s00216-019-02241-y
3. Rustam YH, Reid GE (2018) Analytical challenges and recent advances in mass spectrometry based lipidomics. Anal Chem 90(1):374–397. https://doi.org/10.1021/acs.analchem.7b04836
4. Cabezas-Cruz A, Espinosa P, Alberdi P, de la Fuente J (2019) Tick–pathogen interactions: the metabolic perspective. Trends Parasitol 35(4):316–328. https://doi.org/10.1016/j.pt.2019.01.006
5. Chakrapani N, Fischer J, Swiontek K, Codreanu-Morel F, Hannachi F, Morisset M, Mugemana C, Bulaev D, Blank S, Bindslev-Jensen C, Biedermann T, Ollertand M, Hilger C (2022) α-Gal present on both glycolipids and glycoproteins contributes to immune response in meat-allergic patients. J Allergy Clin Immunol 150(2):396–405. https://doi.org/10.1016/j.jaci.2022.02.030

Part III

Food Allergy Models

Chapter 13

Predictive Functional Profiling of Allergy-Associated Microbiota with PICRUSt2

Elianne Piloto-Sardiñas, Alejandra Wu-Chuang and Alejandro Cabezas-Cruz

Abstract

Functional profiling of host-associated microbiota is essential for elucidating its role in immune dysregulation, such as in allergic diseases. Predictive bioinformatics tools like PICRUSt2 (Phylogenetic Investigation of Communities by Reconstruction of Unobserved States, version 2) enable the inference of microbial metabolic potential from 16S rRNA gene amplicon data. This chapter describes the computational workflow for using PICRUSt2 to analyze allergy-associated microbiota. The protocol begins with a standardized amplicon sequence variant (ASV) table and its corresponding taxonomy and sequence files. Users execute the core PICRUSt2 pipeline to predict gene family abundances (EC numbers, KO categories) and reconstruct MetaCyc metabolic pathways. Predicted pathway abundances are then analyzed for differential abundance between sample groups using statistical methods in R, and metabolic interaction networks are constructed for visualization. The workflow is executed in a standard Unix command-line environment and is compatible with output from upstream pipelines such as QIIME 2. This protocol provides a standardized method to generate testable hypotheses about microbiome function in allergy research from widely available 16S rRNA data.

Keywords 16S rRNA gene, Amplicon sequence variant (ASV), Functional prediction, PICRUSt2, Bioinformatics pipeline, Metabolic pathways, Microbiome, Allergy

1 Introduction

The functional contribution of host-associated microbial communities is a fundamental determinant of physiological and immunological homeostasis. These communities can be viewed as dynamic ecosystems, where keystone taxa and their interactions govern stability, resilience, and functional output [1–3]. Specific microbial metabolic outcomes, such as the production of short-chain fatty acids (SCFAs), vitamins, and anti-inflammatory metabolites, directly influence host immune tolerance, barrier integrity, and competitive exclusion of pathogens [4–6]. In allergic diseases, dysbiosis and this altered microbial function have been consistently linked to disease susceptibility

José de la Fuente and Alejandro Cabezas-Cruz (eds.), *Allergic Disease Models*, Methods in Molecular Biology, vol. 3092, https://doi.org/10.1007/978-1-0716-5635-8_13,

and severity, highlighting the critical need to move beyond taxonomic characterization to profile functional potential [7, 8].

However, direct functional characterization of microbial communities via metagenomics or metabolomics can be cost-prohibitive and computationally demanding for large-scale studies. PICRUSt2 (Phylogenetic Investigation of Communities by Reconstruction of Unobserved States, version 2) addresses this gap by providing a predictive, phylogenetically aware framework to infer functional potential from widely available 16S rRNA gene amplicon data [9]. The tool maps amplicon sequence variants (ASVs) to a curated genome database to predict the abundance of gene families (*e.g.*, Enzyme Commission numbers, KEGG Orthologs) and reconstruct metabolic pathways (*e.g.*, MetaCyc). While predictive, this approach has been validated in dedicated studies where its functional inferences showed reliable correlation with experimental data (*e.g.*, qPCR), demonstrating its utility for generating robust hypotheses [10, 11].

Beyond generating lists of differentially abundant pathways, a systemic, ecological analysis of predicted functions offers deeper insight. Functional network analysis can reveal interactions between metabolic pathways, assess the robustness of the functional profile to perturbations, and link taxonomic shifts to emergent functional stability or vulnerability [12]. This perspective is crucial for interpreting how microbial communities maintain or lose their functional contributions in conditions like allergic inflammation.

This chapter provides a comprehensive, step-by-step protocol for applying PICRUSt2 to allergy-associated microbiota and for conducting advanced downstream ecological analysis. The workflow begins with standardized ASV tables, executes the core prediction pipeline, and guides statistical analysis of differential pathway abundance, functional metabolic network construction, and robustness assessment in R. This protocol offers a reproducible computational method to systematically profile and interrogate the functional landscape of microbiomes in allergic inflammation, transforming 16S rRNA data into testable, mechanistically informed hypotheses.

2 Materials

2.1 Input data

- FASTQ files (16S rRNA gene sequencing data)
- FASTA file (representative sequences)
- BIOM file (feature/ASV table)
- Tab-delimited text file (metadata) (*see* **Note 1** for file naming conventions)

2.2 Software and computational resources

- QIIME2 (v.2021.4)
- PICRUSt2 (latest stable release)
- Python 3.x

- RStudio ($\geq$ v.1.4)
- Gephi (v.0.9.2)

2.3 Databases

- SILVA 138 (99% OTUs, 515 F/806 R region)
- KEGG Orthology (KO)
- MetaCyc
- Enzyme Commission (EC)

2.4 R Packages

- phyloseq, ggplot2, tidyverse, Heatplus, gplots, RColorBrewer, htmltools.
- DESeq2, ALDEx2, vegan, matrixStats
- Decontam
- igraph, SpiecEasi, NetSwan
- openxlsx, devtools, metagMisc

2.5 Computational requirements

- Unix/Linux or macOS operating system (Windows Subsystem for Linux for Windows users)
- Conda package manager
- 4 CPUs
- 8-16 GB RAM
- Sufficient disk space for large sequence files and analysis artifacts

3 Methods

3.1 Overview

The following protocol performs predictive functional profiling of allergy-associated microbiota using PICRUSt2. The workflow consists of three main stages: (i) preparing input data from QIIME 2, (ii) executing the PICRUSt2 pipeline, and (iii) conducting downstream statistical analysis and network construction in R (*see* **Notes 1&2**).

3.2 Export QIIME 2 Files for PICRUSt2

PICRUSt2 requires FASTA and BIOM format files. Export the necessary data from QIIME 2 artifacts. Before proceeding, ensure sample identifiers are consistent (*see* **Note 3**).

1. Export the denoised feature table:

```
qiime tools export \
--input-path dada2/filtered_table.qza \
--output-path Exported-filtered_table
```

2. Export the representative sequences:

```
qiime tools export \
--input-path dada2/filtered_repseqs.qza \
```

```
--output-path Exported-filtered_repseq
```

3.3 Execute the PICRUSt2 Pipeline

Run the following commands in a terminal with the PICRUSt2 conda environment activated (conda activate picrust2).

1. Place the representative sequences into a reference phylogenetic tree:

```
place_seqs.py -s Exported-filtered_repseqs/dna-
sequences.fasta -o out.tre -p 10 --intermediate
intermediate-place_seqs
```

2. Predict marker gene families:

```
hsp.py -i 16S -t out.tre -p 10 -n -o marker_
predicted_and_nsti.tsv.gz
```

3. Predict gene families using Enzyme Commission (EC) numbers:

```
hsp.py -i EC -t out.tre -o EC_predicted.tsv.gz -p 10
```

4. Predict gene families using KEGG Orthology (KO) identifiers:

```
hsp.py -i KO -t out.tre -o KO_predicted.tsv.gz -p
10
```

5. Run the metagenome prediction pipeline (using EC predictions as an example):

```
metagenome_pipeline.py -i Exported-filtered_table/
                       feature-table.biom \
                       -m marker_predicted_and_
                       nsti.tsv.gz \
                       -f EC_predicted.tsv.gz \
                       -o EC_metagenome_out
                       --strat_out
```

6. Decompress and rename the unstratified output for downstream analysis:

```
gunzip -c EC_metagenome_out/pred_metagenome_
unstrat.tsv.gz > input_table.tsv
```

3.4 Downstream statistical analysis in R

3.4.1 Import and filter the data

1. Load the required R packages.

```
library(htmltools)
library(DESeq2)
library(ggplot2)
```

2. Import the predicted metagenome table (`input_table.tsv`) and the metadata file (`input_metadata.tsv`) (*see* **Note 1**).

```
countData <- read.csv("input_table.tsv", header =
TRUE, sep = "\t", row.names = 1)
metaData <- read.csv("input_metadata.tsv", header
= TRUE, sep = "\t", row.names = 1)
```

3. Filter low-abundance features to reduce noise.

```
countData <- countData[rowSums(countData) >= 20,]
countData <- countData[rowCounts(countData > 0) >
= 3,]
```

3.4.2 Verify data consistency

1. Ensure sample identifiers match between the metadata and count matrix.

```
all(rownames(metaData) %in% colnames(countData))
all(rownames(metaData) == colnames(countData))
```

2. Reorder the count matrix columns to match the metadata row order if necessary.

```
countData <- countData[, rownames(metaData)]
all(rownames(metaData) == colnames(countData))
```

3.4.3 Differential abundance analysis with DESeq2

1. Construct a DESeqDataSet object. Replace Condition with your specific group variable (*see* **Note 2**).

```
dds <- DESeqDataSetFromMatrix(countData = countData,
                              colData = metaData,
                              design = ~ Condition)
```

2. Set the reference level for the Condition factor (*e.g.*, "Control") (*see* **Note 2**).

```
dds$Condition <- relevel(dds$Condition, ref =
"Condition 1")
```

3. Run the DESeq2 analysis.

```
dds <- DESeq(dds)
```

4. Extract, summarize, and order the results by adjusted p-value.

```
res <- results(dds)
summary(res)
resOrdered <- res[order(res$padj),]
```

5. Export the complete results table

```
write.csv(as.data.frame(resOrdered), file = "
results_all.csv")
```

3.4.4 Visualization of results

1. Generate a volcano plot to visualize differential abundance

```
with(res, plot(log2FoldChange, -log10(pvalue), pch
= 20,
 main = "Volcano plot of predicted functions", cex
= 1.5))
with(subset(res, padj < 0.05),
```

```
points(log2FoldChange, -log10(pvalue), pch = 20,
col = "blue"))
with(subset(res, padj < 0.05 & abs(log2FoldChange)
> 2),
 points(log2FoldChange, -log10(pvalue), pch = 20,
col = "red"))
```

2. Export a subset of highly significant results.

```
resSig <- subset(res, padj < 0.05 &
abs(log2FoldChange) > 2)
write.csv(as.data.frame(resSig), file = "results_
significant.csv")
```

3.5 Comparative analysis of pathways: Generate Venn Diagrams

1. Prepare separate lists of significantly enriched pathways (*e.g.*, from results_significant.csv) for each experimental condition (*see* **Note 2**).
2. Access the online Venn diagram tool at the Bioinformatics & Evolutionary Genomics platform: http://bioinformatics.psb.ugent.be/webtools/Venn/.
3. Input the lists of pathways for each condition into the corresponding fields on the web tool.
4. Generate the diagram to visualize shared and condition-specific functional pathways.
5. Download the resulting image in the desired format (*e.g.*,.png or.svg) for publication.

3.6 Functional network construction

3.6.1 Import data into R

1. Load the required R packages.

```
library(openxlsx)
library(SpiecEasi)
library(igraph)
```

2. Import the filtered MetaCyc pathway abundance table (Level 3) (*see* **Note 1 & 4**).

```
table <- as.matrix(read.xlsx("input_network_table.
xlsx", sheet="Condition 1", colNames = TRUE,
rowNames = TRUE))
```

3. Remove empty rows and transpose the matrix.

```
table <- table[rowSums(table) > 0,]
table <- t(table)
```

3.6.2 Infer correlations and apply cutoff

1. Calculate pairwise correlations between pathways using the SparCC algorithm.

```
sparcc.table <- sparcc(table, iter = 20, inner_iter
= 10, th = 0.3)
sparcc.graph <- sparcc.table$Cor
```

2. Apply a correlation strength cutoff of 0.75 (*see* **Note 5**).

```
sparcc.cutoff <- 0.75
sparcc.graph <- ifelse(abs(sparcc.graph) >=
sparcc.cutoff, sparcc.graph, 0)
colnames(sparcc.graph) <- colnames(table)
rownames(sparcc.graph) <- colnames(table)
diag(sparcc.graph) <- 0
```

3.6.3 Export network for visualization

1. Convert the adjacency matrix into an igraph object and export it in GraphML format.

```
ig.sparcc <- graph.adjacency(sparcc.graph, mode =
"undirected", weighted = TRUE, diag = FALSE)
write_graph(ig.sparcc, " network_output_Condition1.
graphml", format = "graphml")
```

2. Repeat steps 2-6 for each experimental condition, adjusting the input file and output name accordingly.

3.6.4 Visualize and Analyze Networks in Gephi

1. Open the exported.graphml file in Gephi (v0.9.2 or later).
2. Apply the ForceAtlas2 layout algorithm for visualization (*see* **Note 6**).
3. Calculate key topological metrics (*e.g.*, modularity, network diameter, average degree) using Gephi's statistical tools (*see* **Note 7**).

3.7 Network Robustness Analysis

3.7.1 Node Removal Analysis

1. Load the required R packages.

```
library(igraph)
library(NetSwan)
library(data.table)
```

2. Import the edge list file exported from Gephi (*see* **Note 8 & 9**) and create an igraph object.

```
elec <- read.csv("input_edge_list_Condition1.csv",
header = TRUE)
elec <- as.matrix(elec)
gra <- graph.edgelist(elec, directed = FALSE)
```

3. Calculate connectivity loss using the swan_combinatory function for four removal strategies: random, betweenness centrality, degree centrality, and cascading (*see* **Note 9**).

```
f4 <- swan_combinatory(gra, 10)
f4 <- as.data.frame(f4)
```

4. Rename the columns for clarity.

```
setnames(f4,
        old = c("V1","V2","V3","V4","V5"),
        new = c("fraction_nodes",
              "loss_connec_BNC",
              "loss_connec_DEG",
              "loss_connec_Cascading",
              "loss_connec_Random"))
```

5. Export the results table.

```
write.csv(f4, file = "robustness_results_Condition1.
csv", row.names = FALSE)
```

6. Generate a plot to visualize the connectivity loss curves (*see* **Note 10**).

```
 plot(f4[,1], f4[,5], type = "o", col = "yellow",
     xlab = "Fraction of nodes removed",
     ylab = "Connectivity loss")
lines(f4[,1], f4[,3], type = "o", col = "red")
lines(f4[,1], f4[,4], type = "o", col = "orange")
lines(f4[,1], f4[,2], type = "o", col = "blue")
legend("bottomright",
     c("Random", "Betweenness", "Degree", "Cascading"),
      lty = c[1,1,1,1], pch = c[1,1,1,1], cex = 0.5,
      col = c("yellow","blue","red","orange"))
```

7. Repeat steps 2-6 for each experimental condition, adjusting input and output file names.

3.7.2 Node Addition Analysis

1. Load the required R packages.

```
library(igraph)
library(coin) # Permutation test
library(boot) # Bootstrapping
library(stats) # p.adjust()
```

2. Import the edge list file and create an igraph object.

```
edge_list <- read.table("input_edge_list_
Condition1.csv", header = TRUE, sep = ",")
edge_list <- edge_list[,-1]
edge_list$Source <- as.character(edge_list$Source)
edge_list$Target <- as.character(edge_list$Target)
all.vertices <- unique(c(edge_list$Source,
edge_list$Target))
g <- graph_from_data_frame(edge_list, directed =
FALSE, vertices = all.vertices)
```

3. Set simulation parameters: number of simulations (n.sim) and number of nodes to add (n.add).

```
n.sim <- 10 # Number of simulations
n.add <- 100 # Number of nodes to add per
simulation
```

4. Run the simulation to iteratively add nodes and calculate the Largest Connected Component (LCC) and Average Path Length (APL) at each step (*see* **Note 11**).

```
robustness <- numeric(n.add * n.sim)
path.lengths <- numeric(n.add * n.sim)
new.nodes <- paste0("NewNode", 1:n.add)
for(i in 1:n.add){
  for(j in 1:n.sim){
    g.new <- add.vertices(g, i, name = new.
    nodes[1:i])
    new.edges <- cbind(sample(new.nodes[1:i], i,
    replace = TRUE),
                            sample(all.vertices, i,
replace = TRUE))
    g.new <- add_edges(g.new, new.edges)
        # Largest Connected Component (LCC)
    ccs <- clusters(g.new)
    max.cc <- max(ccs$csize)
    robustness[(i-1)*n.sim+j] <- max.cc
        # Average Path Length (APL)
    path.lengths[(i-1)*n.sim+j] <- average.path.
    length(g.new)
    }
}
```

5. Perform statistical tests (Wilcoxon signed-rank) and apply Benjamini-Hochberg correction (*see* **Note 12**).

```
robustness.wilcox <- wilcox.test(robustness, mu =
0, exact = FALSE)
path.lengths.wilcox <- wilcox.test(path.lengths, mu
= 0, exact = FALSE)
pvalues <- c(robustness.wilcox$p.value, path.
lengths.wilcox$p.value)
adjusted_pvalues <- p.adjust(pvalues, method =
"BH")
```

6. Fit linear regression models to assess the relationship between the number of added nodes and each network metric (LCC, APL). Export the model summaries.

```
robustness.df <- data.frame(nodes_added = rep(1:n.
                           add, each = n.sim),
                           robustness = robustness)
path.lengths.df <- data.frame(nodes_added =
rep(1:n.add, each = n.sim), path_lengths = path.
lengths)
robustness.lm <- lm(robustness ~ nodes_added, data
= robustness.df)
path_lengths.lm <- lm(path_lengths ~ nodes_added,
data = path.lengths.df)
# Export summaries
write.csv(data.frame(Model = "robustness.lm",
```

```
                              R_squared =summary(robustness.
                              lm)$r.squared,
                              Coefficient = coef(robustness.
                              lm)[2],
                              P_value =
coef(summary(robustness.lm))[2, "Pr(>|t|)"]),
              "LCC_lm_info_Condition1.csv.csv", row.
              names = FALSE)
write.csv(data.frame(Model = "path_lengths.lm",
                              R_squared = summary(path_
                              lengths.lm)$r.squared,
                              Coefficient = coef(path_lengths.
                              lm)[2],
                              P_value = coef(summary(path_
                              lengths.lm))[2, "Pr(>|t|)"]),
              "path_lengths_lm_info_Condition1.csv",
              row.names = FALSE)
```

7. (Optional) Export the full simulation results for visualization in external software (*e.g.*, GraphPad Prism) (*see* **Note 13**).

```
write.csv(robustness.df, "robustness_node_addition_
Condition1.csv", row.names = FALSE)
write.csv(path.lengths.df, "path_lengths_node_
addition_Condition1.csv", row.names = FALSE)
```

8. Repeat steps 9-14 for each experimental condition.

3.8 Taxa Contribution Analysis

3.8.1 Import Stratified PICRUSt2 Output

1. Load the required R packages.

```
library(dplyr)
library(tidyr)
library(networkD3)
library(htmlwidgets)
Import the stratified PICRUSt2 output file (functions
× taxa).
stratData <- read.delim("KO_metagenome_out/pred_
                              metagenome_strat.tsv",
                              header = TRUE, sep = "\t",
                              row.names = 1)
```

2. Separate the combined "Function|Taxon" column into two distinct columns.

```
stratData_sep <- stratData %>%
  tibble::rownames_to_column("Function_Taxon") %>%
  tidyr::separate(Function_Taxon, into =
  c("Function","Taxon"), sep = "\\|")
```

3. Filter the data to retain only taxa with a relative abundance greater than 1%.

```
taxa_abundance <- stratData_sep %>%
  group_by(Taxon) %>%
  summarise(total = sum(across(where(is.numeric))))
  %>%
  mutate(rel_abund = total / sum(total))
selected_taxa <- taxa_abundance %>%
  filter(rel_abund > 0.01) %>%
  pull(Taxon)
strat_filtered <- stratData_sep %>%
  filter(Taxon %in% selected_taxa)
```

3.8.2 Build Taxa–Function Network

1. Summarize the contribution values for each function–taxon pair.

```
taxa_contrib <- strat_filtered %>%
  group_by(Function, Taxon) %>%
  summarise(value = sum(across(where(is.numeric))))
  %>%
 ungroup()
```

2. Create node and edge lists for the bipartite network.

```
nodes <- data.frame(name = unique(c(taxa_
contrib$Function, taxa_contrib$Taxon)))
links <- taxa_contrib %>%
  mutate(source = match(Function, nodes$name) - 1,
         target = match(Taxon, nodes$name) - 1,
         value = value)
```

3. Generate an interactive force-directed network and save it as an HTML file.

```
net <- forceNetwork(Links = links, Nodes = nodes,
                    Source = "source", Target =
                    "target",
                    NodeID = "name", Value =
                    "value",
                    opacity = 0.9, zoom = TRUE)
# Save as HTML
saveWidget(net, "Taxa_Function_Network_Condition1.
html", selfcontained = TRUE)
```

4. Repeat steps 2-7 for each experimental condition, adjusting the input and output file names.

4 Notes

1. File naming for reproducibility. Adhere to the following standardized file names throughout the workflow:
   - input_table.tsv (PICRUSt2 functional abundance output)
   - input_metadata.tsv (sample metadata)

- input_network_table.xlsx (filtered MetaCyc Level 3 pathway table)
- network_output.graphml (network for Gephi visualization)
- input_edge_list.csv (edge list for robustness analysis)

2. Variable naming convention. The term Condition is used as a generic placeholder. Replace "Condition 1" and "Condition 2" in all code and file names with your specific experimental group labels (*e.g.*, "Control", "Allergy").
3. Sample ID verification before PICRUSt2. Before running the PICRUSt2 pipeline (Section 3.3), verify that the feature/ASV identifiers in the exported QIIME 2 feature table (feature-table.biom) and representative sequences file (dna-sequences.fasta) are consistent.
4. Pathway database level selection. Perform functional network analysis (Section 3.6) using MetaCyc pathways at hierarchical Level 3 to balance functional specificity with analytical stability.
5. Adjusting the network correlation cutoff. The correlation cutoff of 0.75 in Section 3.6.2 is a recommended starting point for constructing a conservative, high-confidence network using the SparCC algorithm [13]. Increase this threshold (*e.g.*, to 0.8 or 0.85) for more stringent, sparse networks. Decrease it (*e.g.*, to 0.65 or 0.7) to capture more potential associations, acknowledging the increased risk of including spurious edges. The choice should be justified based on network biology expectations.
6. Network layout in Gephi. For functional network visualization (Section 3.6.4), apply the ForceAtlas2 layout algorithm in Gephi [14]. Adjust its repulsion/attraction settings to clearly reveal module (community) structure.
7. Key network metrics for comparison. Calculate and record the following topological metrics in Gephi [14] (Section 3.6.4) for each condition's network to enable comparison: modularity, network diameter, and average degree.
8. Preparing edge list files for robustness analysis. In the R scripts for robustness analysis (Sections 3.7.1 and 3.7.2), replace the generic placeholder <edge_list> with the actual filename of your edge list (*e.g.*, "network_Condition1_edges.csv").
9. Interpreting node removal strategies. The swan_combinatory function (Section 3.7.1) tests four strategies [15]. Compare the resulting connectivity loss curves: Random removal models stochastic failure, while Betweenness, Degree, and Cascading removals test targeted attacks on network hubs.

10. Interpreting robustness from removal curves. A network is more robust if its connectivity loss curve increases more slowly, meaning a larger fraction of nodes must be removed to disrupt the network (Section 3.7.1).
11. Metrics for node addition simulations. In node addition analysis (Section 3.7.2), monitor two key metrics: the Largest Connected Component (LCC), which indicates overall network cohesion, and the Average Path Length (APL), which measures integration efficiency [16].
12. Statistical validation of robustness metrics. Use the integrated bootstrapping and permutation tests (Section 3.7.2) to generate confidence intervals and assess the statistical significance of the LCC and APL trends observed during node addition.
13. Final visualization of results. Import the exported.csv results files (*e.g.*, from robustness analyses) into dedicated graphing software like GraphPad Prism to generate publication-quality figures for comparative visualization.

Competing interests

The authors have no conflicts of interest to declare that are relevant to the content of this chapter.

References

1. RE Ley, CA Lozupone, M Hamady, R Knight, JI Gordon (2008) Worlds within worlds: evolution of the vertebrate gut microbiota. *Nature Reviews Microbiology*, 6(10): 776–788.
2. S Banerjee, K Schlaeppi, MG Van Der Heijden (2018) Keystone taxa as drivers of microbiome structure and functioning. *Nature Reviews Microbiology* 16(9): 567–576.
3. L Röttjers, K Faust (2019) Can we predict keystones? *Nature Reviews Microbiology* 17(3): 193–193.
4. Y Belkaid, TW Hand (2014) Role of the microbiota in immunity and inflammation. *Cell* 157(1), 121–141.
5. M Levy, E Blacher, E Elinav (2017) Microbiome, metabolites and host immunity *Current opinion in microbiology*, 35, 8–15.
6. Hashem ZS (2025) Bacterial Metabolites in Defense: A Crucial Aspect of Microbial Interaction and Host Protection. In Metabolic Dynamics in Host-Microbe Interaction. Singapore: Springer Nature Singapore, 101-120.
7. KE Fujimura, SV Lynch (2015) Microbiota in allergy and asthma and the emerging relationship with the gut microbiome. *Cell host & microbe* 17(5), 592–602.
8. DG Peroni, G Nuzzi, I Trambusti, ME Di Cicco, P Comberiati (2020) Microbiome composition and its impact on the development of allergic diseases. *Frontiers in immunology* 11: 700.
9. GM Douglas, VJ Maffei, JR Zaneveld, SN Yurgel, JR Brown, CM Taylor, C Huttenhower, MGI Langille (2020) PICRUSt2 for prediction of metagenome functions. *Nat Biotechnol* 38: 685–688.
10. L Mateos-Hernández, D Obregón, J Maye, J Borneres, N Versille, J de La Fuente, A Estrada-Peña, A Hodžić, L Šimo, A Cabezas-Cruz (2020) Anti-tick microbiota vaccine impacts Ixodes ricinus performance during feeding. *Vaccines* 8(4): 702.
11. L Mateos-Hernández, D Obregón, A Wu-Chuang, J Maye, J Bornères, N Versillé, J de la Fuente, S Díaz-Sánchez, LG Bermúdez-Humarán, E Torres-Maravilla, A Estrada-Peña,

A Hodžić, L Šimo, A Cabezas-Cruz (2021) Anti-microbiota vaccines modulate the tick microbiome in a taxon-specific manner. *Frontiers in Immunology* 12: 704621.
12. PG Paulino, L Abuin-Denis, A Maitre, E Piloto-Sardiñas, D Obregon, HA Santos, A Cabezas-Cruz (2024) Dissecting the impact of Anaplasma phagocytophilum infection on functional networks and community stability of the tick microbiome. *International Microbiology* 27(4): 1205–1218.
13. J Friedman, EJ Alm (2012). Inferring correlation networks from genomic survey data. *PLoS computational biology*, 8(9): e1002687.
14. Bastian M, Heymann S, Jacomy M (2009) Gephi: An open source software for exploring and manipulating networks. WebAtlas.
15. Lhomme S (2015) Analyse spatiale de la structure des réseaux tech niques dans un contexte de risques. Cybergeo: Eur J Geography. https://doi.org/10.4000/cybergeo.26763.
16. Freitas S, Yang D, Kumar S, et al (2021) Evaluating graph vulnerability and robustness using TIGER. In: Proceedings of the 30th ACM International Conference on Information & Knowledge Manage ment. ACM, Virtual Event Queensland Australia, 4495–4503.

Check for updates

Chapter 14

Experimental Food Allergy: From Induction to Anaphylaxis Assessment in a Rat Model

Diana Reyes-Pavón, Mariela Jiménez, Luis G. Bermúdez-Humarán and Eva Salinas

Abstract

This chapter describes the experimental methodology used to establish a rat model of oral food allergy. The final protocol has been optimized for the Wistar strain. The model mimics chronic oral allergen re-exposure and allows the study of effector-phase allergic responses. It encompasses all experimental stages, including preliminary procedures, intramuscular sensitization, the use of pro–type 2 immune response adjuvants, allergen re-sensitization, and the final oral challenge, using ovalbumin (OVA). Key outcomes include clinical parameters such as sign scoring, diarrhea classification, and body temperature changes during oral anaphylaxis. In addition, immunological and inflammatory markers such as OVA-specific IgE titer, mast cell protease II, or even histamine can be assessed.

Keywords Allergic sensitization, Experimental allergy, Food allergy, Anaphylaxis

1 Introduction

Animal models of food allergy have become indispensable tools to elucidate the mechanisms underlying allergic sensitization [1, 2] and gut-barrier responses [3, 4] to food allergens [5]. They provide a controlled environment to reproduce key immunological and physiological features of the pathology [6, 7]. A variety of experimental models have been developed, differing in species, allergen type, sensitization route, and adjuvant use [8, 9]. This is valuable but requires careful standardization. This chapter provides an overview of the main strategies used for the effective induction of experimental food allergy in rats [10], detailing the immunological principles, methodological considerations, and the assessment of anaphylaxis to assess physiological modifications in the animals [11].

In the present model, adapted from Jiménez and Abril-Gil and collaborators [12, 13], sensitizing variables such as IgE levels and intestinal inflammation were considered key indicators of successful

José de la Fuente and Alejandro Cabezas-Cruz (eds.), *Allergic Disease Models*, Methods in Molecular Biology, vol. 3092,
https://doi.org/10.1007/978-1-0716-5635-8_14,

oral allergy induction, simulating repeated oral re-exposure during a chronic phase. In addition, other serum measures, including mast cell protease II levels and histamine concentration, were satisfactorily validated.

2 Materials

2.1 Rat Housing

1. Male Wistar rats (8 weeks old, 100–120 g) were obtained from the institutional bioterium and certified as specific pathogen–free. Animals were maintained under controlled environmental conditions, at an ambient temperature of 20–24 °C, with relative humidity maintained at 40%–60%, under a 12 h/12 h light/dark cycle. Experimental protocol, animal housing and handling conditions need the approval and supervision of the Institutional Ethics Committee and compliance with the institutional guidelines for experimental animal care and the national regulatory norm (e.g., Animal Welfare Mexican Law NOM-062-ZOO-1999).
2. Egg-protein-free rodent chow diet.
3. Drinking troughs.
4. Drinking water.
5. Analytical balance for animals (precision: ± 0.1 g).
6. Cages with sawdust and shelters.

2.2 Systemic Sensitization

1. Sterile saline solution (0.9%).
2. Aluminum hydroxide (aluminum hydroxide (40 mg/mL) and magnesium hydroxide (40 mg/mL)).
3. OVA (45 kDa).
4. 4 IU/0.5 mL of inactivated *Bordetella pertussis.*
5. Sterile syringes (0.5 mL, 31 G×8 mm and 1 mL, 27 G×13 mm).
6. Mini cotton balls.
7. Antiseptic solution (ethanol 70%).
8. Nitrile gloves.
9. Sharps disposal container.
10. Anesthesia chamber.
11. Animal restraint board.
12. Conical microtubes with caps of 1.5 mL.

2.3 Oral Challenges

1. Intraesophageal cannula.
2. Syringe(s) compatible with the feeder (3 mL).

3. $NaHCO_3$ 0.2 M solution (vehicle).
4. OVA.
5. Pipettes and tips for solution preparation.
6. Nitrile gloves.

2.4 Anaphylaxis Challenge

1. Intraesophageal cannula.
2. Syringe(s) compatible with the feeder (3 mL).
3. $NaHCO_3$ 0.2 M solution (vehicle).
4. OVA.
5. Pipettes and tips for solution preparation.
6. Nitrile gloves.

2.5 Assessment of the Anaphylactic Response

1. Clean metallic or plastic observation chambers.
2. Digital thermometer.
3. Clinical scoring sheets for anaphylaxis signs.
4. Timer (or timing device).
5. Pre-warmed cushion.

3 Methods

3.1 Rat Handling and Maintenance

1. Acclimate animals to the experimental facility during 7 days prior to the beginning of the food allergy protocol.
2. Wistar rats ($n = 10$) should be randomly assigned to the sham control group (SH) and food allergy group (FA).
3. House the animals (five animals per cage) in individually ventilated cages provided with sawdust and shelter and with the controlled environmental conditions, as mentioned in Sect. 2.1.
4. Keep the animals with *ad libitum* access to filtered water and an egg-protein-free rodent chow diet (*see* **Note 1**).
5. Change sawdust from cages once per week or according to institutional standard operating procedures (*see* **Note 2**).
6. Label the cages with the group, date of animal arrival to the experimental facility, date of any treatments, type of food, number or animals, and any other special need.
7. Routinely assess cage hygiene and animal health status according to facility standard operating procedures.

3.2 Systemic Sensitization

1. To perform this food allergy model, use OVA as the allergen. The complete protocol is detailed in Fig. 1.

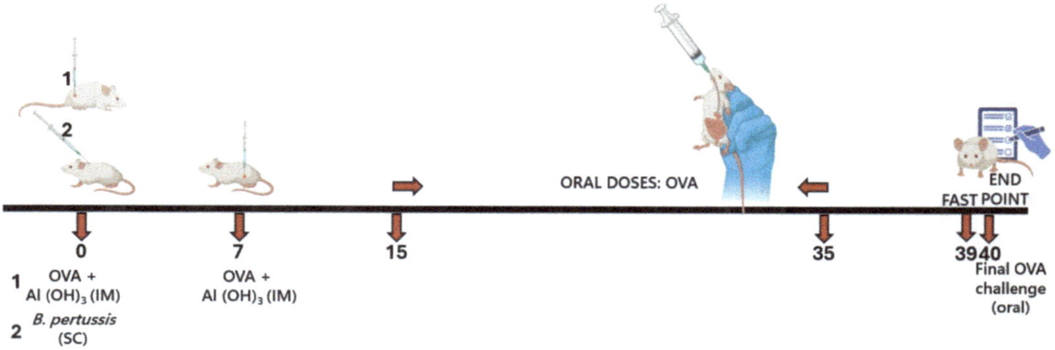

Fig. 1 Complete food allergy model: (**1**) Allergen injection and (**2**) *Bordetella pertussis* injection. *IM: Intramuscular, SC: Subcutaneous*

2. Prepare the animals by identification, health check, and weight on day 0 (*see* **Note 3**).
3. Prepare a dose of 1 mg of OVA (*see* **Note 4**) adsorbed onto 7.8 mg of aluminum hydroxide in 1 mL of saline solution into Eppendorf tubes (*see* **Note 5**).
4. Introduce the rat to the anesthesia chamber until it is asleep (*see* **Note 6**).
5. Put the rat in a dorsal position and identify the large hindlimb muscle.
6. Perform an aseptic practice, disinfecting skin at the injection site.
7. Intramuscular sensitization: Inject the contents of the OVA-aluminum hydroxide syringe at a 90° position (*see* **Note 6**) [12]. The immunological mechanisms underlying sensitization are included in Fig. 2.
8. In the same position, identify the dorsal midline or dorsolateral region, in the zone along the lower back/flanks, lateral to the spine.
9. Prepare your dose, destroying any air bubbles.
10. Perform an aseptic practice, disinfecting skin at the injection site.
11. Inject *B. pertussis* dose subcutaneously, divided into two punctures in the lower back.
12. Wait until the animal is conscious and observe for adverse reactions, assessing mobility, injection site condition, and general welfare. A simple evaluation should also include body weight, food and water intake, behavior (specifically after immunization and challenges), and any signs of injury, inflammation, or infection (*see* **Note 7**).

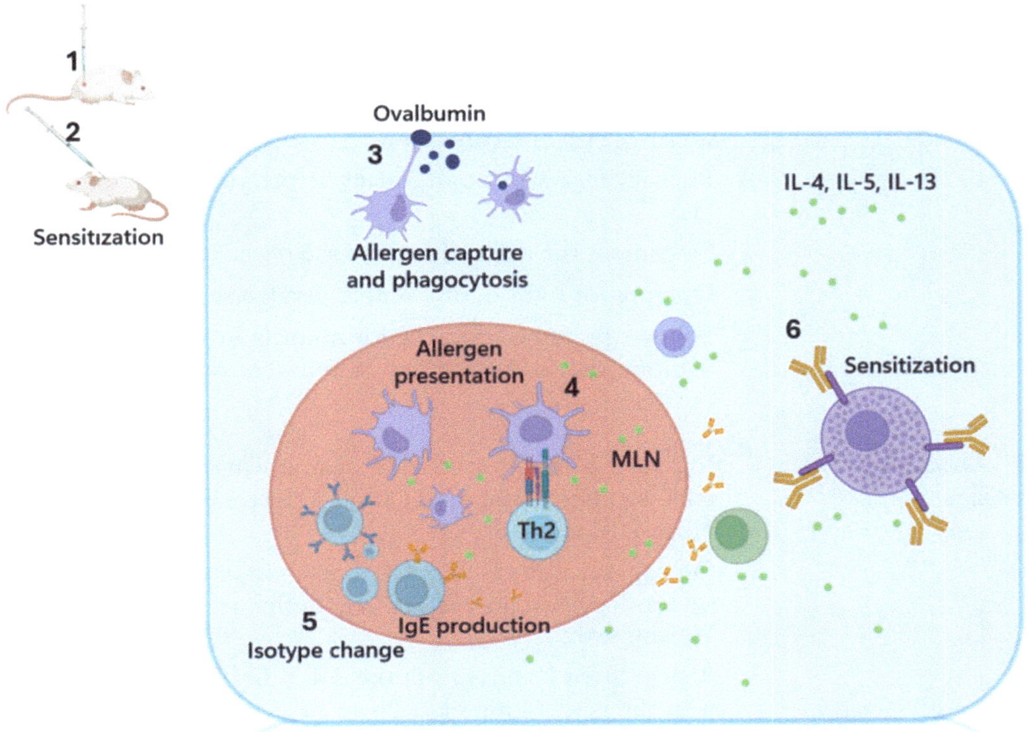

Fig. 2 Uptake of the food antigen: (**1**) allergen injection, **(2)** *Bordetella pertussis* injection and its interaction with antigen-presenting cells, such as dendritic cells **(3)**, lead to the secretion of cytokines. These cytokines initiate the polarization and maturation of dendritic cells in the lymph node, enabling them to process the antigen and present it to T lymphocytes for their differentiation into Th2, Th9, and follicular T cells **(4)**. The Th2, Th9, and follicular T-cell subpopulations, together with other cells of the intestinal microenvironment, produce cytokines that contribute to B-lymphocyte maturation and immunoglobulin class switching toward IgE production **(5)**. The environment rapidly becomes enriched in type 2 profile cytokines (IL-4, IL-5, IL-13). IgE then binds to its high-affinity receptor (FcεRI) on the surface of mast cells, thereby sensitizing these cells. MNL, mesenteric node lymph; IL, interleukin

13. Intramuscular re-sensitization: On day 7, repeat Steps 3–7, detailed above, for the administration of the intramuscular booster dose to the animals in group FA (*see* **Note 8**), following the same preparation (*see* **Note 4**).
14. Follow the same protocol for animals in the SH group, but without the OVA; instead, administer only the aluminum hydroxide and *B. pertussis* dose in sensitization step and aluminum hydroxide in re-sensitization step.

3.3 Oral Challenges

1. On day 15 after the first immunization, prepare the OVA solution in $NaH(CO)_3$ buffer at a concentration of 1 mg/mL per rat (*see* **Note 9**).

2. Place each dose in the 3 mL syringe and insert the esophageal cannula.
3. Hold each rat in the FA group with one hand and immobilize it. Insert the cannula intraesophageally and slowly and continuously empty the contents.
4. Repeat Steps 1 through 3 daily at the same time until day 35 [13].
5. Document the day and time of the dose.
6. Observe for signs of discomfort or adverse effects.
7. Follow the same protocol for animals in the SH group, but without the OVA; instead, administer only the $NaH(CO)_3$ buffer.

3.4 Anaphylaxis Challenge

1. Initiate a 24-h fast on day 39 by removing food from the containers in both groups, SH and FA. Leave water *ad libitum* (*see* **Note 10**).
2. On day 40, after the first immunization, prepare 2 mL of OVA solution per rat in $NaH(CO)_3$ buffer at a concentration of 100 mg/mL.
3. Repeat Steps 2 and 3 from the 3.4 indicator (Oral challenges), with the new solution (*see* **Note 11**).

The pathophysiological mechanisms of the challenge are reviewed in Fig. 3.

3.5 Assessment of the Anaphylactic Response

1. Three blinded observers assess clinical signs every 10 min during the first hour and at 1.5, 2, and 24 h after administration of the last allergen provocation test.
   Rate the demonstrations using the following rating scale: 0: No signs; 1: Scratching and rubbing around the nose and head; 2: Puffiness around the eyes and mouth, diarrhea, piloerection, reduced activity, and/or increased respiratory rate; 3: Wheezing, labored respiration, and cyanosis around the mouth and tail; 4: No activity after prodding, or presence of tremors and convulsions; 5: Death [14] (*see* **Notes 12–14**).
2. Along with recording clinical signs, record rectal temperature with a digital thermometer and calculate the temperature drop.
3. Finally, at the times indicated above, analyze the features of diarrhea in all animals by using the following score to grade changes in stool appearance: 0, no changes; 1, soft but well formed; 2, soft, non-formed; 3, one episode of liquid diarrhea; 4, at least two episodes of liquid diarrhea [15].

Figure 4 shows representative images of the clinical manifestations.

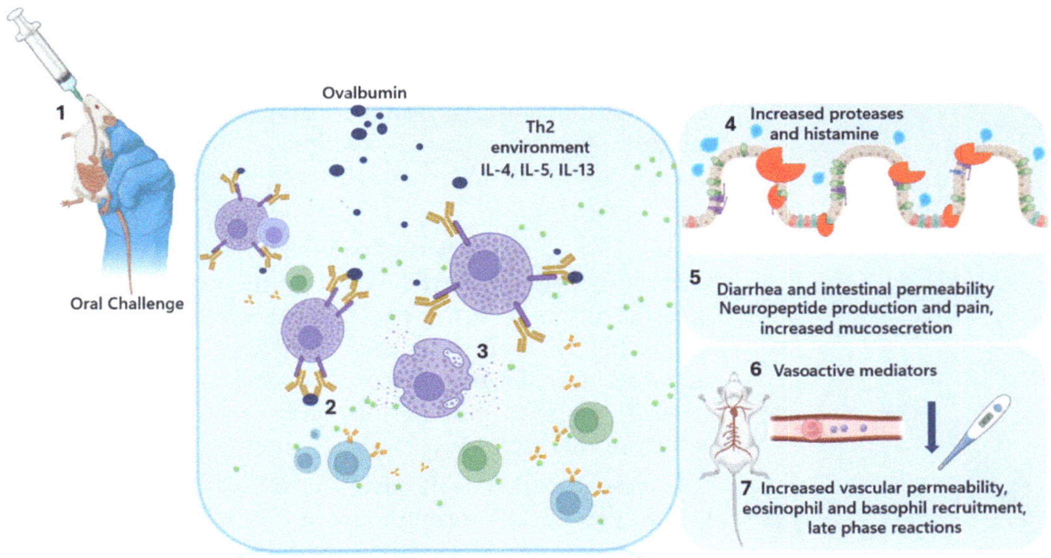

Fig. 3 Upon subsequent exposures to the allergen **(1)**, particularly at high doses, cross-linking of the FcεRI–IgE complexes **(2)** activates multiple phosphorylation cascades and induces changes in intracellular calcium concentrations in mast cells, leading to cellular activation and the release of preformed and newly produced vasoactive mediators **(3)**. Physiological manifestations then appear as a consequence of mast cell activation by the allergen. The most significant effects generated by the released mediators include: increased mucus secretion due to the action of neuropeptides on intestinal goblet cells **(4)**; increased intestinal permeability through effects on tight junctions **(5)**; vasodilation, which allows the allergen to enter the bloodstream and bind to previously sensitized basophils, thereby amplifying systemic effects through the release of additional vasoactive mediators **(6)** and finally, the recruitment of numerous leukocytes that drive the late-phase reaction, perpetuate the inflammatory process, and cause tissue damage, exacerbating disease manifestations **(7)**. Severe reactions may also occur, including alterations in blood pressure or body temperature

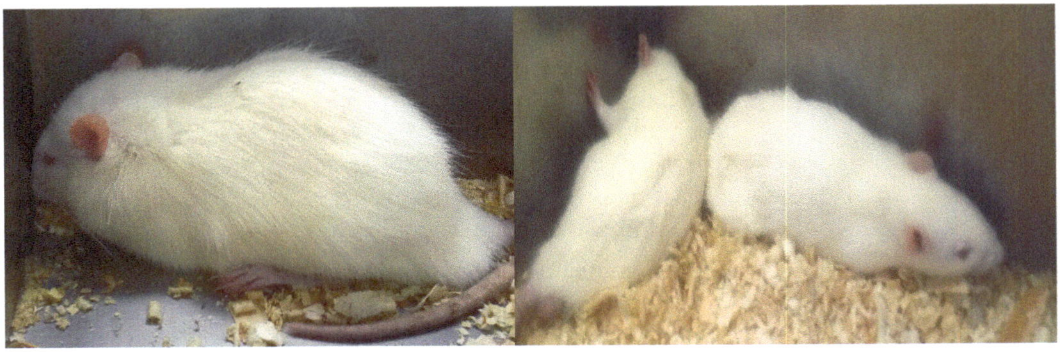

Fig. 4 The left image shows piloerection, partial eyelid closure, and signs of discomfort, with the ears angled backward. The right image illustrates an antalgic posture, characterized by a markedly arched back, tucked abdomen, and reduced locomotion, with minimal or very slow movements observed

4 Notes

1. Preparing the animals in advance and maintaining the parental diets free of allergens minimizes unintended allergen pre-exposure, thereby enhancing the internal validity of the experimental model.
2. Remember to check with the animal facility or the specialized bioterium to determine whether the animals have already received any antiparasitic and/or antibiotic treatment. If not, a specific antiparasitic and/or antibiotic treatment should be administered to minimize infectious diseases that could enhance the Th2 immune response or otherwise affect the results of this long-term model.
3. It is recommended to monitor the animals' health status and weight periodically throughout the entire protocol.
4. OVA is highly hygroscopic. Use a dry, low-humidity room or desiccator cabinet for weighing, ideally under 40% relative humidity. If possible, minimize air exposure by weighing quickly and efficiently: prepare all materials (balance, weighing paper, spatula, container) before opening the vial. Keep the OVA container sealed until the moment of weighing and recap it immediately afterward.
5. When preparing the doses for OVA administration, remember to protect them from heat and light. The same applies for the *B. pertussis* doses.
6. Please refer to these documents if you have any questions about the inhaled anesthetic choice or injection methodology or hints [16–18].
7. In order to systematically evaluate your animals, ensure that they are monitored immediately after procedures and at defined intervals thereafter. The evaluation should be conducted in an organized, reproducible, and predefined manner, and should include observable signs such as activity level, posture, grooming, social interaction, nesting behavior, and facial expressions. Whenever possible, multiple indicators should be assessed in combination.
8. Alternate from one leg to another for choosing injection sites between one intramuscular sensitization and the other (7 days between them). This will minimize cumulative tissue damage, reduce discomfort and hyperalgesia, and prevent confounding inflammatory responses from the injection site. Finally, it will preserve correct locomotion and normal access to food and water after the injections [19].

9. To prepare 1 L of the $NaHCO_3$ solution, weigh 16.8 g $NaHCO_3$ (≥99.7% reagent, 0.2 mol/L × 1 L × 84.01 g/mol) and dissolve in 900 mL of sterile deionized Water (DI). Add the weighed $NaHCO_3$ and stir until fully dissolved. Finally, bring the solution up to the final volume of 1 L with sterile DI water. Store at 4 °C in a tightly capped bottle to prevent CO_2 exchange with air, and label with concentration and date.

   Preparing a stock solution of $NaHCO_3$ may save you some time. To do so, prepare 1.0 M $NaHCO_3$ stock solution: dissolve 84.01 g $NaHCO_3$ in 1 L water (or 8.401 g per 100 mL). Cap to limit CO_2 exchange. Dilute to 1 mL of 0.2 M: Pipette 200 μL of the 1.0 M stock; add water to a final 1 mL and mix well. The calculation can be adjusted to ensure sufficient volume for all doses (for example, preparing 11 mL for 10 doses to allow for one additional dose). Keep it under refrigeration and open it under sterile conditions, as it will be administered to animals. Use Milli-Q ultrapure or deionized water, calibrated pipettes, and a tightly capped tube or sterile flasks ($NaHCO_3$ solutions exchange CO_2 with air over time). Heating should be avoided because bicarbonate is thermally unstable and decomposes at elevated temperatures.

10. Fasting is important before the anaphylactic challenge, enhancing the absorption of the challenge antigens and improving detection of allergic responses [20].

11. To prepare for the challenge, place the animals in a clean observation chamber containing freshly replaced substrate. This will be used both to assess clinical manifestations and to quantify and evaluate diarrheic depositions. A pre-warmed cushion under the chamber is a refinement measure that will ensure that their body temperature is maintained during the anaphylactic challenge and hypothermia is only provoked by anaphylaxis in all cases. Remember, small rodents are particularly susceptible to procedure-induced hypothermia during anesthesia, restraint, or stress [21, 22].

12. Other evaluation scales also appropriate for anaphylactic challenge are:

    Evaluate within 0–45 min from the allergen administration: 0: No symptoms; 1: Reduced activity; 2: Scratching and rubbing around the nose, the ears and eyes, partial immobility; 3: Prostration, pilar erection, total immobility; 4: Edema around the mouth and the eyes, puffiness around the eyes; 5: No activity after prodding, convulsion, and death [23].

    Evaluate at 0, 20, 40, 60 min from the allergen administration: 0: Normal mobility; 1: Decreased spontaneous movement; 2: Spontaneous movement only after a pinch; 3: Slow or absent spontaneous movement after a pinch; 4: Inability to maintain upright posture; 5: Death [24].

In order to evaluate and report discomfort and pain, the Grimace Scale can be used. It is a validated, non-invasive behavioral assessment tool used to quantify pain in laboratory animals based on changes in facial expression. It relies on the systematic evaluation of specific facial action units (FAUs) that are correlated with pain states. In rats, FAUs typically include orbital tightening, nose and cheek bulge, ear position, and whisker change. Animals can be observed either directly or via still images or video recordings, and each FAU is scored independently on an ordinal scale (0 = not present, 1 = moderately present, 2 = obviously present). The individual scores are then combined to generate an overall grimace score, which reflects the level of pain experienced by the animal [25].

13. If using the Grimace Scale, remember that it requires standardized conditions to ensure reliability and reproducibility. Animals should be assessed in a calm environment, free from external stressors, and observations should be conducted at defined time points, such as before and after the allergen challenge. Whenever possible, scoring should be performed by at least two trained observers blinded to treatment groups to minimize subjective bias. The use of photographic or video-based analysis allows for repeated evaluation and inter-observer reliability testing. Importantly, baseline scores must be established prior to experimental manipulation to account for individual variability.

14. Remember to always ensure animal welfare, as all animal techniques explained here (oral gavage, injections) are invasive. In accordance with the principles of Replacement, Reduction, and Refinement (3Rs) [26], all interventions should be carefully justified. Animals should be handled exclusively by trained and experienced personnel and acclimatized to both the housing conditions and handling procedures prior to the initiation of the study. Procedures should be performed in a calm and controlled environment, using appropriate restraint techniques to minimize stress and discomfort. For oral gavage, suitably sized and flexible gavage needles will have to be employed, administration volumes will adhere to established guidelines, and solutions will be slowly delivered to reduce the risk of injury or aspiration. Injections should be carried out using sterile equipment and small-gauge needles, with careful selection of the route of administration and rotation of injection sites to prevent local tissue damage. Animals will be closely monitored during and after all procedures for signs of pain, distress, or adverse effects, and endpoints will

be predefined and strictly applied. All experimental protocols shall receive prior approval from the relevant institutional ethics committee, and animal welfare considerations should also be continuously evaluated throughout the study to ensure both ethical compliance and the reliability of experimental outcomes.

Funding

This work was supported by Autonomous University of Aguascalientes, Grant Nos. PIBB 15-9 N and PIBB 18-3.

References

1. Knippels LM, vanWijk F, Penninks AH (2004) Food allergy: what do we learn from animal models?. Curr Opin Allergy Clin Immunol 4(3):205–209
2. Dolence JJ, Krempski JW, O'Konek JJ (2023) Editorial: food allergy: advances in basic and translational animal models. Front Allergy 4:1299113
3. Drønen EK, Namork E, Dirven H, Nygaard UC (2022) Suspected gut barrier disruptors and development of food allergy: adjuvant effects and early immune responses. Front Allergy 3:1029125
4. Poto R, Fusco W, Rinninella E, Cintoni M, Kaitsas F, Raoul P et al (2023) The role of gut microbiota and leaky gut in the pathogenesis of food allergy. Nutrients 16(1):92
5. Van Gramberg JL, De Veer MJ, O'Hehir RE, Meeusen ENT, Bischof RJ (2013) Use of animal models to investigate major allergens associated with food allergy. J Allergy 2013:1–10
6. Yu W, Freeland DMH, Nadeau KC (2016) Food allergy: immune mechanisms, diagnosis and immunotherapy. Nat Rev Immunol 16(12):751–765
7. Reyes-Pavón D, Jiménez M, Salinas E (2020) Fisiopatología de la alergia alimentaria. Rev Alerg Mex 67(1):34–53
8. Jensen-Jarolim E, Pali-Schöll I, Roth-Walter F (2017) Outstanding animal studies in allergy I. From asthma to food allergy and anaphylaxis. Curr Opin Allergy Clin Immunol 17(3):169–179
9. Kazemi S, Danisman E, Epstein MM (2023) Animal models for the study of food allergies. Curr Protoc 3(3):e685
10. Reyes-Pavón D, Cervantes-García D, Bermúdez-Humarán LG, Córdova-Dávalos LE, Quintanar-Stephano A, Jiménez M et al (2020) Protective effect of glycomacropeptide on food allergy with gastrointestinal manifestations in a rat model through down-regulation of Type 2 immune response. Nutrients 12(10):2942
11. Zuurveld M, Kiliaan PCJ, Van Grinsven SEL, Folkerts G, Garssen J, Van'T LB et al (2023) Ovalbumin-Induced epithelial activation directs monocyte-derived dendritic cells to instruct type 2 inflammation in T cells which is differentially modulated by 2′-fucosyllactose and 3-fucosyllactose. J Innate Immun 15(1):222–239
12. Li S, Liang Z, Mazhar M, Liao Y, Kwan KM, Yang PC, Chu KH, Liu X, Xiao X (2026) Mouse Models of Food Allergies: Where Do They Take Us?. Clin Rev Allergy Immunol 69(1):23.
13. Abril-Gil M, Garcia-Just A, Pérez-Cano FJ, Franch À, Castell M (2015) Development and characterization of an effective food allergy model in brown norway rats. PLoS ONE 10(4):e0125314
14. Li XM, Serebrisky D, Lee SY, Huang CK, Bardina L, Schofield BH et al (2000) A murine model of peanut anaphylaxis: T- and B-cell responses to a major peanut allergen mimic human responses. J Allergy Clin Immunol 106(1):150–158
15. Duncker SC, Philippe D, Martin-Paschoud C, Moser M, Mercenier A, Nutten S (2012) Nigella sativa (Black Cumin) seed extract alleviates symptoms of allergic diarrhea in mice, iInvolving opioid receptors. PLoS ONE 7(6):e39841
16. Norwegian University of Science and Technology, Faculty of Medicine, Department of Comparative Medicine. Guide to administration of substances to rodents: guidelines for the administration of substances [Internet]. Faculty

of Medicine, Department of Comparative Medicine; 2023. Available from: https://www.ntnu.edu/documents/22063822/0/Guide%2Bto%2Badministration%2Bof%2Bsubstances%2B%281%29.pdf?utm_source=chatgpt.com
17. Kay S, Valerie AS (2015) Compound administration in rodents: injection techniques. Visualized Experiments [Internet]. Available from: https://app.jove.com/v/10198/compound-administration-in-rodents-injection-techniques?utm_source=chatgpt.com
18. Oh SS, Narver HL (2024) Mouse and rat anesthesia and analgesia. Curr Protoc 4(2):e995
19. Thuilliez C, Dorso L, Howroyd P, Gould S, Chanut F, Burnett R (2009) Histopathological lesions following intramuscular administration of saline in laboratory rodents and rabbits. Exp Toxicol Pathol 61(1):13–21
20. Bird JA, Leonard S, Groetch M, Assa'ad A, Cianferoni A, Clark A et al (2020) Conducting an oral food challenge: an update to the 2009 adverse reactions to foods committee work group report. J Allergy Clin Immunol: Practice 1:75–90
21. Guidelines for the care and use of mammals in neuroscience and behavioral research (2003), National Academies Press, Washington, D.C. Available from: https://www.nationalacademies.org/publications/10732
22. Caro AC, Hankenson FC, Marx JO (2013) Comparison of thermoregulatory devices used during anesthesia of C57BL/6 mice and correlations between body temperature and physiologic parameters. J Am Assoc Lab Anim Sci 52(5):577–583.
23. Proust B, Astier C, Renaudin JM, Magueur E, Maurice D, Belcourt C, Yen FT, Kanny G, Bihain BE, Jacquenet S (2009) A murine model of cow's milk protein-induced allergic reaction: use for safety assessment of hidden milk allergens. Eur Ann Allergy Clin Immunol 41:85–94
24. Khodoun MV, Morris SC, Shao WH, Potter C, Angerman E, Kiselev A et al (2021) Suppression of IgE-mediated anaphylaxis and food allergy with monovalent anti-FcεRIα mAbs. J Allergy Clin Immunol 147(5):1838–1854
25. Sotocina SG, Sorge RE, Zaloum A, Tuttle AH, Martin LJ, Wieskopf JS et al (2011) The rat grimace scale: a partially automated method for quantifying pain in the laboratory rat via facial expressions. Mol Pain 7:1744–8069
26. Hubrecht RC, Carter E (2019) The 3Rs and humane experimental technique: implementing change. Animals 9(10):754

Part IV

Microbiota and Host–Microbiome Interactions in Allergy

Chapter 15

Integrative Analysis of Gut Microbiota in Alpha-Gal Research

Alejandra Wu-Chuang, Apolline Maitre, Lourdes Mateos-Hernández and Alejandro Cabezas-Cruz

Abstract

Studying the gut microbiota in α-gal murine models can help uncover host–microbiota–immune interactions. In this protocol, we present a comprehensive workflow for the analysis of gut microbiota in α-gal murine models. We describe the procedures for fecal sample collection, 16S rRNA gene sequencing, and construction of microbial co-occurrence networks.

Keywords Microbiota analysis, 16S rRNA gene sequencing, Co-occurrence networks, α-Gal model

1 Introduction

Exposure of α1,3-galactosyltransferase knockout (α1,3GT KO) mice, which lack endogenous α-Gal expression, to α-Gal-expressing microbes such as *Escherichia coli* O86:B7, offers a physiologically relevant model for studying α-Gal-specific immune responses [1, 2]. However, microbial exposure can reshape the gut microbiome, which has a major role in immune responses [3, 4]. Characterizing the gut microbiota in α-Gal murine models can reveal important microbial interactions related to α-Gal-related immune responses. This protocol outlines the major steps in microbiota analysis, from fecal sample collection to microbial network construction. It describes bioinformatic workflows for the study of 16S rRNA-based datasets, including procedures for microbial community profiling and construction of microbial co-occurrence networks for the study of microbial interactions.

2 Materials

2.1 Fecal Sample Collection

1. Clean cages.
2. Sterile forceps.

José de la Fuente and Alejandro Cabezas-Cruz (eds.), *Allergic Disease Models*, Methods in Molecular Biology, vol. 3092, https://doi.org/10.1007/978-1-0716-5635-8_15,

3. Sterile 1.5 mL DNA-free microtubes.
4. Ice bucket.

2.2 DNA Extraction

1. NucleoSpin tissue DNA extraction kit (*see* **Note 1**).
2. Centrifuge.
3. Thermomixer.
4. Vortex.
5. Pipette.
6. Sterile pipette tips with filters.
7. Sterile 1.5 mL DNA-free microtubes.
8. Nuclease-free water.

2.3 Post-extraction Quality Control

1. NanoDrop One Microvolume UV-Vis Spectrophotometer.
2. Pipette.
3. Sterile pipette tips with filters.

2.4 16S rRNA Gene Sequencing

1. 16S rRNA amplicon sequencing service.
2. Dry ice.

2.5 Computational Resources

1. Linux-based server.
2. Storage: ≥100 GB free space.

3 Methods

3.1 Collection of Fecal Samples

1. Collect fecal samples from transgenic mice immunized with *E. coli* O86:B7 and the mock control group (*see* **Note 2**).
2. Place the individual mouse in a clean, empty cage with no bedding or food and wait until natural defecation occurs.
3. Collect one to three fresh pellets into pre-labeled 1.5 mL microtubes (*see* **Note 3**) using sterile forceps.
4. Place collected fecal samples on ice and store them at −80 °C until DNA extraction.

3.2 DNA Extraction of Fecal Samples

1. Thaw fecal samples on ice to minimize DNA degradation.
2. Include at least three blank controls where water is used as template.
3. Prepare Buffer B5 and Proteinase K according to NucleoSpin tissue DNA extraction kit instructions.

4. Add approximately 25 mg of feces to 180 μL Buffer T1 and 25 μL Proteinase K for each sample.
5. Vortex to mix and incubate samples at 56 °C in a shaking incubator (*see* **Note 4**) for 1–3 h or overnight until complete lysis is obtained.
6. Add 200 μL Buffer B3, mix by vortexing, and incubate at 70 °C for 10 min to complete the lysis process.
7. Add 210 μL of 96%–100% ethanol to each sample and vortex vigorously.
8. Load the entire sample onto a NucleoSpin Tissue Column previously placed in a collection tube.
9. Centrifuge for 1 min at 11 000 × *g* and discard the flow-through.
10. Wash silica membrane with 500 μL Buffer BW, centrifuge for 1 min at 11 000 x *g*, and discard the flow-through.
11. Perform a second wash by adding 600 μL Buffer B5 to the column, centrifuge for 1 min at 11 000 × *g*, and discard flow-through.
12. Dry silica membrane by centrifuging the column for 1 min at 11 000 × *g*.
13. Place column into a clean 1.5 mL microcentrifuge tube.
14. Elute DNA by adding 100 μL of nuclease-free water directly to the membrane.
15. Incubate for 1 min at room temperature.
16. Centrifuge for 1 min at 11,000 × *g*.

3.3 Post-extraction Quality Control

1. Quantify DNA concentration with NanoDrop (*see* **Note 5**).
2. Check DNA purity ratios using NanoDrop ($OD_{260/280} \approx 1.8$, $OD_{260/230} \approx 2.0$).
3. Store DNA samples at −20 °C for short-term and −80 °C for long-term storage.

3.4 16S rRNA Gene Sequencing

1. Prepare DNA samples in 1.5 mL safe-lock microtubes. Samples should have between 20 and 50 ng/μL, ≥20 μL volume per sample, and should meet the purity requirements (see Sect. 3.3).
2. Ship samples and blank controls on dry ice in a sealed secondary container to a sequencing service
3. Choose a specific hypervariable region of 16S rRNA gene to sequence, such as the V4 region, using universal bar-coded primers 515F/806R.

3.5 Bioinformatic Analysis

3.5.1 Installation Instructions

Install Quantitative Insights Into Microbial Ecology (QIIME) 2 [5] within a conda environment, following the corresponding installation instructions.

Install R and RStudio [6] following the instructions provided here. Once installed, open RStudio and install the different libraries [7–20] (*see* **Note 6**):

```
> BiocManager::install("decontam")
> BiocManager::install("phyloseq")
> install.packages("ggplot2")
> install.packages("openxlsx")
> BiocManager::install("SpiecEasi")
> install.packages("igraph")
> BiocManager::install("ALDEx2")
> BiocManager::install("Heatplus")
> install.packages("RColorBrewer")
> install.packages("dplyr")
> install.packages("stringr")
> install.packages("circlize")
> install.packages("vegan")
> install.packages("tidyverse")
> install.packages("gplots")
```

3.5.2 Processing of 16S rRNA Raw Sequencing Reads

1. Activate QIIME 2 Environment and set up the working directory (*see* **Note 7**):

```
> conda activate qiime2-<version>
> cd /your/working/directory/
```

(*see* **Note 8**).

2. Import raw sequencing data into QIIME 2 (*see* **Notes 9 and 10**).

```
> qiime tools import \
  --type
'SampleData[PairedEndSequencesWithQuality]' \
  --input-path manifest.csv \
  --output-path paired-end-demux.qza \
  --input-format PairedEndFastqManifestPhred33
```

3. Demultiplex and assess read quality (*see* **Note 11**).

```
> qiime demux summarize \
--i-data paired-end-demux.qza \
--o-visualization demux.qzv
Visualize output files using https://view.qiime2.
org/ (QIIME2 visualizer) by uploading the 'demux.
qzv'file.
```

4. Denoise sequences using DADA2 [21] (*see* **Note 12**).

```
> qiime dada2 denoise-paired \
  --i-demultiplexed-seqs paired-end-demux.qza \
--p-trim-left-f 5 \
--p-trim-left-r 5 \
  --p-trunc-len-f 180 \
  --p-trunc-len-r 210 \
  --output-dir dada2
> qiime metadata tabulate \
  --m-input-file dada2/denoising_stats.qza \
  --o-visualization denoising_stats.qzv
```

5. Create a metadata table that contains information about your samples, such as experimental treatment, timepoints, and/or other information considered important (for more information, see here).
6. Create visual statistical summaries and the file containing representative sequences of sequencing data

```
> qiime feature-table summarize \
  --i-table dada2/table.qza \
  --o-visualization table.qzv \
  --m-sample-metadata-file metadata.tsv
>qiime feature-table tabulate-seqs \
  --i-data dada2/rep-seqs.qza \
  --o-visualization rep-seqs.qzv
```

7. Download the taxonomy classifier to be used for taxa identification here [22, 23]. Select the Silva 138 99% OTUs from 515F/806R region of sequences (*see* **Note 13**).
8. Perform taxonomic classification using a Naïve Bayes classifier pre-trained on the SILVA 138 database for the 515F–806R (V4) region [24, 25]:

```
> qiime feature-classifier classify-sklearn \
  --i-classifier silva-138-515f-806r-nb-classifier.
qza \
  --i-reads dada2/rep-seqs.qza \
  --o-classification taxonomy.qza
>qiime metadata tabulate \
  --m-input-file taxonomy.qza \
  --o-visualization taxonomy.qzv
```

9. Export feature table and convert to a format compatible with R:

```
> qiime tools export \
  --input-path dada2/table.qza \
  --output-path Exported-table
```

```
> biom convert \
  --i Exported-table/feature-table.biom \
  --o Exported-table/Table.tsv \
  --to-tsv
```

3.5.3 Identify Contaminants from Sequencing Data

1. Prepare input files:
   a. "Table.tsv" generated in Step 9,
   b. Open "taxonomy.qzv" from Step 8 in QIIME 2 visualizer and download as.tsv.
   c. Prepare "metadata1.tsv" similar to Step 5. Add a column indicating which samples are negative controls.
2. Load required libraries and import data using R:

```
library(phyloseq); packageVersion("phyloseq")
library(ggplot2); packageVersion("ggplot2")
library(decontam); packageVersion("decontam")
feature <- read.csv('Table.tsv', header = TRUE, sep
= "\t", row.names = 1)
taxonomy <- as.matrix(read.csv('Taxonomy.tsv',
header = TRUE, sep = "\t", row.names = 1))
metadata <- read.csv('metadata1.tsv', header =
TRUE, sep = "\t", row.names = 1)
```

3. Create a Phyloseq object [8]:

```
OTU <- otu_table(feature, taxa_are_rows = TRUE)
TAX <- tax_table(taxonomy)
metadata <- sample_data(data.frame(metadata))
ps <- phyloseq(OTU, TAX, metadata)
```

4. Inspect library size in each sample:

```
df <- as.data.frame(sample_data(ps))
df$LibrarySize <- sample_sums(ps)
df <- df[order(df$LibrarySize),]
df$Index <- seq(nrow(df))
number_of_reads <- ggplot(data = df, aes(x
= Index, y = LibrarySize, color = Stage)) +
geom_point()
png("number_of_reads.png")
print(number_of_reads)
dev.off()
```

5. Identify contaminants using the "prevalence" method:

```
sample_data(ps)$is.neg <- sample_data(ps)$Sample_
or_Control == "Control Sample"
contamdf.prev <- isContaminant(ps, method =
"prevalence", neg = "is.neg")
write.csv(contamdf.prev, "contamdf.csv", row.names
= FALSE)
```

6. Remove contaminants and create a final feature table:

```
ps.noncontamprev <- prune_taxa(!contamdf.
prev$contaminant, ps)
                                 feature_noncontam <
- as.data.frame(as(otu_table(ps.noncontamprev),
"matrix"))
                                 feature_noncontam <- data.
frame("#OTU ID" = rownames(feature_noncontam)
                                 feature_noncontam, check.
names = FALSE)
                                 write.table(feature_
noncontam, "Prevalence.txt", sep = "\t", row.names
= FALSE, quote = FALSE)
```

3.5.4 Alpha and Beta Diversity Analysis

1. Eliminate the negative control columns from the table generated "Prevalence.txt" before subsequent analysis.
2. Import the generated "Prevalence.txt" table into QIIME 2 for downstream analysis:

```
> biom convert -i Prevalence.txt -o table.from_txt_
hdf52.biom --table-type="OTU table" --to-hdf5
> qiime tools import \
  --input-path table.from_txt_hdf52.biom \
  --type 'FeatureTable[Frequency]' \
  --input-format BIOMV210Format \
  --output-path feature-table-decontam.qza
```

3. Create visual summaries:

```
> qiime feature-table summarize \
  --i-table feature-table-decontam.qza \
  --o-visualization table_decontam.qzv \
  --m-sample-metadata-file metadata.tsv
```

4. Construct a phylogenetic tree:

```
> qiime phylogeny align-to-tree-mafft-fasttree \
  --i-sequences dada2/ rep-seqs.qza \
  --p-n-threads 15 \
  --output-dir tree
```

5. Perform alpha-rarefaction (*see* **Note 14**):

```
> qiime diversity alpha-rarefaction \
   --i-table feature-table-decontam.qza \
  --i-phylogeny tree/rooted_tree.qza \
  --m-metadata-file metadata.tsv \
  --p-max-depth <max_depth> \
  --o-visualization alpha-rarefaction.qzv
```

6. Perform alpha and beta diversity analyses (*see* **Note 15**):

```
>qiime diversity core-metrics-phylogenetic \
  --i-phylogeny tree/rooted_tree.qza \
  --i-table feature-table-decontam.qza \
  --p-sampling-depth <sampling_depth> \
  --m-metadata-file metadata.tsv\
  --output-dir core-metrics-results
```

7. Determine alpha and beta diversity significance (*see* **Note 16**):

```
>qiime diversity alpha-group-significance \
  --i-alpha-diversity core-metrics-results/shannon_
vector.qza \
  --m-metadata-file metadata.tsv \
  --o-visualization core-metrics-results/shannon-
group-significance.qzv
>qiime diversity beta-group-significance \
  --i-distance-matrix core-metrics-results/
unweighted_unifrac_distance_matrix.qza \
  --m-metadata-file metadata.tsv \
  --m-metadata-column Group \
  --o-visualization core-metrics-results/
unweighted-unifrac-treatment-significance.qzv \
```

8. Create principal coordinates plots for visualizing microbial community composition:

```
>qiime emperor plot \
  --i-pcoa diversity-core-metrics-results/
unweighted_unifrac_pcoa_results.qza \
  --m-metadata-file metadata.tsv \
  --o-visualization unweighted-unifrac-emperor-
days-since-experiment-start.qzv
```

3.5.5 Differential Abundance Analysis

1. Collapse the feature table at the taxonomic level of interest (*see* **Note 17**):

```
> qiime taxa collapse \
  --i-table feature-table-decontam.qza \
  --i-taxonomy taxonomy.qza \
  --p-level 6 \
  --o-collapsed-table table-level6.qza
```

2. Eliminate low-abundance taxa by filtering taxa with fewer than 10 total reads or present in fewer than 30% of samples, and export the table for further analysis. Set to the minimum number of samples corresponding to 30% of the total sample number (e.g., 3 for 10 samples):

```
> qiime feature-table filter-features \
  --i-table table-level6.qza \
  --p-min-frequency 10 \
  --p-min-samples <min_samples> \
  --o-filtered-table table-level6-filtered.qza
> qiime tools export \
  --input-path table-level6-filtered.qza \
  --output-path Exported-table-level6-filtered
> biom convert -i Exported-table-level6-filtered/
feature-table.biom -o Exported-table-level6-
filtered/Table-level6-filtered.tsv --to-tsv.
```

3. Identify differentially abundant taxa using ALDEx2 [13] in R, which accounts for compositional data:

   a. Prepare input files.

   b. Load required packages and import data:

```
library(ALDEx2)
library(dplyr)
library(stringr)
asv<- read.csv("Table-level6-filtered.tsv", header=
T, row.names = 1, sep = '\t')
metadata <- read.csv("metadata2.tsv", header=T,sep
= '\t')
summary(metadata)
```

   c. Define the experimental condition or grouping variable for the differential abundance test before calculating centered log-ratio (CLR) values:

```
conds <- as.character(metadata$Condition)
a.clr <- aldex.clr(asv, conds, mc.samples=128,
denom="all", verbose=F)
clr.values <- aldex.effect(a.clr, include.sample.
summary=TRUE)
write.csv(clr.values,"clr-values.csv")
```

   d. Run statistical analysis:

```
aldexo <- aldex(asv, conds, mc.samples=128, test=
"t", denom="all", verbose=FALSE)
```

   e. Create and export a final table containing only features that are significantly different with their respective CLR values:

```
aldexo.f <- filter(aldexo, we.eBH < 0.05)
aldexo.f$asvID <- rownames(aldexo.f)
clr.values$asvID <- rownames(clr.values)
aldexo.clr.f <- merge(aldexo.f, clr.values, by=
c("asvID"="asvID"))
```

```
aldexo_subset <- aldexo.clr.f[, c(1,
which(grepl("rab.sample", names(aldexo.clr.f))))]
colnames(aldexo_subset) <- str_
remove(colnames(aldexo_subset), "rab.sample.")
```

```
write.csv(aldexo_subset,"aldexo_subset.
csv",row.names = FALSE)
```

Visualize Differentially Abundant Taxa Using a Heatmap:

1. Prepare input files "aldexo_subset.csv" and "metadata2.tsv" previously created in Step 19.
2. Load required libraries and import data:

```
library(Heatplus)
library(RColorBrewer)
library(vegan)
library(gplots)
library(tidyverse)
data1 <- read.csv("aldexo_subset.csv", header =
TRUE, row.names = 1)
data1 <- t(data1)
colnames(data1) <- str_remove(colnames(data1),
"g__")
metadata <- read.csv("metadata2.tsv", header =
TRUE, sep='\t')
data <- data1[metadata$sample.id,]
```

3. Compute distance and clustering:

```
Matrix.dist.g <- vegdist(t(data), method =
"euclidean")
col.clus <- hclust(Matrix.dist.g, "aver")
```

4. Choose and assign color annotations for different treatment conditions:

```
metadata$Condition <- factor(metadata$Condition)
condition_levels <- levels(metadata$Condition)
condition_colors <- c("#FFB266", "#FF8000")
col_side_colors <
- condition_colors[metadata$Condition]
```

5. Define heatmap color palette

```
scalecolor <-colorRampPalette(c("deeppink4","whit-
esmoke","darkslategray4"), space = "rgb")(100)
```

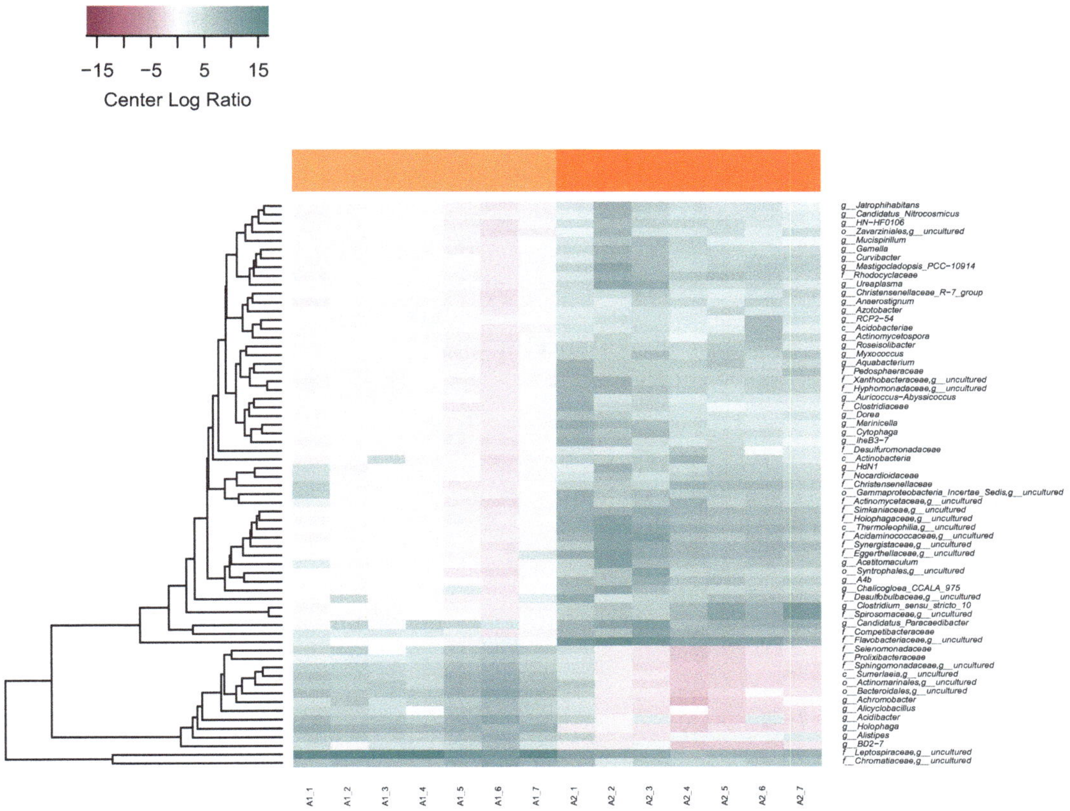

Fig. 1 Example of a heatmap representing centered log-ratio (CLR) values of differentially abundant bacterial taxa

6. Generate clustered heatmap and save as PDF (see Fig. 1 and **Note 18**):

```
pdf("ALDEx2_heatmap.pdf", width = 7, height = 8)
heatmap.2(t(as.matrix(data)), Rowv
= as.dendrogram(col.clus), labRow=
as.expression(lapply(colnames(data), function(a)
bquote(italic(.(a))))), Colv = FALSE, na.rm = TRUE,
col = scalecolor, dendrogram = "row", trace =
"none", density.info = "none", cexRow = 0.5, cexCol
= 0.5, lhei = c(2, 2.5), key.title = NA, key.xlab
= "Centered log-ratio (CLR)", margins = c(4, 13),
ColSideColors = col_side_colors)
dev.off()
```

3.5.6 Microbial Co-occurrence Network

(A) Construction of Microbial Co-occurrence Networks

1. Prepare input table "Table-level6-filtered.xlsx" (*see* **Note 19**).

2. Load required packages, import input table, and define important parameters:

```
library(openxlsx)
library(SpiecEasi)
library(igraph)
>input_file <- "Table-level6-filtered.xlsx"
>sparcc_cutoff <- 0.75
>sparcc_iter <- 20
>sparcc_inner_iter <- 10
```

3. List all worksheet names in the input file and select the worksheet to analyze:

```
>sheet_names <- getSheetNames(input_file)
>cat("Detected sheets:\n")
>print(sheet_names)
                                          >sheet_name <-
sheet_names[1] # Select the worksheet to analyze
```

4. Remove taxa with zero total abundance and transpose table:

```
prepare_sparcc_data <- function(input_file, sheet_
name) {
  table_raw <- read.xlsx(input_file, sheet = sheet_
name, rowNames = TRUE, colNames = TRUE)
  nonzero_rows <- rowSums(table_raw) > 0
  table_filtered <- table_raw[nonzero_rows,]
  table_t <- t(table_filtered)
  return(table_t)
}
table_t <- prepare_sparcc_data(input_file,
sheet_name)
```

5. Run SPARCC [26]:

```
run_sparcc <- function(table_t, sparcc_iter,
sparcc_inner_iter, threshold = 0.3) {
sparcc.res <- sparcc(table_t, iter = sparcc_iter,
inner_iter = sparcc_inner_iter, th = threshold)
  return(sparcc.res$Cor)
}
cor.mat <- run_sparcc(table_t, sparcc_iter,
sparcc_inner_iter)
```

6. Apply correlation cutoff and build igraph object:

```
build_and_export_network <- function(cor.mat, taxa_
names, sheet_name, sparcc_cutoff) {
  cor.filtered <- ifelse(abs(cor.mat) >= sparcc_
cutoff, cor.mat, 0)
```

```
  diag(cor.filtered) <- 0
  colnames(cor.filtered) <- taxa_names
  rownames(cor.filtered) <- taxa_names
  ig.sparcc <- graph.adjacency(cor.filtered, mode =
"undirected", weighted = TRUE, diag = FALSE)
  output_file <- paste0(sheet_name, "-Cutoff",
sparcc_cutoff * 100, ".graphml")
  write_graph(ig.sparcc, output_file, format =
"graphml")
  return(ig.sparcc)
}
```

```
  ig <- build_and_export_network(cor.mat,
taxa_names = colnames(table_t), sheet_name,
sparcc_cutoff)
```

(B) Visualization of Microbial Co-occurrence Networks

1. Open Gephi [27] and import the file created in Step 20.
2. Run "Overview" panel and choose the preferred layout (Ex: Fruchterman Reingold), and run until clusters form, then stop it.
3. Use filters to remove low-degree nodes or low-weight edges as desired.
4. Use the "Statistics" panel and compute topological features such as degree, modularity, network diameter, average degree, weighted degree, and clustering coefficient.

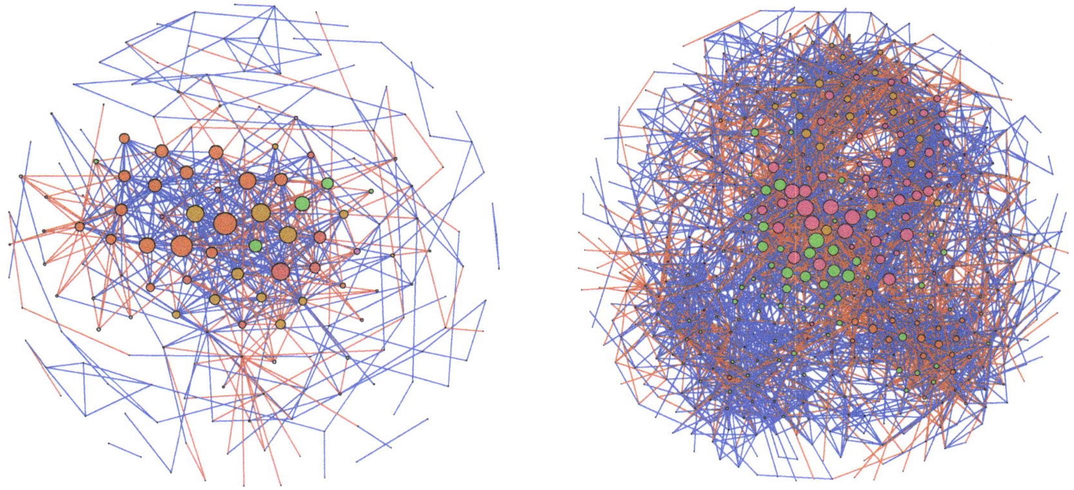

Fig. 2 Examples of bacterial co-occurrence networks inferred from 16S rRNA gene sequences. Nodes represent bacterial taxa and edges stand for co-occurrence correlation (SparCC > 0.75 or < −0.75). Node size in co-occurrence networks is proportional to the eigenvector centrality value, and node color is based on the modularity class. Positive and negative interactions between co-occurring bacteria are represented by the blue and red edges, respectively. Only nodes with at least one connecting edge are displayed

5. Change appearance of networks in the "appearance" panel (*see* **Note 20**).
6. Export high-resolution PNG/PDF/SVG from the "Preview" panel (see Fig. 2).

4 Notes

1. Alternative commercial kits for DNA extraction can be used. Some examples of suitable kits include QIAamp DNA Stool Mini Kit (QIAGEN, Dublin, Ireland) or Quick-DNA Fecal/Soil Microbe Miniprep Kit (Zymo Research, USA).
2. Wear and exchange gloves frequently and, importantly, between cages to avoid cross-contamination.
3. Clean and disinfect forceps between mice or use new disposable forceps for each collection.
4. Alternatively, use a still heating block and vortex occasionally during incubation.
5. For accurate quantification of DNA concentration, use Qubit dsDNA HS Assay or Quant-iT™ PicoGreen™ dsDNA Assay Kits.
6. Installation of R libraries is required only once. Likewise, QIIME 2 and PICRUSt2 [28] need to be installed only once in their respective conda environments.
7. All input files to be used in subsequent analysis should be in the working directory.
8. Replace <version> with the version of QIIME2 installed. For instance, if your QIIME2 version is 2021.4, use > conda activate qiime2-2021.4
9. The pipeline shown here is for raw paired-end FASTQ files. For importing different types of data, see here.
10. The manifest.csv is a file that contains the absolute paths of the paired-end sequence files. For more information, see here.
11. Quality profiles are visualized to determine optimal trimming parameters to be used for the denoising step.
12. Parameters such as "--p-trunc-len-f" and "--p-trunc-len-r" should be adjusted based on the quality of the sequences.
13. Naive Bayes classifiers should be chosen based on the hypervariable region that was sequenced. Taxonomic classifiers perform better when trained based on sequencing parameters, including the primers that were used for sequencing.

For more information about training, classifiers can be found here.

14. The "--p-max-depth" refers to the maximum rarefaction depth and should be set based on the "table_decontam.qzv" file generated in Step 10. Based on the read counts per sample, choose a cutoff that is slightly below the lowest sequencing depth among the majority of samples.
15. The "--p-sampling-depth" should be chosen based on the same criteria as in **Note 14**.
16. Statistical analysis can be performed for other metrics of alpha and beta diversity analyses by using as input the .qza file of the desired metric.
17. Levels refer to the taxonomic rank and can be chosen from 1 to 7, where 1 represents the Domain, and 7 represents the species.
18. Variables of the heatmap.2 function can be modified in order to change the appearance of the plot. For more information, see here.
19. This table should contain different sheets, each one for one experimental condition (Ex: Control vs. Treated). In each sheet, the rows should represent the bacterial taxa and the columns the samples with the reads per bacterial taxon.
20. Nodes size can be changed by their degree, nodes color by modularity. Edge colors can be changed by type of correlation (positive–negative correlation). Additionally, labels for each node can be added or removed.

Competing Interests

The authors declare no competing interests.

References

1. Posekany KJ, Pittman HK, Bradfield JF, Haisch CE, Verbanac KM (2002) Induction of cytolytic anti-Gal antibodies in alpha-1,3-galactosyltransferase gene knockout mice by oral inoculation with Escherichia coli O86:B7 bacteria. Infect Immun 70(11):6215–6222. https://doi.org/10.1128/IAI.70.11.6215-6222.2002
2. Yilmaz B, Portugal S, Tran TM, Gozzelino R, Ramos S, Gomes J, Regalado A, Cowan PJ, d'Apice AJ, Chong AS, Doumbo OK, Traore B, Crompton PD, Silveira H, Soares MP (2014) Gut microbiota elicits a protective immune response against malaria transmission. Cell 159(6):1277–1289. https://doi.org/10.1016/j.cell.2014.10.053
3. Zheng D, Liwinski T, Elinav E (2020) Interaction between microbiota and immunity in health and disease. Cell Res 30:492–506. https://doi.org/10.1038/s41422-020-0332-7
4. Wu HJ, Wu E (2012) The role of gut microbiota in immune homeostasis and autoimmunity. Gut Microbes 3(1):4–14. https://doi.org/10.4161/gmic.19320
5. Bolyen E, Rideout JR, Dillon MR, Bokulich NA, Abnet CC, Al-Ghalith GA, Alexander H, Alm EJ, Arumugam M, Asnicar F et al (2019)

Reproducible, interactive, scalable and extensible microbiome data science using QIIME 2. Nat Biotechnol 37:852–857. https://doi.org/10.1038/s41587-019-0209-9
6. Posit team (2025) RStudio: integrated development environment for R. Posit Software, PBC, Boston, MA. http://www.posit.co/
7. Davis NM, Proctor DM, Holmes SP, Relman DA, Callahan BJ (2018) Simple statistical identification and removal of contaminant sequences in marker-gene and metagenomics data. Microbiome 6:226. https://doi.org/10.1186/s40168-018-0605-2
8. McMurdie PJ, Holmes S (2013) phyloseq: an R Package for reproducible interactive analysis and graphics of microbiome census data. PLoS One 8:e61217. https://doi.org/10.1371/JOURNAL.PONE.0061217
9. Wickham H (2016) ggplot2: Elegant graphics for data analysis. Springer, New York. https://ggplot2.tidyverse.org
10. Schauberger P, Walker A (2025) openxlsx: read, write and edit xlsx files. R package version 4.2.8. https://ycphs.github.io/openxlsx/index.html
11. Kurtz ZD, Müller CL, Miraldi ER, Littman DR, Blaser MJ, Bonneau RA (2015) Sparse and compositionally robust inference of microbial ecological networks. PLoS Comput Biol 11:e1004226. https://doi.org/10.1371/JOURNAL.PCBI.1004226
12. Csardi G, Nepusz T (2006) The igraph software package for complex network research. Int J Complex Syst 1695:1–9. https://igraph.org
13. Fernandes AD, Reid JN, Macklaim JM, McMurrough TA, Edgell DR, Gloor GB (2014) Unifying the analysis of high-throughput sequencing datasets: characterizing RNA-seq, 16S rRNA gene sequencing and selective growth experiments by compositional data analysis. Microbiome 2:1–13. https://doi.org/10.1186/2049-2618-2-15
14. Alexander P (2024) Heatplus: heatmaps with row and/or column covariates and colored clusters. R package version 3.12.0
15. Neuwirth E (2022) ColorBrewer Palettes. R Package RColorBrewer Version 1:1–3. https://CRAN.R-project.org/package=RColorBrewer
16. Wickham H, François R, Henry L, Müller K, Vaughan D (2025) dplyr: a grammar of data manipulation. R package version 1.1.4. https://dplyr.tidyverse.org
17. Wickham H (2025) stringr: simple, consistent wrappers for common string operations. R package version 1.6.0. https://stringr.tidyverse.org
18. Gu Z, Gu L, Eils R, Schlesner M, Brors B (2014) circlize implements and enhances circular visualization in R. Bioinformatics 30:2811–2812
19. Oksanen J, Simpson GL, Guillaume Blanchet F, Kindt R, Legendre P, Minchin PR, O'Hara RB, Solymos P, Stevens MHH, Szoecs E et al (2024) vegan: Community Ecology Package
20. Wickham H, Averick M, Bryan J, Chang W, McGowan LD, François R, Grolemund G, Hayes A, Henry L, Hester J et al (2019) Welcome to the Tidyverse. J Open Source Softw 4(43):1686. https://doi.org/10.21105/JOSS.01686
21. Callahan BJ, McMurdie PJ, Rosen MJ, Han AW, Johnson AJA, Holmes SP (2016) DADA2: high-resolution sample inference from Illumina amplicon data. Nat Methods 13:581–583. https://doi.org/10.1038/nmeth.3869
22. Quast C, Pruesse E, Yilmaz P, Gerken J, Schweer T, Yarza P, Peplies J, Glockner FO (2013) The SILVA ribosomal RNA gene database project: improved data processing and web-based tools. Nucleic Acids Res 41:D590–D596
23. Yilmaz P, Parfrey LW, Yarza P, Gerken J, Pruesse E, Quast C, Schweer T, Peplies J, Ludwig W, Glockner FO (2014) The SILVA and "All-species Living Tree Project (LTP)" taxonomic frameworks. Nucleic Acid Res 42:D643–D648
24. Robeson MS II, O'Rourke DR, Kaehler BD, Ziemski M, Dillon MR, Foster JT, Bokulich NA (2021) RESCRIPt: reproducible sequence taxonomy reference database management. PLoS Comput Biol 17(11):e1009581. https://doi.org/10.1371/journal.pcbi.1009581
25. Bokulich NA, Kaehler BD, Rideout JR, Dillon M, Bolyen E, Knight R, Huttley GA, Gregory Caporaso J (2018) Optimizing taxonomic classification of marker-gene amplicon sequences with QIIME 2's q2-feature-classifier plugin. Microbiome 6(1):90
26. Friedman J, Alm EJ (2012) Inferring correlation networks from genomic survey data. PLoS Comput Biol 8(9):e1002687. https://doi.org/10.1371/journal.pcbi.1002687
27. Bastian M, Heymann S, Jacomy M (2009) Gephi: an open source software for exploring and manipulating networks. In: Cohen WW, Nicolov N, Glance N, Hurst M (eds) Proceedings of the 3rd International AAAI Conference on Weblogs and Social Media (ICWSM-09). AAAI Press, Menlo Park, CA, USA, pp 361–362. https://doi.org/10.1609/icwsm.v3i1.13937
28. Douglas GM, Maffei VJ, Zaneveld JR, Yurgel SN, Brown JR, Taylor CM, Huttenhower C, Langille MGI (2020) PICRUSt2 for prediction of metagenome functions. Nat Biotechnol 38(6):685–688. https://doi.org/10.1038/s41587-020-0548-6

Check for updates

Chapter 16

Experimental Model to Study Microbiota Perturbation Role in Alpha-Gal Syndrome

Lourdes Mateos-Hernández, Alejandra Wu-Chuang, Apolline Maitre and Alejandro Cabezas-Cruz

Abstract

This chapter describes a standardized experimental protocol to induce and analyze galactose-α-1,3-galactose (α-Gal)–specific IgE responses in α-Gal–deficient (α1,3-galactosyltransferase knockout) mice. The absence of endogenous α-Gal in this model enables controlled sensitization to exogenous α-Gal–containing antigens from microbial, arthropod, and mammalian sources. The protocol integrates sequential oral exposure to α-Gal–expressing *Escherichia coli* O86:B7, capsule-based infestation with *Ixodes ricinus* nymphs, oral administration of porcine tissue homogenate, and subcutaneous injection of tick-derived protein extracts. Step-by-step procedures are provided for antigen preparation, animal handling, tick infestation, blood sampling, and indirect ELISA-based quantification of α-Gal–specific IgE. The protocol is designed for repeated sampling over time following defined antigen exposures and can be applied as an in vivo experimental framework for studies of α-Gal–associated IgE responses.

Keywords α-Gal syndrome, α-Gal knockout mice, *Escherichia coli* O86:B7, *Ixodes ricinus*, Gut microbiota

1 Introduction

Allergy to the carbohydrate galactose-α-1,3-galactose (α-Gal) is a tick-bite–associated IgE-mediated hypersensitivity to mammalian meat and related products, characterized in humans by delayed allergic reactions following consumption of non-primate mammalian tissues [1–3]. Although its clinical relevance is increasingly recognized, the mechanisms and exposure pathways driving α-Gal sensitization remain incompletely understood [2, 4].

α-Gal knockout (KO) mice constitute a suitable experimental model, as the absence of endogenous α-Gal allows the induction and analysis of immune responses to exogenous α-Gal–containing antigens [4]. Environmental sources of α-Gal include Gram-negative bacteria, mammalian tissues, and salivary components of hematophagous ticks such as *I. ricinus*

José de la Fuente and Alejandro Cabezas-Cruz (eds.), *Allergic Disease Models*, Methods in Molecular Biology, vol. 3092, https://doi.org/10.1007/978-1-0716-5635-8_16,

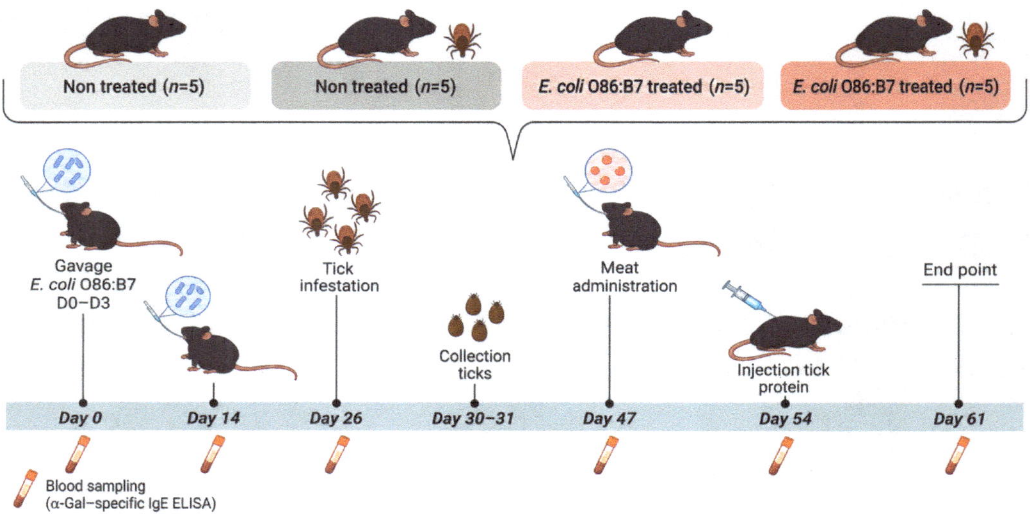

Fig. 1 Experimental design and exposure timeline in α-Gal knockout mice. Created with BioRender.com. Schematic representation of the experimental groups and exposure schedule in α-Gal–deficient mice. Mice were assigned to four groups ($n = 5$ per group): untreated controls, untreated mice subjected to tick infestation, mice treated with *Escherichia coli* 086:B7, and *E. coli* 086:B7–treated mice subjected to tick infestation. Oral gavage with *E. coli* 086:B7 or PBS was performed on days 0, 1, 3, and 14. Capsule-based infestation with *I. ricinus* nymphs was carried out on day 26, followed by tick recovery on days 30–31. Oral administration of porcine tissue homogenate was performed on day 47, and subcutaneous injection of tick protein extract was performed on day 54. Blood samples were collected at the indicated time points for indirect ELISA-based quantification of α-Gal–specific IgE. The experimental endpoint was reached on day 61

[3–5]. In particular, *E. coli* O86:B7 expresses high levels of α-Gal and provides a standardized means of controlled intestinal exposure via oral gavage, with documented effects on anti-α-Gal antibody responses and immune modulation in α-Gal–deficient mice [4, 6].

Here, we present a standardized protocol using α-Gal KO mice that integrates defined bacterial, tick, and mammalian-tissue exposures to enable reproducible induction and analysis of α-Gal–specific IgE responses, addressing current methodological gaps in experimental models of α-Gal sensitization and providing a platform for mechanistic and interventional studies [4, 6]. The overall experimental design and exposure timeline are shown in Fig. 1.

2 Materials

2.1 Animals and Housing

2.1.1 Animals

1. α-Gal–deficient (α1,3GT knockout) mice, C57BL/6 background

2.1.2 Animal Handling and Anesthesia

1. Isoflurane vaporizer system
2. Isoflurane anesthetic
3. Heating surface
4. Ophthalmic ointment (e.g., antibiotic)
5. Proparacaine hydrochloride
6. Animal hair clippers or depilatory cream
7. Sterile cotton swabs or gauze
8. Alcohol swabs

2.2 General Laboratory Equipment

1. Micropipettes and sterile pipette tips
2. Bench-top centrifuge
3. Microcentrifuge
4. Vortex mixer
5. Incubators (37 °C)
6. Spectrophotometer
7. Lysis-cycler
8. 250 mL baffled flask
9. Protective plastic band
10. Biosafety cabinet

2.3 Reagents and Consumables

1. Phosphate-buffered saline (PBS) 137 mM NaCl, 2.7 mM KCl, 10 mM Na_2HPO_4, 1.8 mM KH_2PO_4, pH 7.4
2. PBS supplemented with 0.05% (v/v) Tween-20 (PBST)

2.4 Antigen Preparation and Exposure Materials

2.4.1 Bacterial Culture and Oral Gavage

1. *E. coli* O86:B7 (ATCC 12701)
2. Luria–Bertani (LB) broth
3. LB agar plates
4. Feeding needles suitable for adult mice

2.4.2 Porcine Tissue Homogenate

1. Porcine kidney tissue
2. Homogenization equipment
3. Sterile tubes

2.4.3 Tick Protein Extract

1. *I. ricinus* nymphs
2. Protein quantification kit (e.g., BCA assay)
3. Triton X-100

2.4.4 Capsule-Based Tick Infestation

1. EVA foam
2. Double-sided adhesive foam
3. Latex or veterinary-grade adhesive
4. Transparent plastic lids
5. Fine forceps

2.5 Injection and Blood Sampling Materials

1. Sterile syringes
2. Fine-gauge needles
3. Glass Pasteur pipettes with rounded tips
4. Collection tubes

2.6 ELISA Reagents and Equipment

2.6.1 Equipment

1. High-binding 96-well ELISA plates
2. Plate washer (manual or automated)
3. Microplate reader (450 nm)

2.6.2 Reagents

1. α-Gal–GlcNAc–HSA conjugate
2. Blocking buffer (1% human serum albumin, HSA)
3. HRP-conjugated anti-mouse IgE (ε-chain specific)
4. TMB substrate
5. Sulfuric acid (H_2SO_4)

3 Methods

3.1 Preparation of *E. coli* 086:B7 Inoculum for Oral Gavage (Fig. 2) (see Note 2)

3.1.1 Thawing and Culture Initiation

1. Retrieve a −80 °C glycerol stock of *E. coli* O86:B7 (ATCC 12701).
2. Thaw the stock briefly on ice.
3. Inoculate 100 μL of thawed stock into 50 mL Luria Broth (LB) in a sterile 250 mL baffled flask.

3.1.2 Bacterial Growth

1. Incubate cultures overnight at 37 °C with shaking at 180–190 rpm.

3.1.3 OD600–CFU Standard Curve (Performed 1 Day Before Dosing)

1. Blank a spectrophotometer using sterile LB.
2. At approximately six growth points spanning OD600 = 0.05–0.8, withdraw fresh aliquots and measure OD600 immediately.

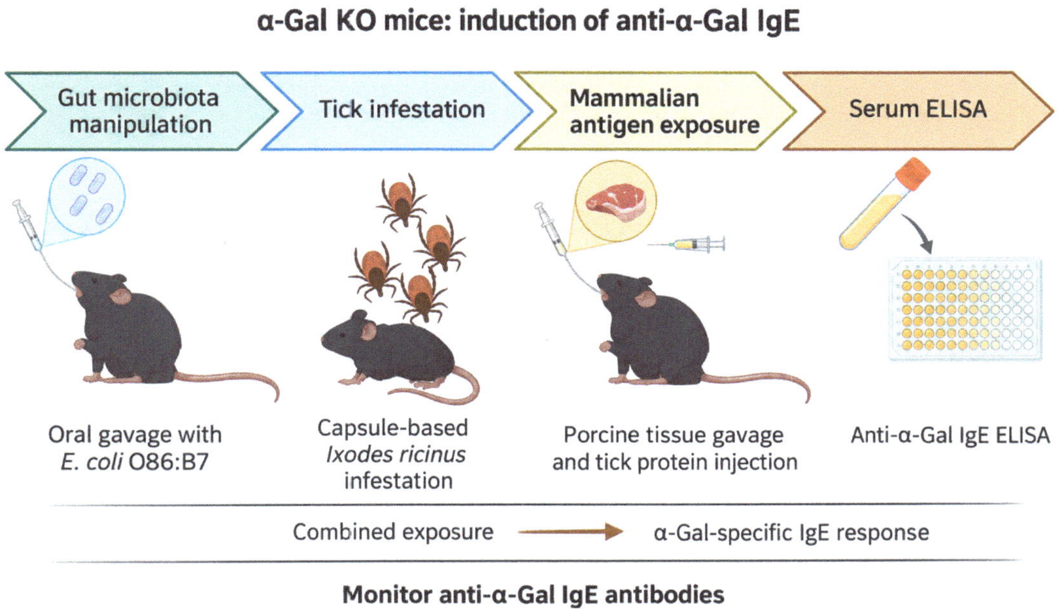

Fig. 2 Workflow for combined microbial, arthropod, and mammalian antigen exposure in α-Gal knockout mice. Created with BioRender.com. Overview of the sequential experimental workflow in α-Gal–deficient mice. The protocol consists of four main procedural components: repeated oral gavage with α-Gal–expressing *E. coli* O86:B7, capsule-based infestation with *I. ricinus* nymphs, oral administration of porcine tissue homogenate combined with subcutaneous injection of tick protein extracts, and serological assessment of α-Gal–specific IgE using an indirect ELISA. The workflow illustrates the order of antigen exposures and the timing of sample collection during the experimental protocol

3. For samples with OD600 >0.8, dilute into LB, record dilution factors, and back-calculate OD600.
4. For each time point, prepare tenfold serial dilutions (10^{-1}–10^{-7}) in sterile PBS.
5. Plate 100 μL of each dilution in duplicate onto LB agar plates.
6. Incubate plates inverted at 37 °C overnight.
7. Select plates containing 30–300 colonies.
8. Calculate CFU/mL using:
   CFU/mL = (colony counts ÷ volume plated [mL]) × dilution factor.
9. Pair OD600 values with corresponding CFU/mL.
10. Fit $\log_{10}$(CFU/mL) versus OD600 and record the regression equation, coefficient of determination (R^2), and validated OD range.

3.1.4 Quantification of the Dosing Culture

1. Measure OD600 of the overnight culture (blanked with LB).
2. Convert OD600 to CFU/mL using the established standard curve.

3.1.5 Preparation of Oral Gavage Dose

1. Calculate the total CFU required for all mice plus one additional dose.
2. Centrifuge the calculated culture volume at 1000 × *g* for 5 min at 4 °C.
3. Discard the supernatant and wash the pellet twice with cold sterile PBS.
4. Resuspend the pellet to a final concentration of 1×10^7 CFU per 100 μL PBS per mouse.
5. Perform oral gavage on days 0, 1, 3, and 14.
6. Use only pyrogen-free reagents throughout preparation and administration.

3.1.6 Oral Gavage Technique (see Note 3)

1. Use a 22 G × 25 mm feeding sonde for adult C57BL/6 mice.
2. Measure the distance from the mouth to the last rib and mark the sonde.
3. Insert the sonde along the palate into the esophagus without force.
4. Advance to the marked depth and deliver the dose at 0.2–0.5 mL/s.
5. Withdraw the sonde and monitor animals for 2–5 min.

3.2 Porcine Tissue Homogenate for Oral Gavage

3.2.1 Homogenate Preparation

1. Weigh 0.5 g porcine kidney tissue.
2. Add 1 mL sterile PBS.
3. Homogenize using a lysis-cycler at 6000 rpm for 30 s, repeated three times.
4. Place samples on ice for 1 min between cycles.
5. Centrifuge at 200 × *g* for 5 min and transfer the supernatant.
6. Further centrifuge the supernatant at 1000 × *g* for 5 min.
7. Retain the clarified supernatant.
8. Pool homogenates to the required final volume.

3.3 Tick Protein Extract Preparation (see Note 4)

3.3.1 Tick Homogenization

1. Combine *I. ricinus* nymphs (approximately 50% fed and 50% unfed; total ≈94).
2. Add 3 mL PBS containing 1% Triton X-100 (approximately 500 μL per 15 ticks).

3. Homogenize using a lysis-cycler at 6000 rpm for 30 s, repeated twice.
4. Place samples on ice for 1 min between cycles.

3.3.2 Clarification and Quantification

1. Centrifuge at 200 × *g* for 5 min.
2. Collect the supernatant.
3. Determine protein concentration using a BCA assay.
4. Prepare aliquots to allow delivery of 50 μg protein in 100 μL PBS per mouse.

3.4 Animals and Experimental Design (see Note 1)

3.4.1 Experimental Groups (Fig. 1)

1. Assign mice randomly to four experimental groups ($n = 5$ per group):
   - Untreated mice
   - Untreated mice subjected to tick infestation
   - Mice treated with *E. coli* O86:B7
   - *E. coli* O86:B7–treated mice subjected to tick infestation

3.4.2 Exposure and Sampling Timeline

1. Oral gavage with *E. coli* O86:B7 or PBS was performed on days 0, 1, 3, and 14.
2. Capsule-based *I. ricinus* nymph infestation was carried out on day 26, followed by recovery of engorged ticks on days 30–31.
3. Oral administration of porcine tissue homogenate was performed on day 47.
4. Subcutaneous injection of tick protein extract was conducted on day 54.
5. Collect blood samples on days 0, 14, 26, 47, 54, and 61 for serological analysis.

3.5 Capsule-Based Infestation of Mice with I. ricinus Nymphs (Day 26)

3.5.1 Capsule Construction and Attachment (see Note 5)

1. Construct a capsule using EVA foam laminated to double-sided adhesive foam.
2. Punch an outer diameter of 20 mm and an inner diameter of 12 mm.
3. Affix a transparent plastic lid (20 mm diameter; total capsule height ≈4 mm).
4. Anesthetize mice with isoflurane (<80 breaths/min).

5. Shave and clean the dorsal area between scapular and occipital regions.
6. Apply latex glue to the EVA surface and wait 1 min.
7. Press capsule onto skin for 3 min, ensuring complete edge adhesion.

3.5.2 Tick Placement

1. Introduce up to 20 nymphs per capsule using fine forceps.
2. Reinforce capsules with a protective plastic band if needed.

3.5.3 Tick Recovery (see Note 6)

1. Re-anesthetize animals for tick collection.
2. Cut a cross in the capsule lid to retrieve ticks.
3. House animals individually.
4. Allow natural capsule detachment (typically 8–9 days post-placement).

3.6 Porcine Tissue Administration (Day 47)

1. Administer porcine tissue homogenate by oral gavage.
2. Use the same gavage technique described in Sect. 3.1.6.

3.7 Tick Protein Injection (Day 54)

3.7.1 Subcutaneous Administration

1. Prepare 50 μg tick protein in 100 μL PBS per mouse.
2. Maintain total injection volume ≤10 mL/kg.
3. Induce anesthesia with 3%–4% isoflurane and maintain at 1%–2%.
4. Place the mouse in sternal recumbency and extend the neck.
5. Create a skin tent at the interscapular region.
6. Insert a 22 G needle bevel-up at a 10°–15° angle into the subcutaneous space.
7. Inject slowly until a visible bleb forms.
8. Withdraw the needle and apply gentle pressure for 2–3 s.
9. Monitor recovery for 5–10 min.
10. Record mouse ID, injection site, volume, time, and observations.

3.8 Blood Sampling (Retro-Orbital)

3.8.1 Sample Collection (see Note 7)

1. Anesthetize mice with isoflurane (3%–4% induction, 1%–2% maintenance).
2. Place animals on a warming surface.
3. Apply one drop of 0.5% proparacaine to the sampled eye.

4. Wait 5 min and remove excess liquid.
5. Use a sterile glass Pasteur pipette with a rounded tip.
6. Position the mouse and retract eyelids at the medial canthus.
7. Insert the pipette at a 30°–45° angle toward the opposite ear.
8. Collect blood and withdraw along the same path.
9. Apply pressure for 10–20 s then apply ophthalmic antibiotic ointment.
10. Monitor recovery for 10–15 min.
11. Alternate eyes across sessions.
12. Record all sampling parameters.

3.9 Indirect ELISA for Anti-α-Gal IgE (see Note 8)

3.9.1 Plate Coating

1. Dilute α-Gal–GlcNAc–HSA to 0.5 μg/mL in PBS.
2. Add 100 μL per well.
3. Incubate 2 h at room temperature with gentle rotation, then overnight at 4 °C.

3.9.2 Blocking

1. Wash plates 3× with PBST (PBS + 0.05% Tween-20).
2. Add 100–200 μL of 1% HSA in PBS per well.
3. Incubate 1 h at room temperature.

3.9.3 Serum Incubation

1. Wash plates 3× with PBST.
2. Dilute serum 1:50 in PBS.
3. Add 100 μL per well in duplicates or triplicates.
4. Incubate 1 h at 37 °C.

3.9.4 Detection (see Note 9)

1. Wash plates 5× with PBST.
2. Add anti-mouse IgE (ε-chain specific) diluted 1:1000 in blocking buffer.
3. Incubate 1 h at room temperature.

3.9.5 Development and Reading

1. Wash plates 5× with PBST.
2. Add 100 μL TMB substrate per well.
3. Incubate 5–20 min at room temperature, protected from light.
4. Stop the reaction with 50 μL 1 M H_2SO_4.
5. Read absorbance at 450 nm within 15 min.

3.9.6 Data Processing (see Note 10)

1. Calculate background-corrected absorbance values and summarize technical replicates according to the analytical needs of the study.

4 Notes

1. Ethical approval and regulatory compliance. All animal experiments were performed in authorized facilities by trained personnel and were approved by the Institutional Animal Care and Use Committee (IACUC; protocol #16914-2018091117221496). The approved protocol covered repeated blood sampling, parasite infestation, and delayed hypersensitivity outcomes.
2. Aseptic technique and bacterial inoculum consistency. Strict aseptic technique is required during bacterial culture and gavage preparation. Variability in CFU delivery between batches is a common source of inconsistent sensitization and should be minimized by careful culture standardization.
3. Variability in gut colonization. Inter-individual differences in gut colonization by *E. coli* O86:B7 are expected even with identical CFU doses. Lack of sensitization in individual animals does not necessarily indicate technical failure
4. Tick feeding status and extract variability. Protein yield and composition vary with tick feeding status. Feeding state should be documented, and ticks should be frozen promptly if immediate processing is not possible.
5. Capsule-based tick infestation. Insufficient capsule adhesion is a frequent cause of tick loss. Prevent glue from entering the capsule chamber and correct edge gaps immediately after placement.
6. Monitoring during late experimental stages. Animals may show increased sensitivity during late stages of the protocol. Enhanced monitoring is recommended when IgE-mediated responses are expected to peak.
7. Blood sampling constraints. Cumulative blood volume limits and alternation of eyes between sessions must be respected. Monitor animals for ocular irritation or delayed recovery.
8. Low or absent anti-α-Gal IgE signals. Weak or absent anti-α -Gal IgE signals may reflect insufficient cumulative antigen exposure rather than technical failure of the ELISA. Baseline serum samples collected at day 0, prior to any experimental manipulation, should be included as internal negative controls for each animal to assess background reactivity. In addition, wells incubated without serum and wells lacking α

-Gal–conjugated antigen should be included on each plate to monitor nonspecific binding of detection antibodies and substrate background. Verification of antigen coating efficiency, serum dilution, development time, and completion of all scheduled exposure steps is recommended when low or borderline signals are observed.

9. Delayed IgE response kinetics. The α-Gal–specific IgE responses are delayed. Early serum samples may yield low or negative ELISA signals despite successful sensitization and should be interpreted cautiously.
10. Expected outcome. Following completion of the full exposure schedule, α-Gal–specific IgE can typically be detected by indirect ELISA in a subset of α-Gal KO mice. The magnitude and timing of the response may vary between animals and experimental batches.

Competing interests

The authors have no conflicts of interest to declare that are relevant to the content of this chapter.

References

1. Commins SP, Satinover SM, Hosen J, Mozena J, Borish L, Lewis BD, Woodfolk JA, Platts-Mills TA (2009) Delayed anaphylaxis, angioedema, or urticaria after consumption of red meat in patients with IgE antibodies specific for galactose-alpha-1,3-galactose. J Allergy Clin Immunol 123(2):426–433. https://doi.org/10.1016/j.jaci.2008.10.052
2. van Nunen S (2015) Tick-induced allergies: mammalian meat allergy, tick anaphylaxis and their significance. Asia Pac Allergy 5(1):3–16. https://doi.org/10.5415/apallergy.2015.5.1.3
3. Commins SP, James HR, Kelly LA, Pochan SL, Workman LJ, Perzanowski MS, Kocan KM, Fahy JV, Nganga LW, Ronmark E, Cooper PJ, Platts-Mills TA (2011) The relevance of tick bites to the production of IgE antibodies to the mammalian oligosaccharide galactose-α-1,3-galactose. J Allergy Clin Immunol 127(5):1286–1293. https://doi.org/10.1016/j.jaci.2011.02.019
4. Galili U (2005) The alpha-gal epitope and the anti-Gal antibody in xenotransplantation and in cancer immunotherapy. Immunol Cell Biol 83(6):674–886. https://doi.org/10.1111/j.1440-1711.2005.01366.x
5. Commins SP (2020) Diagnosis & management of alpha-gal syndrome: lessons from 2,500 patients. Expert Rev Clin Immunol 16(7):667–677. https://doi.org/10.1080/1744666X.2020.1782745
6. Yilmaz B, Portugal S, Tran TM, Gozzelino R, Ramos S, Gomes J, Regalado A, Cowan PJ, d'Apice AJ, Chong AS, Doumbo OK, Traore B, Crompton PD, Silveira H, Soares MP (2014) Gut microbiota elicits a protective immune response against malaria transmission. Cell 159(6):1277–1289. https://doi.org/10.1016/j.cell.2014.10.053

Part V

Multiomics Approaches, AI, and Data Integration

Chapter 17

Integration of Multiomics Datasets to Identify Allergy Biomarkers and Protective Signatures

Isidro Sobrino and José de la Fuente

Abstract

The identification of biomarkers for diseases such as allergies requires analytical strategies capable of handling high-dimensional and heterogeneous biological data. Advances in omics technologies have enabled the systematic exploration of molecular features across multiple biological levels. However, their effective integration and interpretation remain challenging. In this chapter, we present a data-driven bioinformatic framework for biomarker discovery based on single-omics and multi-omics datasets, with a particular focus on allergy-related applications. We describe complementary strategies for biomarker prioritization, including differential analysis, ranking-based approaches, machine learning–driven feature selection, network-based methods, and supervised multi-omics integration using DIABLO. Special attention is given to experimental design, batch correction, data leakage prevention, and validation principles. The proposed workflow adapts methodological approaches to the structure and availability of the data, supporting horizontal, vertical, and exploratory diagonal integration scenarios. This chapter provides practical guidance and reproducible examples to assist researchers in extracting biologically meaningful and robust biomarkers from omics and multi-omics studies.

Keywords Allergy, Biodata, Biomarker, Machine learning, Network, Omics

1 Introduction

The U.S. Food and Drug Administration (FDA) and the National Center for Biotechnology Information (NCBI) define a biomarker as "A defined characteristic that is measured as an indicator of normal biological processes, pathogenic processes or responses to an exposure or intervention." [1]. In the context of allergy, a typical biomarker could be elevated IgE levels, as well as specific gene expression patterns that characterize allergy subtypes, disease progression, and severity [2].

Traditionally, biomarkers have been tested by molecular techniques based on prior knowledge of the disease, searching for molecules within a known biological process. This is time-consuming and reduces the success rate [3].

José de la Fuente and Alejandro Cabezas-Cruz (eds.), *Allergic Disease Models*, Methods in Molecular Biology, vol. 3092,
https://doi.org/10.1007/978-1-0716-5635-8_17,

With the appearance of omics sciences, the biomarker discovery process has been revolutionized. Omics data is high-dimensional, usually quantitative, and allows us to examine thousands of biological features (genes, proteins, etc.) simultaneously. With a well-designed experiment and appropriate statistical analysis, we can test many potential biomarkers simultaneously without prior knowledge of the disease [4]. Also, it brings the possibility of discovering multi-feature biomarkers, groups of molecules that act together [5].

Another important aspect of this methodology is that correlation does not imply causality, meaning that most of our findings will have an indirect effect on the pathology of interest and are not necessarily the causal mechanism. Integration of multiple omics can help with the interpretability of discovered biomarkers, as it gives us a panel of correlated biomarkers from different biological layers, for example, the presence of methylation on genes that correlate with differential expression of genes or proteins, but it will always be correlative knowledge. Because of this, statistical methods must always be accompanied by biological knowledge and downstream experiments to move from correlational knowledge to causality [6].

In this chapter, we present a complete data-driven bioinformatic workflow that extracts the full potential of omics datasets for biomarker discovery, accounting for their characteristics and addressing different integration options based on data type, with a focus on Allergy biomarker discovery and experimental design.

2 Materials

2.1 Hardware Requirements

No special computational resources are required, as the packages used do not rely on Graphics Processing Units (GPUs). However, some functions can be time-consuming on standard computers or laptops, so having a high-performance Central Processing Unit (CPU) is recommended to improve efficiency.

2.2 Software Requirements

All packages described here work under the R programming language. It is recommended to use RStudio, an Integrated Development Environment (IDE), to facilitate visualization, code management, and interactive data analysis. The workflow relies on several R packages tailored to specific tasks:

1. Limma [7] for preprocessing, normalization, and differential expression analysis of transcriptomic data.
2. sva [8] for batch effect correction using methods like ComBat [9].
3. Caret [10] for biomarker prioritization and model training using classification metrics (e.g., AUC).

4. glmnet [11] for penalized regression models such as Lasso and Elastic Net that enable embedded feature selection.
5. WGCNA [12] for network-based analysis and identification of co-expression modules and hub genes.
6. mixOmics, particularly its DIABLO framework [13], for supervised multi-omics integration and biomarker discovery. Visualization tools such as ggplot2, pheatmap, and factoextra are also employed to explore data structure and quality throughout the analysis. These packages collectively provide a robust and flexible foundation for high-dimensional, multilayer biological data analysis.

2.3 Databases and Datasets

The omics data required for biomarker discovery can be obtained either from in-house experimental studies or by taking advantage of publicly available datasets. Among the main public repositories, the Gene Expression Omnibus (GEO) from NCBI stands out as one of the most comprehensive databases, hosting thousands of transcriptomic, epigenomic, proteomic, and other high-throughput datasets across a wide range of organisms and experimental conditions [14]. Other relevant repositories include ArrayExpress (EMBL-EBI) [15], SRA (Sequence Read Archive) [16], and EGA (European Genome-phenome Archive) [17] for controlled-access data. Specialized repositories also exist, such as ImmPort, a repository of experiments on immunology, which is very useful for finding allergy-related studies [18].

When searching for data in GEO or similar platforms, it is important to clearly define the biological context of interest (e.g., disease, tissue, treatment, phenotype) and use relevant keywords and filters (such as organism, study type, or data availability). GEO allows searching by study accession numbers (e.g., GSE123456), platforms (e.g., GPL IDs), or specific terms like "asthma RNA-seq" or "multi-omics allergy."

Different types of datasets are available depending on their design:

1. Single-omic datasets, containing one molecular layer (e.g., only transcriptomics).
2. Multi-omics paired datasets, where multiple omics (e.g., transcriptomics + methylomics) are measured on the same set of samples.

Another option is to mix datasets from different cohorts, which means data taken from different biological individuals with the aim of augmenting the biological replicates (*see* **Note 1**).

2.4 Experimental Designs

Defining appropriate biological groups is crucial, as this decision directly shapes the interpretation of biomarkers. For example, to identify markers of allergic disease, a comparison between allergic patients and healthy controls is essential. In contrast, identifying protective or tolerance-associated signatures may require comparing individuals with different degrees of clinical severity or exposure but with similar allergic risk backgrounds (*see* **Note 2**).

3 Methods

3.1 General Overview of the Pipeline

1. The pipeline presented here is based on data-driven decision-making, where different strategies are selected depending on the nature and structure of the available data to apply the most appropriate method for biomarker discovery.
2. If the goal is to identify a single biomarker, traditional single-omic analyses can be applied—for example, differential expression analysis in transcriptomics, selecting candidate genes based on *p*-values or fold change. Alternatively, validation metrics such as the area under the curve (AUC) can be used to assess the discriminative power of individual features between two phenotypic conditions.
3. If the aim is to identify a panel of biomarkers, predictive models are built and evaluated, and the features selected by the best-performing models are interpreted as the biomarker panel.
4. If the focus is on biological interpretability—often at the expense of discriminative power—network-based analyses can be used to identify modules of co-expressed genes or molecular features, which are considered as functional biomarker units.
5. To increase the number of biological replicates, a common strategy is to merge datasets from independent studies measuring the same omics layer. This approach, known as horizontal integration, requires harmonization procedures to remove technical batch effects, such as ComBat or similar methods.
6. When multiple omics layers are measured on the same set of samples (i.e., paired multi-omics), vertical integration can be applied. This strategy enhances biomarker robustness by capturing complementary molecular layers of the biological system. For this purpose, we recommend the use of DIABLO, a supervised multi-omics integration framework specifically designed for biomarker discovery. In this context, a high number of biological replicates is particularly important due to the complexity of the models.
7. Finally, a more exploratory but less-established approach is diagonal integration, which involves combining omics datasets from different cohorts. This strategy is still under active research and depends heavily on prior knowledge, as cross-cohort variability

and noise can be substantial. Therefore, we only outline general methodological recommendations, leaving specific choices to the researcher's discretion based on their data and objectives. Figure 1 shows a simple scheme of the proposed pipeline.

3.2 Biomarker Discovery Based on Single-Omics Strategies

1. Horizontal integration of same-omics datasets. If we are working with datasets from different studies or cohorts but measuring the same omic layer, these datasets can be merged into a single dataset to increase the number of samples and, consequently, the statistical power. However, this merging process introduces non-biological variability, such as batch effects, differences in sample processing, or sequencing platforms, which can obscure real biological signals. To address this, it is essential to apply data harmonization techniques that remove technical noise while preserving underlying biological variation, enabling the joint analysis of the datasets [9].
2. One of the most widely used methods for this purpose is ComBat, an empirical Bayes approach for batch effect correction. The original ComBat implementation assumes that the data follow a Gaussian distribution and therefore must be applied after a log2 transformation of the data [9]. However, to handle raw count data directly—especially in RNA-seq

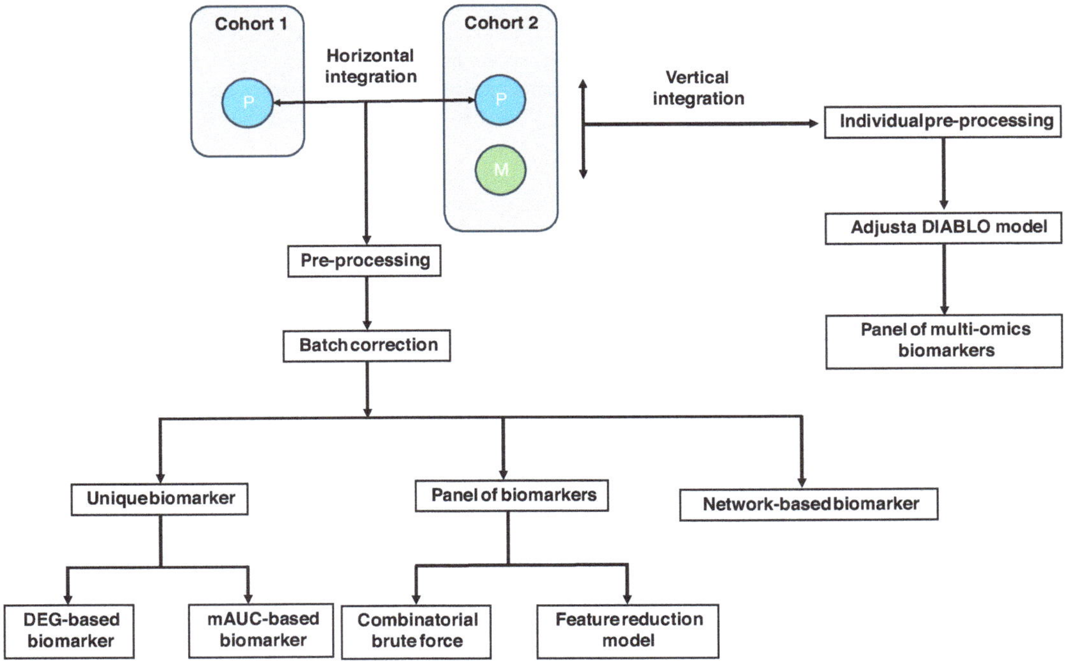

Fig. 1 Decision-making framework for multi-omics biomarker discovery tailored to heterogeneous data structures. Cohort 1 represents an independent dataset comprising a single omics layer, whereas Cohort 2 represents a dataset with paired multi-omics measurements

transcriptomics—more recent adaptations such as ComBat-seq [19] (for count-based RNA-seq data) and ComBat-met [20] (for methylation data) have been developed. These versions allow correction of batch effects while preserving the discrete nature of count data or the structure of epigenomic profiles, respectively. This section describes a pipeline using the empirical Bayes method ComBat, following normalization and transformation steps compatible with GEO-style expression data.

3.3 Step 1: Preprocess and Normalize Each Dataset Separately

To harmonize GEO-style expression matrices, we first check whether log2 transformation is necessary and normalize each dataset independently using quantile normalization (*see* **Note 3**):

```
# GEO-style log2 transformation function
geo_log2_if_needed <- function(ex) {
 qx <- quantile(ex, c(0, 0.25, 0.5, 0.75, 0.99,
1.0), na.rm = TRUE)
 LogC <- (qx[5] > 100) || ((qx[6] - qx[1] > 50)
&& (qx[2] > 0))
 if (LogC) {
  ex[ex <= 0] <- NaN
  ex <- log2(ex)
}
 ex
}
# Apply transformation and normalization
expr1 <- normalizeBetweenArrays(geo_log2_if_
needed(exprs(dataset1)), method = "quantile")
expr2 <- normalizeBetweenArrays(geo_log2_if_
needed(exprs(dataset2)), method = "quantile")
```

3.4 Step 2: Merge Expression Matrices and Align Metadata

```
# Merge expression matrices by columns
expr_norm <- cbind(expr1, expr2)
# Align metadata
meta <- meta[colnames(expr_norm),, drop = FALSE]
meta$dataset <- factor(meta$dataset)
meta$group <- factor(meta$group)
```

3.5 Step 3: Remove Incomplete Rows Before Correction

ComBat requires complete data, so rows with NA or NaN values must be removed:

```
keep <- complete.cases(expr_norm)
expr_norm <- expr_norm[keep,]
```

3.6 Step 4: Apply ComBat for Batch Correction

```
library(sva)
expr_bc <- ComBat(
 dat = expr_norm,
 batch = meta$dataset,
 mod = model.matrix(~ group, data = meta),
 par.prior = TRUE,
 prior.plots = FALSE
)
```

3.7 Step 5: Visualize Batch Effect Correction Using PCA

Before and after ComBat correction, PCA plots are useful to assess batch effect removal:

```
# PCA before and after correction
pca1 <- prcomp(t(expr_norm), scale. = TRUE)
pca2 <- prcomp(t(expr_bc), scale. = TRUE)
# Plotting
library(ggplot2)
df_before <- data.frame(PC1 = pca1$x[,1], PC2 =
pca1$x[,2],
               dataset = meta$dataset, group =
meta$group)
df_after <- data.frame(PC1 = pca2$x[,1], PC2 =
pca2$x[,2],
               dataset = meta$dataset, group =
meta$group)
ggplot(df_before, aes(PC1, PC2, color = dataset,
shape = group)) +
 geom_point(size = 3) +
 labs(title = "PCA BEFORE ComBat") +
 theme_bw()
ggplot(df_after, aes(PC1, PC2, color = dataset,
shape = group)) +
 geom_point(size = 3) +
 labs(title = "PCA AFTER ComBat") +
 theme_bw()
```

After correction, the samples should no longer cluster primarily by dataset, indicating successful removal of batch effect. In Fig. 2, an example of PCA before and after batch correction is shown. This enables downstream analysis of the integrated data, such as differential expression or biomarker discovery.

3.8 Differential Data Analysis

Once the datasets have been integrated and batch effects removed, the next step is to apply omics-specific transformations and perform differential data analysis. It is also crucial to detect potential outliers and apply additional filtering, such as removing lowly expressed genes or features with zero variance. For transcriptomic data, a popular approach is the Linear Models for Microarray Data (limma) package, which uses linear modeling and empirical Bayes moderation to compute differentially expressed features. Below is an example pipeline for differential analysis:

3.9 Step 1: Perform Differential Analysis (Transcriptomic Data Example with Limma)

```
library(limma)
library(ggplot2)
# Use batch-corrected expression matrix
expr <- expr_bc
# Align and define metadata
meta <- meta[colnames(expr),, drop = FALSE]
meta$group <- factor(meta$group)
# Design matrix for group comparison
design <- model.matrix(~ group, data = meta)
```

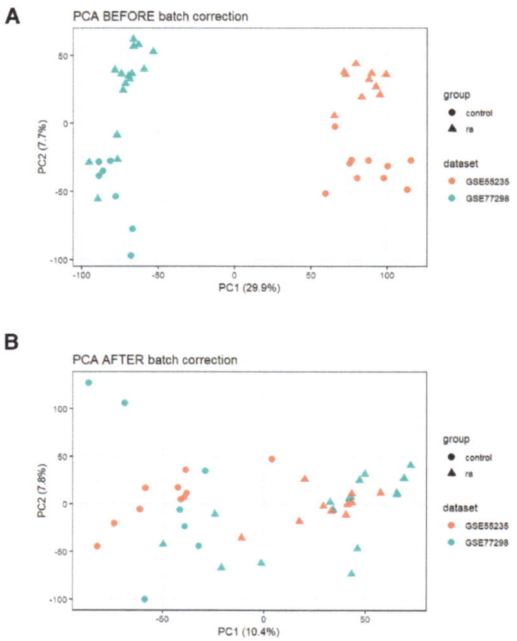

Fig. 2 Example of Batch correction effect. (**A**) PCA pre-batch correction with ComBat, showing a clear separation between batches. (**B**) PCA post-ACTH correction showing elimination of the separation between batches. Examples are based on in-house analysis derived from data obtained from the Gene Expression Omnibus (GEO) under accession numbers GSE77298 and GSE55235 on rheumatoid asthma

```
# Fit linear model and apply empirical Bayes
moderation
fit <- eBayes(lmFit(expr, design))
# Check available coefficients
colnames(coef(fit))
# Extract differential features (e.g., for group
comparison)
res <- topTable(fit, coef = "groupa", number =
Inf, sort.by = "P")
res$feature_id <- rownames(res)
```

3.10 Step 2: Define p-adj and FoldChange Thresholds

```
# Subset significant results based on adjusted
p-value threshold
res_sig <- subset(res, adj.P.Val < 0.05)
# Volcano plot annotation
res_df <- na.omit(as.data.frame(res))
res_df$significance <- "NS"
res_df$significance[res_df$adj.P.Val < 0.05 &
abs(res_df$logFC) > 1] <- "Significant"
# Optional: Volcano plot visualization
```

```
ggplot(res_df, aes(x = logFC, y =
-log10(adj.P.Val), color = significance)) +
 geom_point(alpha = 0.7) +
  scale_color_manual(values = c("grey", "red")) +
  theme_bw() +
  labs(title = "Volcano Plot", x = "log2 Fold
Change", y = "-log10 Adjusted P-Value")
```

In this example, differentially expressed features are selected based on an adjusted *p*-value threshold (FDR < 0.05) and a log2 fold change.

3.11 Unique Biomarker Discovery

After differential analysis, a typical step is performing enrichment analyses (e.g., GO or KEGG) to extract functional meaning. However, for biomarker discovery, the objective is different: to prioritize individual genes, proteins, or other features (depending on the omics datasets we are working with) that best separate biological conditions with different strategies.

3.12 Strategy 1: Ranking Differential Features Using Statistical Metrics Such as Adjusted p -Value or log2 Fold Change (logFC), or a Combination of Both

```
res_df <- na.omit(as.data.frame(res))
res_df$feature <- rownames(res_df)
# Ranking: FDR + |logFC|
res_df <- res_df[order(res_df$adj.P.Val,
-abs(res_df$logFC)),]
top_de <- head(res_df, 50)
top_de$score_mix <- abs(top_de$logFC) *
(-log10(top_de$adj.P.Val))
head(top_de[, c("feature", "logFC", "adj.P.Val",
"score_mix")], 10)
```

3.13 Strategy 2: Ranking Features by Discriminative Power

We can define the area under the curve in a case-control model (classical AUC) as the probability of a case sample having higher expression than a control. Here, we used the modified AUC (mAUC) metric, as proposed in [21]. Unlike the classic AUC, mAUC introduces a margin (δ) to favor more confident separation between conditions. The mAUC formula in Eq. 1:

$$\mathrm{mAUC}(\lambda) = (1-\lambda)\Pr(Y > X) + \lambda \Pr(Y > X + \delta)$$

where:

- Y and X represent the expression values of the case (disease) and control samples, respectively.
- λ is the weighting parameter ($0 \leq \lambda \leq 1$) that balances the standard AUC and the shifted AUC.
- δ denotes the margin threshold, defined as $\delta = 1.96\sigma_0$ (where σ_0 is the standard deviation of the control group), which ensures the measure is scale-independent.

mAUC is basically a weighted sum of the classical AUC and a shifted AUC considering the displacement parameter δ. This will give us a more robust metric of how well the biomarker separates the groups. The R implementation is as follows:

3.14 Step 1: Compute Delta (δ) Based on Control Group Variability

The standard deviation σ_0 is strictly calculated from the control group samples for each gene. Then, we apply Eq. 2 to define the margin:

$$\delta = 1.96 \times \sigma_0$$

```
# Indices for Control (0) and Case (1)
idx_ctrl <- which(y == 0)
idx_case <- which(y == 1)
# Calculate Sigma (SD of control) and Delta (1.96
* Sigma) per gene
sigma0_vec <- apply(expr_bc[, idx_ctrl], 1, sd,
na.rm = TRUE)
delta_vec <- 1.96 * sigma0_vec
```

3.15 Step 2: Calculate mAUC per Gene Using the Formula

Iterate through each gene to calculate the standard AUC and the shifted AUC using the specific δ derived in Step 1. The final mAUC is the weighted sum of both.

```
# Helper function to calculate probability
P(Y > X)
calc_prob <- function(vals_case, vals_ctrl) {
 mean(outer(vals_case, vals_ctrl, FUN =
function(y, x) {
  ifelse(y > x, 1, ifelse(y == x, 0.5, 0))
}))
}
# Iterate over genes to calculate mAUC
mauc_scores <- sapply(seq_len(nrow(expr_bc)),
function(i) {
 val_case <- expr_bc[i, idx_case]
 val_ctrl <- expr_bc[i, idx_ctrl]
 delta <- delta_vec[i]
 # Calculate Original AUC and Shifted AUC
 auc_orig <- calc_prob(val_case, val_ctrl)
 auc_shifted <- calc_prob(val_case, val_ctrl +
delta)
 # Apply formula: mAUC = (1 - lambda) * AUC +
lambda * Shifted_AUC
 (1 - lambda) * auc_orig + lambda * auc_shifted
})
```

3.16 Step 3: Rank Genes and Organize Results

Finally, we consolidate the scores and parameters into a dataframe and sort the genes in descending order (highest mAUC first) to identify the best markers.

```
# Create results table
results <- data.frame(
 gene = rownames(expr_bc),
 mauc = mauc_scores,
 delta = delta_vec,
 sigma0 = sigma0_vec)
# Sort by mAUC in descending order
results_ranked <- results[order(results$mauc,
decreasing = TRUE),]
# Display top 10 genes
head(results_ranked, 10)
```

This method provides an interpretable and quantitative way to prioritize biomarkers by their ability to distinguish phenotypes. It is especially useful when looking for robust single biomarkers. You can also plot mAUC scores to assess global separation (*see* **Note 4**).

3.17 Panel of Biomarkers

1. If we aim to identify more than one biomarker, either to increase classification accuracy or to capture a broader biological signal, we can select a panel of biomarkers. This is typically done by training a classification model (such as logistic regression, K-nearest neighbors, or support vector machines) on a training set and evaluating its performance on an independent test set. The biomarkers included in the most predictive model are considered the most informative features.
2. However, as previously mentioned, these selected biomarkers are those with the highest predictive power, which does not necessarily mean they are causative or the only ones associated with the phenotype. They represent a predictive signature rather than a mechanistic explanation.
3. Step 0 (common to all methods): Split the original dataset into test and train sets. Here, expr_bc is the expression matrix after batch correction and normalization, and meta$group is the phenotype label. We ensure that the rownames (samples) align between the expression data and the metadata.

```
# Transpose gene expression matrix: samples x
features
X <- t(expr_bc)
y <- factor(meta$group, levels = c("control",
"ra"))
stopifnot(rownames(X) == rownames(meta))
Then, the dataset is split into training and test
sets for model evaluation.
#reserve ~20% of samples from each class as a test
set for model evaluation.
set.seed(123)
idx_ra <- which(y == "ra")
```

```
idx_c <- which(y == "control")
test_ra <- sample(idx_ra, ceiling(0.2 *
length(idx_ra)))
test_c <- sample(idx_c, ceiling(0.2 *
length(idx_c)))
test_idx <- sort(c(test_ra, test_c))
train_idx <- setdiff(seq_along(y), test_idx)
Xtr <- X[train_idx,, drop=FALSE]
Xte <- X[test_idx,, drop=FALSE]
ytr <- y[train_idx]
yte <- y[test_idx]
```

This stratified split maintains class balance in both training and test sets.

3.18 Panel of Biomarkers by Brute Force

1. When aiming to identify a small, highly discriminative panel of biomarkers (e.g., two to four genes), brute-force combinatorial model testing can be a practical strategy. This approach systematically evaluates every possible combination of candidate features (within a manageable size limit) using a classification model and ranks them by discriminative power on an independent test set.
2. Step 1: Select candidate features. Use the top-ranked genes (e.g., based on mAUC or DE) as potential biomarker candidates. For computational feasibility, we select a small set (e.g., top 12) (*see* **Note 5**)
   cand <– head(results_ranked$gene, 12)
   kmax <– 4 # Max combination size (e.g., 2 to 4)
3. Step 2: Define evaluation function. A helper function is defined to (a) fit a logistic regression model on training data using selected features, (b) scale test data using training mean/SD, and (c) predict probabilities on test data.
4. Compute AUC using the pROC package.

```
eval_panel <- function(features) {
 sc_tr <- scale(Xtr[, features, drop=FALSE])
 enter <- attr(sc_tr, "scaled:center")
 scalev <- attr(sc_tr, "scaled:scale")
 dftr <- data.frame(y = ytr, sc_tr)
 m <- glm(y ~., data = dftr, family = binomial())
 dfte <- data.frame(scale(Xte[, features, drop=
FALSE], center=center, scale=scalev))
 p <- as.numeric(predict(m, newdata = dfte, type
= "response"))
 roc_obj <- pROC::roc(yte, p, quiet = TRUE,
direction = "auto")
 list(auc = as.numeric(pROC::auc(roc_obj)), roc =
roc_obj, model = m,
     features = features, p = p)
}
```

5. Step 3: Brute-Force feature combination search. Iterate through all combinations of 2 to kmax features from the candidate list and evaluate each model.

```
best <- list(auc = -Inf)
for (k in 2:kmax) {
 combs <- combn(cand, k, simplify = FALSE)
 for (fs in combs) {
 out <- suppressWarnings(suppressMessages(eval_
panel(fs)))
 if (is.finite(out$auc) && out$auc > best$auc)
best <- out
 }
}
```

 This is a brute-force grid search through the combinatorial space of feature sets. The best model is selected based on the maximum AUC on the test set.

6. Step 4: Report and visualize the best panel.

```
best$auc
best$features
Retrieve the features selected in the best panel
and the AUC obtained.
# Plot ROC for best model
plot(best$roc,
   main = paste0("Best panel ROC (k=",
length(best$features),
            ") AUC=", round(best$auc, 3)),
   legacy.axes = TRUE)
abline(a = 0, b = 1, lty = 2)
```

3.19 Feature Selection

In high-dimensional omics datasets, where the number of features far exceeds the number of samples, the best approach to select an optimal panel of biomarkers is to use classifiers that perform feature selection. This reduces overfitting and facilitates biological interpretation.

1. Step 1: Choose a classifier that allows feature selection. Some examples of common models to perform feature selection are (a) LASSO / Elastic Net logistic regression, which selects features by shrinking some coefficients to exactly zero, (b) Support Vector Machine with Recursive Feature Elimination (SVM-RFE), a wrapper method that iteratively removes the least informative variables, and (c) Random Forest, which provides an internal measure of variable importance based on impurity reduction. Caret is used as the primary package for applying these models.
2. Step 2: Define the outcome variable and split the data. The response variable must be binary. When using ROC-based

metrics in caret, the first factor level is treated as the positive class; therefore, levels should be set explicitly.

Priorly, the dataset was split into training and test sets using stratified sampling to preserve class proportions. Additionally, we will split the training dataset into folds for cross-validation to optimize hyperparameters.

```
ctrl5 <- trainControl(
 method = "cv",
 number = 5,
 classProbs = TRUE,
 summaryFunction = twoClassSummary,
 savePredictions = "final"
)
```

3. Step 3: Train a LASSO / Elastic Net model. Regularized logistic regression is trained using the glmnet algorithm. Gene expression values are *z*-score standardized (centering and scaling) within each resampling fold. The optimal regularization parameter (lambda) is selected automatically by cross-validation.

```
m_lasso <- train(
 x = Xtr, y = ytr,
 method = "glmnet",
 preProcess = c("center", "scale"),
 metric = "ROC",
 trControl = ctrl5,
 tuneLength = 50)
 m_lasso$bestTune
```

Genes selected by the LASSO model correspond to those with non-zero regression coefficients at the optimal value of lambda. Figure 3 shows a schematic illustration of the LASSO-based feature selection process for biomarker identification.

```
beta <- coef(m_lasso$finalModel, s =
m_lasso$bestTune$lambda)
beta <- as.matrix(beta)
lasso_genes <- rownames(beta)[beta[, 1]!= 0]
lasso_genes <- setdiff(lasso_genes, "(Intercept)")
length(lasso_genes)
head(lasso_genes)
```

4. Step 4: Rank genes using Random Forest importance. Random Forest classifier is trained on the training data. Gene importance is quantified using the Mean Decrease in Gini index, and the top-ranked genes are retained.

```
rf <- randomForest(
```

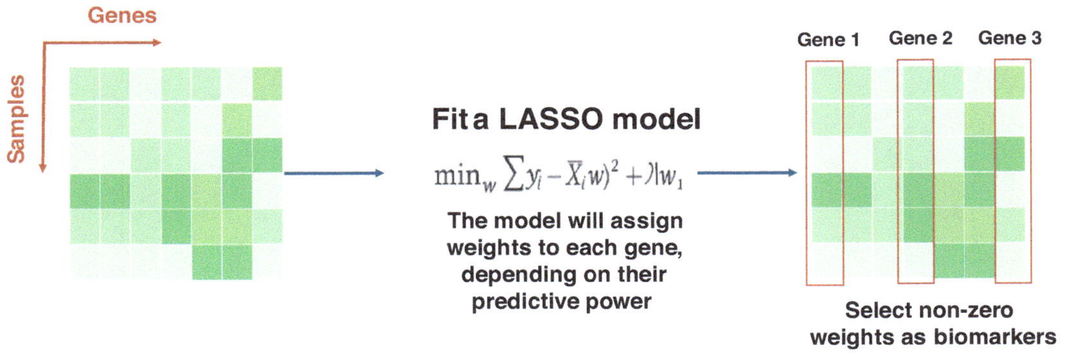

Fig. 3 Schematic illustration of LASSO-based biomarker selection. A LASSO model is fitted to the gene-expression matrix, assigning coefficients to each gene based on its predictive contribution. Genes with non-zero coefficients are retained as candidate biomarkers

```
  x = Xtr,
  y = ytr,
  ntree = 1000,
  importance = TRUE
)
rf_imp <- sort(importance(rf, type = 2)[, 1],
decreasing = TRUE)
rf_genes <- names(rf_imp)[1:min(50,
length(rf_imp))]
length(rf_genes)
head(rf_genes)
```

5. Step 5: Define a consensus biomarker panel. To prioritize robust candidate biomarkers, a consensus gene panel is defined as the intersection of genes selected by LASSO and Random Forest importance ranking. This conservative approach favors genes consistently identified across methods with different assumptions.

```
common_genes <- Reduce(intersect, list(lasso_
genes, rf_genes))
common_genes
```

6. Step 6: Train the final model and evaluate performance on the test set. A final logistic regression model is trained using only the selected gene panel. Scaling parameters are estimated exclusively from the training data and then applied to the test data to avoid information leakage (*see* **Note 6**).

```
panel <- common_genes
sc_tr <- scale(Xtr[, panel, drop = FALSE])
# Store centering and scaling parameters
```

```
cen <- attr(sc_tr, "scaled:center")
sds <- attr(sc_tr, "scaled:scale")
dftr <- data.frame(y = ytr, sc_tr)
m_final <- glm(y ~., data = dftr, family =
binomial())
#scale test data with train scaling.
dfte <- data.frame(
 scale(Xte[, panel, drop = FALSE], center = cen,
scale = sds))
p <- predict(m_final, newdata = dfte, type =
"response")
```

Finally, model performance is evaluated on the test set using the area under the ROC curve (AUC).

```
roc_obj <- roc(
 response = yte,
 predictor = p,
 levels = c("control", "case"),
 direction = "<",
 quiet = TRUE
)
auc_val <- as.numeric(auc(roc_obj))
auc_val
```

3.20 Network-Based Biomarker Discovery

Network-based analysis has been widely applied in systems biology to provide a better understanding of molecular mechanisms. Most of the time, complex traits such as allergies do not depend on the dysregulation of a single gene, but rather on coordinated changes affecting key genes that propagate their effects in cascades, ultimately perturbing entire biological functions or pathways.

Because of this, another effective strategy to prioritize biomarkers is the use of network-based approaches, where genes are analyzed in the context of their relationships with other genes. In this tutorial, Weighted Gene Co-expression Network Analysis (WGCNA) is used to identify gene modules associated with the phenotype of interest (Fig. 4).

1. Step 1: Start from a previously normalized expression matrix. The analysis begins with a gene expression matrix that has already been normalized and batch-corrected if necessary. The matrix must be organized as samples × genes, as required by WGCNA.

```
library(WGCNA)
options(stringsAsFactors = FALSE)
enableWGCNAThreads()
X <- t(expr_bc) # samples x genes
y <- factor(meta$group, levels =
c("control","ra"))
stopifnot(rownames(X) == rownames(meta))
```

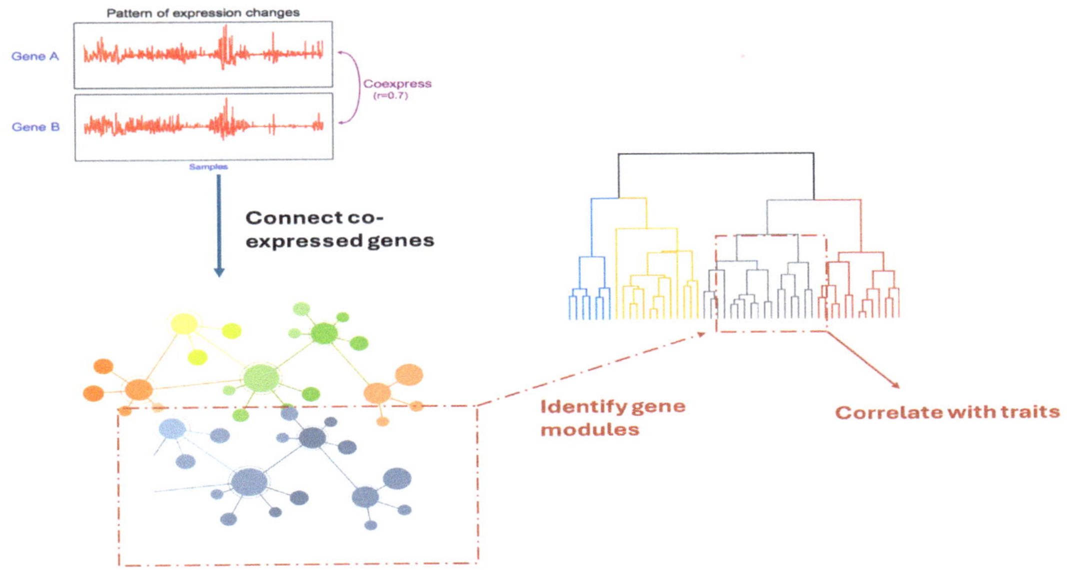

Fig. 4 Overview of a co-expression network–based biomarker discovery approach. Genes with similar expression patterns are grouped into co-expression modules, which are subsequently correlated with phenotypic traits to identify biologically relevant gene modules

```
datExpr0 <- as.data.frame(X)
```

2. Step 2: Define trait dataframe and perform quality control to detect outliers. Before network construction, a quality control step is performed to identify problematic samples or genes. The function goodSamplesGenes() flags samples or genes with too many missing or zero-variance values.

```
# Trait dataframe (0 = control, 1 = RA)
trait <- data.frame(RA = as.numeric(y == "ra"))
rownames(trait) <- rownames(datExpr0)
gsg <- goodSamplesGenes(datExpr0, verbose = 0)
datExpr <- datExpr0[gsg$goodSamples,
gsg$goodGenes]
trait <- trait[rownames(datExpr),, drop = FALSE]
```

3. Step 3: Filter the expression matrix and optionally inspect sample clustering. After removing outliers, hierarchical clustering of samples can be used to visually inspect potential remaining outliers.

```
sampleTree <- hclust(dist(datExpr), method =
"average")
plot(sampleTree, main = "Sample clustering", xlab
= "", sub = "")
```

4. Step 4: Choose the soft-thresholding power. The soft-thresholding power (β) determines how strongly correlations between genes are weighted. It is selected by iteratively evaluating different values and choosing the one that best approximates a scale-free network topology, which is commonly observed in biological networks.

```
sft <- pickSoftThreshold(
 datExpr,
 powerVector = powers,
 networkType = "signed",
 verbose = 0
)
beta <- ifelse(is.na(sft$powerEstimate), 8,
sft$powerEstimate)
beta
```

5. Step 5: Construct the co-expression network and identify modules. The network is constructed using the blockwiseModules function. Modules are defined as clusters of genes showing similar expression profiles across samples. Several parameters can be tuned to obtain more or fewer modules.

```
net <- blockwiseModules(
 datExpr,
 power = beta,
 networkType = "signed",
 TOMType = "signed",
 minModuleSize = 30,
 deepSplit = 2,
 pamRespectsDendro = FALSE,
 mergeCutHeight = 0.25,
 reassignThreshold = 0,
 verbose = 3
)
moduleColors <- labels2colors(net$colors)
table(moduleColors)
```

The resulting gene dendrogram and module assignments can be visualized as follows:

```
plotDendroAndColors(
 net$dendrograms[[1]],
 moduleColors[net$blockGenes[[1]]],
 "Module colors",
 dendroLabels = FALSE,
 hang = 0.03,
 addGuide = TRUE,
 guideHang = 0.05
)
```

6. Step 6: Calculate module eigengenes. Module eigengenes represent the first principal component of each module and summarize the expression of all genes within the module for each sample.

```
MEs <- moduleEigengenes(datExpr, colors =
moduleColors)$eigengenes
MEs <- orderMEs(MEs)
```

7. Step 7: Correlate module eigengenes with the trait of interest. Module eigengenes are correlated with the trait using Pearson correlation. *P*-values are computed using a Student's *t*-test.

```
moduleTraitCor <- cor(MEs, trait$RA, use = "p")
moduleTraitP        <
-  corPvalueStudent(moduleTraitCor, nSamples =
nrow(datExpr))
moduleTraitCor
moduleTraitP
```

A heatmap is commonly used to visualize these associations.

```
textMatrix <- paste0(
 signif(moduleTraitCor, 2), "\n(",
 signif(moduleTraitP, 2), ")"
)
labeledHeatmap(
 Matrix = moduleTraitCor,
 xLabels = "RA (1) vs control (0)",
 yLabels = names(MEs),
 ySymbols = names(MEs),
 colorLabels = FALSE,
 textMatrix = textMatrix,
 setStdMargins = FALSE,
 cex.text = 0.6,
 zlim = c(-1, 1),
 main = "Module-trait relationships"
)
```

8. Step 8: Select biomarker modules. Modules whose eigengenes show the strongest correlation (absolute value) with the trait are selected as candidate biomarker modules.

```
bestME <- rownames(moduleTraitCor)[which.
max(abs(moduleTraitCor[,1]))]
bestME
```

To understand the direction of the association, eigengene values can be visualized by group.

```
par(mar = c(4,4,2,1))
boxplot(
 MEs[[bestME]] ~ y,
 main = paste0(bestME, " eigengene by group"),
 xlab = "group",
 ylab = "Module eigengene (PC1)"
)
```

9. Step 9: Optional functional enrichment analysis. Optionally, genes belonging to trait-associated modules can be subjected to ORA or GSEA (ranking genes by module membership) to identify enriched biological processes or pathways.
10. Step 10: Identify hub genes within the selected module. Hub genes are genes that are highly connected within a module and strongly associated with the trait. Genes can be selected based on (a) Module Membership (MM or kME): correlation between gene expression and module eigengene, (b) Gene Significance (GS): correlation between gene expression and the trait, and (c) topology metrics, with Betweenness centrality for the most connected gene within the module, with KWithin metric or others.

```
# Gene Significance (gene-trait correlation)
GS <- as.numeric(cor(datExpr, trait$RA, use =
"p"))
names(GS) <- colnames(datExpr)
bestColor <- sub("^ME", "", bestME)
genesInMod <- colnames(datExpr)[moduleColors ==
bestColor]
# Module Membership (kME)
MM <- as.numeric(cor(datExpr[, genesInMod, drop =
FALSE],
           MEs[[bestME]], use = "p"))
names(MM) <- genesInMod
```

Hub genes are selected using predefined thresholds on MM and GS.

```
hubGenes <- genesInMod[abs(MM) > 0.8 &
abs(GS[genesInMod]) > 0.5]
length(hubGenes)
head(hubGenes)
```

3.21 Biomarker Discovery Based on Multi-omics Integration Strategies

1. Paired omics. With paired omics, meaning that multiple omics layers are measured in the same samples, we can apply multivariate statistical approaches for biomarker discovery. This is usually called "vertical integration." Among the available methods, DIABLO is one of the most powerful and widely used. DIABLO is a platform designed for multi-omic

biomarker discovery via dimensionality reduction using latent components. It summarizes high-dimensional omics layers into a reduced number of latent components by maximizing the covariance between omics layers and uses a supervised framework to make these components as discriminative as possible between phenotypic traits. As a result, DIABLO identifies groups of features across different biological layers that are correlated, yielding more biologically coherent signatures. These multi-omic signatures can facilitate the inference of regulatory networks, coordinated molecular effects, and cross-layer biological mechanisms. To use DIABLO, the following steps can be followed:

2. Step 1: Pre-process and normalize each omics independently. Each omics layer must be pre-processed following its own statistical assumptions. Normalization and filtering should be performed before integration, and independently for each data type. First, sample ordering is enforced to ensure that all omics are perfectly aligned.

```
stopifnot(all(sample_order %in%
colnames(mrna_mat_feat)))
stopifnot(all(sample_order %in%
colnames(mirna_mat_feat)))
stopifnot(all(sample_order %in%
colnames(meth_mat_feat)))
mrna_mat_feat <- mrna_mat_feat[, sample_order]
mirna_mat_feat <- mirna_mat_feat[, sample_order]
meth_mat_feat <- meth_mat_feat[, sample_order]
```

The phenotype vector is defined as a factor.

```
Y <- factor(c(rep("Control", 10), rep("CFA", 10)),
       levels = c("Control", "CFA"))
summary(Y)
table(Y)
```

3. Step 2: Generate a list of data matrices, one per omics. Each omics must be stored as a separate matrix with dimensions samples × features. All omics are combined into a named list, which is the required input format for DIABLO.

```
data <- list(
 miRNA = X_miRNA,
 mRNA = X_mRNA,
 methylation = X_met
)
all(rownames(data$miRNA) == rownames(data$mRNA))
all(rownames(data$mRNA) ==
rownames(data$methylation))
```

4. Step 3: Quality check using pairwise sparse PLS. Before building the full DIABLO model, pairwise sparse PLS (sPLS) models are fitted to assess the strength of relationships between each pairs of omics.

```
library(mixOmics)
set.seed(123)
list.keepX <- c[25]
list.keepY <- c[25]
pls1 <- spls(data[["miRNA"]], data[["mRNA"]],
        keepX = list.keepX, keepY = list.keepY)
pls2 <- spls(data[["miRNA"]],
data[["methylation"]],
        keepX = list.keepX, keepY = list.keepY)
pls3 <- spls(data[["mRNA"]],
data[["methylation"]],
        keepX = list.keepX, keepY = list.keepY)
```

A circle correlation plot is used to visualize the overlap between omics on the first two latent components.

```
plotVar(pls1, cutoff = 0.5, title = "(a) miRNA vs
mRNA",
    legend = c("miRNA", "mRNA"),
    var.names = FALSE, style = 'graphics',
    pch = c(16, 17), cex = c(2,2),
    col = c('darkorchid', 'lightgreen'))
```

Pairwise correlations between latent components can also be computed.

```
cor(pls1$variates$X, pls1$variates$Y)
```

5. Step 4: Define the DIABLO design matrix. The design matrix controls how strongly each pair of omics is forced to converge. Values range from 0 to 1.

- Values closer to **0** prioritize phenotype discrimination
- Values closer to **1** prioritize omics integration

In Table 1, we can see an example of a design matrix.

```
design <- matrix(
 0.6,
 ncol = length(data),
 nrow = length(data),
 dimnames = list(names(data), names(data))
)
diag(design) <- 0
design
```

Table 1
Design matrix of DIABLO's model

	miRNA	Methylation	mRNA
miRNA	0	0.6	0.6
Methylation	0.6	0	0.6
mRNA	0.6	0.6	0

6. Step 5: Construct the initial DIABLO model. An initial DIABLO model is built using an arbitrarily high number of components.

```
basic.diablo.model <- block.splsda(
 X = data,
 Y = Y,
 ncomp = 5,
 design = design
)
```

7. Step 6: Choose the optimal number of components. The number of components is selected by cross-validation using the perf( ) function. Then generate the classification error plot (Fig. 5A). This plot is generated by the perf function and can be visualized by plot( ). It plots how good the model is on the classification task depending on the distance metric chosen. You should choose a number of components that balances simplicity and classification power.

```
perf.diablo <- perf(
 basic.diablo.model,
 validation = 'Mfold',
 folds = 5,
 nrepeat = 5
)
plot(perf.diablo)
```

The optimal number of components can be automatically extracted.

```
ncomp <- perf.diablo$choice.ncomp$WeightedVote[
 "Overall.BER", "centroids.dist"
]
perf.diablo$choice.ncomp$WeightedVote
```

8. Step 7: Tune the number of features per omics. The number of features retained for each omic is optimized by cross-validation.

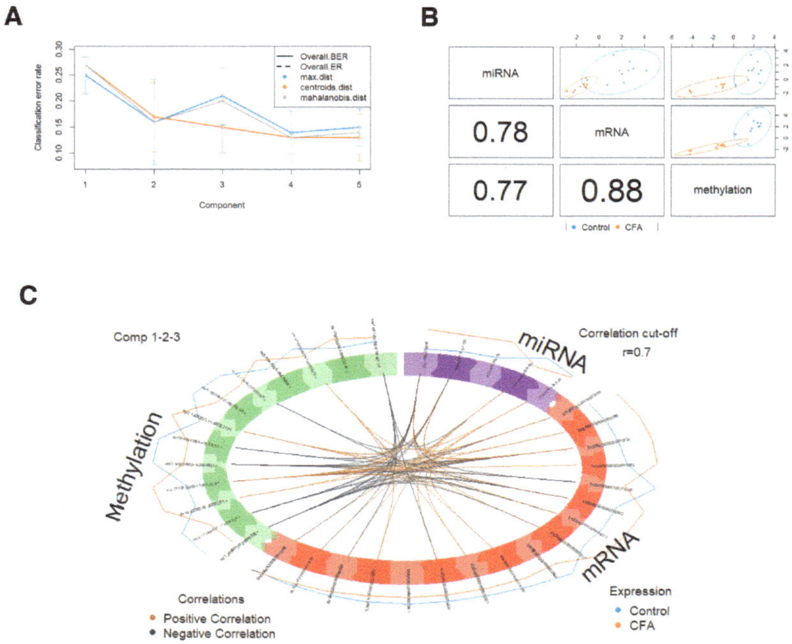

Fig. 5 Example of DIABLO-based multi-omics integration and biomarker identification. (**A**) Cross-validated classification error rates used to select the optimal number of components (*see* **Note 7**). (**B**) Sample projections and correlations between latent components across omics layers, illustrating phenotypic separation and cross-omics integration (*see* **Note 8**). (**C**) Circos plot showing correlations between selected features across omics layers, highlighting putative cross-omics associations (*see* **Note 9**). Plots are derived from in-house analysis of the GEO dataset with accession GSE174631, a dataset of miRNA, mRNA, and methylation measured from rats' model of inflammation-associated chronic pain

First, we need to create a grid of numbers of features for each omics to try:

```
test.keepX <- list(
 mRNA = c(seq(1, 10, 2)),
 miRNA = c(seq(1, 10, 2)),
 proteomics = c(seq(1, 10, 2))
)
```

Then, we are going to use that grid to try different combinations of features:

```
tune <- tune.block.splsda(
 X = data,
 Y = Y,
 ncomp = ncomp,
 test.keepX = test.keepX,
 design = design,
 validation = 'Mfold',
 folds = 5,
```

```
 nrepeat = 1,
 dist = "centroids.dist"
)
list.keepX <- tune$choice.keepX
list.keepX
```

9. Step 8: Fit the final DIABLO model. Fit the final DIABLO model by using the number of components and the list of variables chosen in previous steps.

```
final.diablo.model <- block.splsda(
 X = data,
 Y = Y,
 ncomp = ncomp,
 keepX = list.keepX,
 design = design
)
```

   Selected variables can be extracted per block.

```
selectVar(final.diablo.model, block = 'mRNA', comp
= 1)$mRNA$name
```

10. Step 9: Visualize the DIABLO results. Once the final DIABLO model has been fitted, several graphical outputs can be used to explore the structure of the latent components and assess how well the samples are separated by phenotype.
   The plotDiablo() function provides an overview of the multi-omics integration by displaying the projections of the samples onto the selected latent component(s). This plot summarizes how the different omics blocks jointly contribute to discrimination between classes (Fig. 5B).

```
plotDiablo(final.diablo.model, ncomp = 1)
plotIndiv(final.diablo.model,
      ind.names = FALSE,
      legend = TRUE,
      title = 'DIABLO Sample Plots')
```

11. Step 10: Network and Circos visualizations. To explore the relationships between selected features across omics layers, DIABLO provides network-based visualizations that highlight correlated features across blocks. The circosPlot function displays correlations between selected features from different omics layers in a circular layout. Only correlations above a specified cutoff are shown, allowing the user to focus on the strongest cross-omics associations. This visualization is particularly useful to identify groups of features that are strongly correlated across omics and may participate in shared biological processes. An example of this graph is in Fig. 5C.

```
circosPlot(final.diablo.model, cutoff = 0.7, line
= TRUE,
      color.blocks = c('darkorchid', 'brown1',
'lightgreen'),
      color.cor = c("chocolate3","grey20"), size.
labels = 1.5)
```

Similar but more detailed, the network function can be used to generate a graph-based visualization of the selected features and their correlations. In this network, nodes represent molecular features (e.g., genes, miRNAs, CpGs), and edges represent correlations between features across different omics domains. Increasing the correlation cutoff reduces network complexity and highlights the most robust associations. This can help us infer hypothetical regulatory mechanisms associated with the phenotype (Fig. 6A).

```
network(
 final.diablo.model,
 blocks = c(1,2,3),
 cutoff = 0.75,
 graph.scale = 1,
 size.node = 0.01,
 cex.node.name = 0.7,
 color.node = c('darkorchid', 'brown1',
'lightgreen'),
 lwd.edge = 1.5,
 show.color.key = TRUE,
 save = 'png',
 name.save = 'diablo-network1'
)
```

12. Step 11: Final performance assessment. After model construction and visualization, the predictive performance of the DIABLO model is evaluated using repeated cross-validation. The perf( ) function computes classification error rates using the selected distance metric.

```
perf.diablo <- perf(
 final.diablo.model,
 validation = 'Mfold',
 folds = 5,
 nrepeat = 5,
 dist = 'centroids.dist'
)
```

ROC-AUC curves can also be plotted in order to see the separation power of each omic.

```
plot(perf.diablo)
auc.diablo <- auroc(
 final.diablo.model,
```

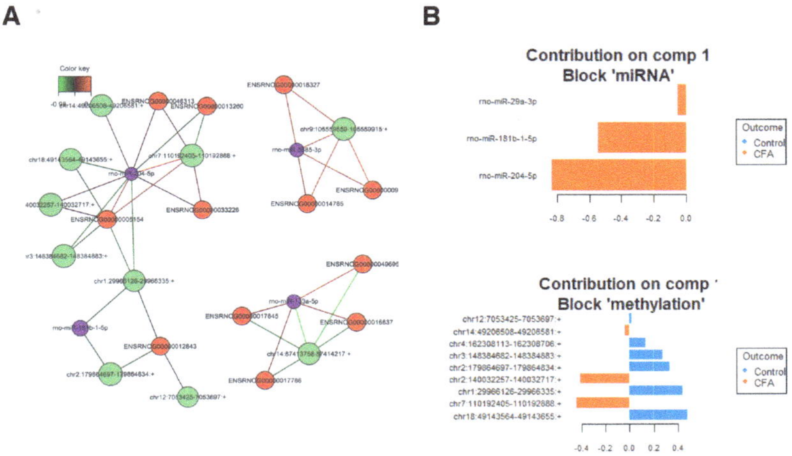

Fig. 6 Example of network-based and feature contribution analysis of selected multi-omics biomarkers. (**A**) correlation-based network linking features across omics layers, highlighting putative cross-omics associations. (**B**) contribution (loadings) of selected miRNA and DNA methylation features to the first latent component, indicating their relative importance in phenotypic discrimination (*see* **Note 10**)

```
  roc.block = "met",
  roc.comp = 3,
  print = FALSE
)
```

Another important plot type is the loading plot, which shows the contribution of each feature to the latent components. This helps us identify the most influential features that drive component separation (Fig. 6B).

```
plotLoadings(final.diablo.model,
       comp = 1,
       contrib = 'max',
       method = 'median')
```

3.22 Independent Cohorts

Integrating datasets from different omics measured in independent cohorts, commonly referred to as diagonal integration, is a highly challenging task and remains an active area of research with no established, standardized methodology. This complexity arises from the presence of both technical variability and biological heterogeneity, rendering conventional batch correction or dimensionality-reduction approaches insufficient.

The most commonly explored strategies rely on incorporating prior biological knowledge, such as known regulatory relationships between molecular entities or curated interaction networks, to guide model construction and integration [22–24]. However, because these approaches strongly depend on the availability and quality of prior knowledge, as well as on the specific omics layers and experimental designs involved, no generalizable methodological framework can currently be recommended

3.23 Expected Outputs and Biomarker Validation

The methodologies described in this chapter are expected to yield candidate biomarkers at multiple molecular levels, including individual features (genes, proteins, miRNAs, or CpG sites), network-derived modules, and integrated multi-omic signatures. Typical outputs include ranked feature lists, latent components, co-expression modules, and predictive biomarker panels accompanied by performance metrics.

A critical step following biomarker discovery is validation, which aims to assess the robustness, generalizability, and biological relevance of the identified candidates. Whenever available, validation should be performed using an independent cohort that was not involved in any stage of model training or feature selection. Independent validation provides the most stringent test of biomarker reproducibility and helps to rule out overfitting or cohort-specific effects.

4 Notes

1. Precautions about mixing datasets. In this case, it is mandatory to carefully look into the experiment metadata and confirm that: (a) the biological conditions are comparable (same model, species, environment), (b) sample-level metadata are available and sufficiently detailed (e.g., treatment, age, sex, batch), (c) consistency of the omics technologies and pre-processing steps is documented. Sometimes, it is better to download the raw files and do the pre-processing ourselvess than rely on pre-processed data matrices, and (d) can sample identifiers be matched across omics in multi-omics datasets.
2. About biological replicates needed for biomarker discovery. Although $n=3$ (biological replicates) for each biological group is a MINIMUM for statistical analysis, most of the models used here use multivariate statistics, using multiple test corrections, fold partition for intra-model validation when machine learning is used, etc., which means that most of the time we need a higher number of samples to correctly apply these methodologies. Consider n between 5 and 10 to start if the biological conditions are highly separable, but sometimes we will need a higher number. The choice of biological replicates has to be done by considering the omics used (e.g., microbiomics can be highly dependent on environmental conditions, introducing high noise rates), the downstream analysis we will do, and the need for independent test validation.

3. Log2 on database datasets. Before starting the analysis, it is essential to determine whether the data are raw or already processed. Some databases provide normalized expression matrices, in which case additional normalization or $\log_2$ transformation should not be applied. When raw data are available, appropriate preprocessing steps must be performed. This can be assessed by consulting the study metadata or by examining simple statistics such as value ranges or distributions.
4. Precaution on using AUC and mAUC. Here, mAUC is used as a ranking criterion rather than as a validation metric. In this context, mAUC is calculated on the training data to prioritize candidate biomarkers. Importantly, mAUC values computed on the training set should not be interpreted as measures of predictive performance. Proper reporting of mAUC as a validation metric requires its calculation on an independent test set to avoid overfitting.
5. Combinatorial problem. Take into account the combinatorial problem: if you choose five genes for your panel, you will have 2^5-1 possible combinations of genes to try, making exhaustive evaluation rapidly infeasible as the number of candidate genes increases. Because of this, this method should be used only if you have a very clear pre-filter of genes.
6. About data leakage. It is important to perform data scaling after splitting the dataset into training and test sets, rather than before. If scaling is applied prior to the split, the scaling parameters (e.g., mean and standard deviation) are computed using information from the entire dataset, including the test samples. This results in data leakage, as information from the test set is inadvertently used during model training, leading to overly optimistic performance estimates. Ideally, we should use an independent test set from a different cohort of samples.
7. Guide to understanding DIABLO plots. This plot shows the classification error as a function of the number of model components. Lower values indicate better predictive performance. Solid and dashed lines represent the overall balanced error rate (BER) and overall error rate (ER), respectively, while different colors correspond to alternative prediction distance metrics. Error bars reflect the variability across cross-validation folds. The optimal number of components is typically selected at the point where the classification error reaches a minimum or stabilizes, balancing predictive accuracy and model complexity, including the metric chosen.
8. Guide to understanding DIABLO plots. DIABLO aims to strike a balance between integrating multiple omics layers and maximizing the discriminative power between phenotypic

groups. Accordingly, the upper diagonal of the figure displays scatter plots showing how well the phenotypes are separated based on the selected variables, summarized by the latent components. Each panel represents sample projections onto the corresponding components, with samples colored according to their phenotypic group. Greater separation between the group ellipses indicates improved discriminative performance of the selected biomarkers. The lower diagonal of the figure illustrates the integrative aspect of the model by reporting the correlations between components from different omics layers. Higher correlation values reflect stronger cross-omics coherence, indicating biologically consistent signals captured across the integrated datasets.

9. Guide to understanding Circos plots. This Circos plot visualizes correlations between features selected from different omics layers across the model components. Links represent pairwise associations above the specified correlation threshold, with colors indicating positive or negative correlations. The circular layout facilitates the identification of coordinated patterns across omics layers and allows the inference of putative regulatory relationships, such as miRNA–mRNA or methylation–expression interactions. However, these associations are purely correlative and do not imply direct regulatory effects or causal relationships. The inferred network (visualized on the network plot) should therefore be interpreted as a hypothesis-generating framework that highlights candidate interactions requiring independent experimental validation.
10. Guide to understanding loading plots. This figure shows the contribution of the selected features to the first latent component for each omics block. Bars represent the loading values of individual variables, indicating both the magnitude and direction of their contribution to the component. Features with larger absolute loadings contribute more strongly to the separation captured by the component. The sign of the loading reflects the direction of the association with the phenotypic groups, as illustrated by the group-wise coloring. Importantly, these contributions indicate association with the latent component rather than direct effect sizes or causal influence. Therefore, the identified features should be interpreted as candidate biomarkers driving group separation within the model, rather than as independently validated regulatory elements.

Competing Interests

The authors have no conflicts of interest to declare that are relevant to the content of this chapter.

References

1. Califf RM (2018) Biomarker definitions and their applications. Exp Biol Med 243(3):213–221. https://doi.org/10.1177/1535370217750088
2. Muraro A, Arasi S (2018) Biomarkers in food allergy. Curr Allergy Asthma Rep 18(11):64. https://doi.org/10.1007/s11882-018-0816-4
3. McDermott JE, Wang J, Mitchell H, Webb-Robertson BJ, Hafen R, Ramey J, Rodland KD (2013) Challenges in biomarker discovery: combining expert insights with statistical analysis of complex omics data. Expert Opin Med Diagn 7(1):37–51. https://doi.org/10.1517/17530059.2012.718329
4. Diaz-Uriarte R, Gómez de Lope E, Giugno R, Fröhlich H, Nazarov PV, Nepomuceno-Chamorro IA, Rauschenberger A, Glaab E (2022) Ten quick tips for biomarker discovery and validation analyses using machine learning. PLoS Comput Biol 18(8):e1010357. https://doi.org/10.1371/journal.pcbi.1010357
5. Zhao Q, Zhang C, Zhang W, Zhang S, Liu Q, Guo Y (2025) Applications and challenges of biomarker-based predictive models in proactive health management. Front Public Health 13:1633487. https://doi.org/10.3389/fpubh.2025.1633487
6. Porcu E, Sjaarda J, Lepik K, Carmeli C, Darrous L, Sulc J, Mounier N, Kutalik Z (2021) Causal inference methods to integrate omics and complex traits. Cold Spring Harb Perspect Med 11(5):a040493. https://doi.org/10.1101/cshperspect.a040493
7. Ritchie ME, Phipson B, Wu D, Hu Y, Law CW, Shi W, Smyth GK (2015) limma powers differential expression analyses for RNA-sequencing and microarray studies. Nucleic Acids Res 43(7):e47. https://doi.org/10.1093/nar/gkv007
8. Leek JT, Johnson WE, Parker HS, Jaffe AE, Storey JD (2012) The sva package for removing batch effects and other unwanted variation in high-throughput experiments. Bioinformatics 28(6):882–883. https://doi.org/10.1093/bioinformatics/bts034
9. Johnson WE, Li C, Rabinovic A (2007) Adjusting batch effects in microarray expression data using empirical Bayes methods. Biostatistics 8(1):118–127. https://doi.org/10.1093/biostatistics/kxj037
10. Kuhn M (2008) Building predictive models in R using the caret package. J Stat Softw 28(5):1–26. https://doi.org/10.18637/jss.v028.i05
11. Friedman J, Hastie T, Tibshirani R (2010) Regularization paths for generalized linear models via coordinate descent. J Stat Softw 33(1):1–22
12. Langfelder P, Horvath S (2008) WGCNA: an R package for weighted correlation network analysis. BMC Bioinformatics 9:559. https://doi.org/10.1186/1471-2105-9-559
13. Singh A, Shannon CP, Gautier B, Rohart F, Vacher M, Tebbutt SJ, Lê Cao KA (2019) DIABLO: an integrative approach for identifying key molecular drivers from multi-omics assays. Bioinformatics 35(17):3055–3062. https://doi.org/10.1093/bioinformatics/bty1054
14. Clough E, Barrett T (2016) The gene expression Omnibus database. Methods Mol Biol 1418:93–110. https://doi.org/10.1007/978-1-4939-3578-9_5
15. Parkinson H, Kapushesky M, Shojatalab M, Abeygunawardena N, Coulson R, Farne A, Holloway E, Kolesnykov N, Lilja P, Lukk M, Mani R, Rayner T, Sharma A, William E, Sarkans U, Brazma A (2007) ArrayExpress—a public database of microarray experiments and gene expression profiles. Nucleic Acids Res 35(Database issue):D747–D750. https://doi.org/10.1093/nar/gkl995
16. Katz K, Shutov O, Lapoint R, Kimelman M, Brister JR, O'Sullivan C (2022) The sequence read archive: a decade more of explosive growth. Nucleic Acids Res 50(D1):D387–D390. https://doi.org/10.1093/nar/gkab1053
17. Lappalainen I, Almeida-King J, Kumanduri V et al (2015) The European Genome-phenome Archive of human data consented for biomedical research. Nat Genet 47(7):692–695. https://doi.org/10.1038/ng.3312
18. Bhattacharya S, Dunn P, Thomas CG, Smith B, Schaefer H, Chen J, Hu Z, Zalocusky KA, Shankar RD, Shen-Orr SS, Thomson E, Wiser J, Butte AJ (2018) ImmPort, toward repurposing of open access immunological assay data for translational and clinical research.

Sci Data 5:180015. https://doi.org/10.1038/sdata.2018.15

19. Zhang Y, Parmigiani G, Johnson WE (2020) *ComBat-seq*: batch effect adjustment for RNA-seq count data. NAR Genom Bioinform 2(3):lqaa078. https://doi.org/10.1093/nargab/lqaa078
20. Wang J (2025) ComBat-met: adjusting batch effects in DNA methylation data. NAR Genom Bioinform 7(2):lqaf062. https://doi.org/10.1093/nargab/lqaf062
21. Yu W, Ivan Chang Yc, Park E (2014) A modified area under the ROC curve and its application to marker selection and classification. J Korean Stat Soc 43:161–175. https://doi.org/10.1016/j.jkss.2013.05.003
22. Li Y, Tan Y, Chen R, Zhou XM, Chen Y, Huo Y, Guo X, Tiwari A, Wen Z, Zhong X, Malin BA, Li B, Yin Z (2025) scCotag: diagonal integration of single-cell multi-omics data via prior-informed co-optimal transport and regularized barycentric mapping. bioRxiv [Preprint] 2025.12.11.693589. https://doi.org/10.64898/2025.12.11.693589
23. Charmpi K, Chokkalingam M, Johnen R, Beyer A (2021) Optimizing network propagation for multi-omics data integration. PLoS Comput Biol 17(11):e1009161. https://doi.org/10.1371/journal.pcbi.1009161
24. Dimitrakopoulos C, Hindupur SK, Häfliger L, Behr J, Montazeri H, Hall MN, Beerenwinkel N (2018) Network-based integration of multi-omics data for prioritizing cancer genes. Bioinformatics 34(14):2441–2448. https://doi.org/10.1093/bioinformatics/bty148

Chapter 18

Modeling Antigen Allergenicity with Bioinformatics Approaches

Isidro Sobrino and José de la Fuente

Abstract

Allergenicity is a multifactorial and context-dependent property that emerges from the interaction between intrinsic protein features and host immune mechanisms. Although regulatory frameworks rely primarily on sequence homology to known allergens, increasing numbers of novel and engineered proteins require more comprehensive computational strategies for preclinical assessment. This chapter describes a modular bioinformatic pipeline for the prediction and characterization of potential protein allergens. The workflow integrates FAO/WHO sequence-based criteria for regulatory compliance, structure-informed homology methods, and machine learning-based classifiers. Proteins identified as potential allergens are subsequently subjected to downstream immunological characterization. This stepwise decision framework enables flexible and biologically informed prescreening of candidate proteins, providing a structured approach prior to experimental validation or systems biology–based analyses.

Keywords Allergy, Biodata, Machine learning, Systems biology, Deep learning, Immunoinformatics

1 Introduction

Allergens are usually proteins capable of inducing an IgE-mediated hypersensitivity reaction. However, allergenicity is not an intrinsic binary property of a molecule but rather an emergent feature resulting from the interaction between the protein and the host immune system [1]. Allergy is a complex immune-mediated disorder influenced by environmental exposure, host genetic predisposition, epithelial barrier integrity, and the physicochemical and structural properties of biomolecules [2].

The acquisition of hypersensitivity involves a prior sensitization phase. In the case of the classic Type-I hypersensitivity during this process, antigen-presenting cells, particularly dendritic cells, process the protein and present derived peptides via MHC class II molecules to naïve $CD4^+$ T cells [3]. Under specific cytokine and costimulatory signals, these T cells differentiate into Th2 cells, producing IL-4, IL-5, and IL-13 [3, 4]. Th2 cells provide help to B cells that have

José de la Fuente and Alejandro Cabezas-Cruz (eds.), *Allergic Disease Models*, Methods in Molecular Biology, vol. 3092,
https://doi.org/10.1007/978-1-0716-5635-8_18,

recognized the same antigen through their B cell receptor, inducing class-switch recombination toward IgE production [4]. The resulting plasma cells secrete allergen-specific IgE, which binds to FcεRI receptors on mast cells and basophils. Upon subsequent exposure, crosslinking of receptor-bound IgE triggers cellular degranulation and the clinical allergic response [3].

In addition to adaptive immune mechanisms, epithelial damage and innate immune sensing through pattern recognition receptors, as well as the release of epithelial-derived alarmins (such as IL-33, TSLP, and IL-25), play key roles in promoting Th2 polarization and allergic sensitization [4].

Organizations such as World Health Organization/International Union of Immunological Societies (WHO/IUIS) Allergen Nomenclature Sub-Committee define a molecule as an allergen if its capacity to bind IgE from allergic patients has been experimentally demonstrated [5]. Typically, this requires positive results in sera from at least five allergic individuals [5]. This means that a protein is considered an allergen if there is experimental evidence from immunoassays, such as ELISA or immunoblot, showing IgE binding in at least five independent samples [5]. However, this does not necessarily imply that the allergen will provoke clinically relevant symptoms in every sensitized individual; rather, it indicates that the molecule has sensitizing potential and should be taken into consideration.

Additional assays may include functional cellular tests, such as basophil activation tests (BAT) [6], as well as in vivo studies in humans or animal models, which help to evaluate the biological relevance of the IgE response [7].

Several studies suggest that features encoded in protein sequence and structure (such as stability, protease resistance, and epitope composition) contribute to allergenicity [8, 9]. This opens a window for computational assessment of allergenicity in silico, as a way of pre-evaluation prior to experimental assessment [10]. This is very important in different fields of science, especially biotechnology, as it allows us to accelerate the vaccine design process, the evaluation of new food protein sources, drugs, and others [11]. However, due to the complexity and non-binary nature of allergenicity, systems biology–based approaches have also been proposed [12].

In this chapter, we present a bioinformatic pipeline composed of different prediction tools to prioritize potential allergens. The pipeline integrates multiple layers of complexity depending on the available input data, with the objective of pre-screening potentially allergenic proteins prior to downstream systems biology–based analyses.

2 Materials

2.1 Hardware

Most of the codes and software described in this workflow can be executed on a conventional personal computer with standard hardware specifications. However, access to a Linux-based operating system is strongly recommended, either through a native distribution such as Ubuntu, or via alternative solutions such as Windows Subsystem for Linux (WSL), a virtual machine, or a remote Linux server.

Homology-based methods (e.g., BLAST searches) and web-based machine learning tools can be run efficiently on standard CPUs. In contrast, deep learning–based approaches may require access to GPU acceleration to achieve reasonable computation times, particularly during model training. GPUs can be accessed through a local CUDA-enabled workstation or via remote high-performance computing (HPC) clusters.

Although deep learning models can technically be executed on CPUs, this may substantially increase execution time and computational cost. In such cases, it is advisable to reduce batch size, limit the number of training epochs, or adjust other model parameters to accommodate hardware constraints.

2.2 Software

All computational analyses are performed in command-line environments using Bash and Python. When standalone implementations are unavailable, publicly accessible web servers are used.

The following software packages and tools were employed in the pipeline:

1. NCBI BLAST+ suite [13] is used to construct local protein databases (makeblastdb) and to perform similarity searches (blastp) according to the FAO/WHO sequence-based criteria.
2. AllerCatPro 2.0 web server (https://allercatpro.bii.a-star.edu.sg/) is used for combined sequence- and structure-based allergenicity assessment [14].
3. AlgPred 2.0 web server (https://webs.iiitd.edu.in/raghava/algpred2/) is used for machine learning–based allergenicity prediction [15].
4. AllerTOP v2.1. web server (https://www.ddg-pharmfac.net/allertop_test/) is used as an additional machine learning–based classifier [16].
5. DeepAlgPro (GitHub repository) is used as an example of a deep learning–based allergenicity prediction framework [17].
6. Custom Python scripts (Python $\geq$ 3.8) are developed for parsing BLAST tabular outputs and processing results from web-based tools. The pandas library is used for structured data manipulation and result aggregation.

All locally installed packages are configured following the dependencies specified in their respective documentation. For deep learning–based analyses, it is essential to create a dedicated Python environment including torch (PyTorch), torchvision, and all additional dependencies required by the DeepAlgPro repository.

Most steps of the workflow can be executed on conventional CPUs. However, GPU acceleration (CUDA-compatible devices) is recommended when training or running deep learning models to reduce computational time.

2.3 Databases and Sequences

Several tools described in this protocol require the use of local reference databases containing experimentally validated allergens. High-quality curation of the reference dataset is critical to ensure reliable homology-based predictions.

The following allergen databases are commonly used and considered in this work:

- AllergenOnline (FARRP) [18].
- Structural Database of Allergenic Proteins (SDAP) [19].
- WHO/IUIS Allergen Nomenclature Database [20].
- COMPARE Database [21].

In the examples presented below, AllergenOnline was selected as the primary reference dataset, as it represents a well-curated resource that provides a reasonable balance between database size and computational efficiency.

When appropriate, database subsets focused on specific allergen categories (e.g., food allergens, plant allergens, inhalant allergens) may be selected to better match the biological context of the study. The choice of database should be aligned with the biological source and intended application of the proteins under investigation.

3 Methods

The pipeline follows a structured decision workflow (Fig. 1) in which an initial list of candidate proteins is systematically filtered using multiple allergenicity prediction approaches, selected according to the specific research objectives. After applying the computational prediction methods, shortlisted candidates are subjected to additional context-dependent analyses. These downstream evaluations integrate biological knowledge of the relevant allergic setting and aim to assess the potential of each protein to activate distinct immunological pathways associated with sensitization and allergic responses.

3.1 Homology-Based Prediction

The first computational method for allergenicity prediction was proposed by the FAO/WHO consultation group in 2001 [22, 23]. This

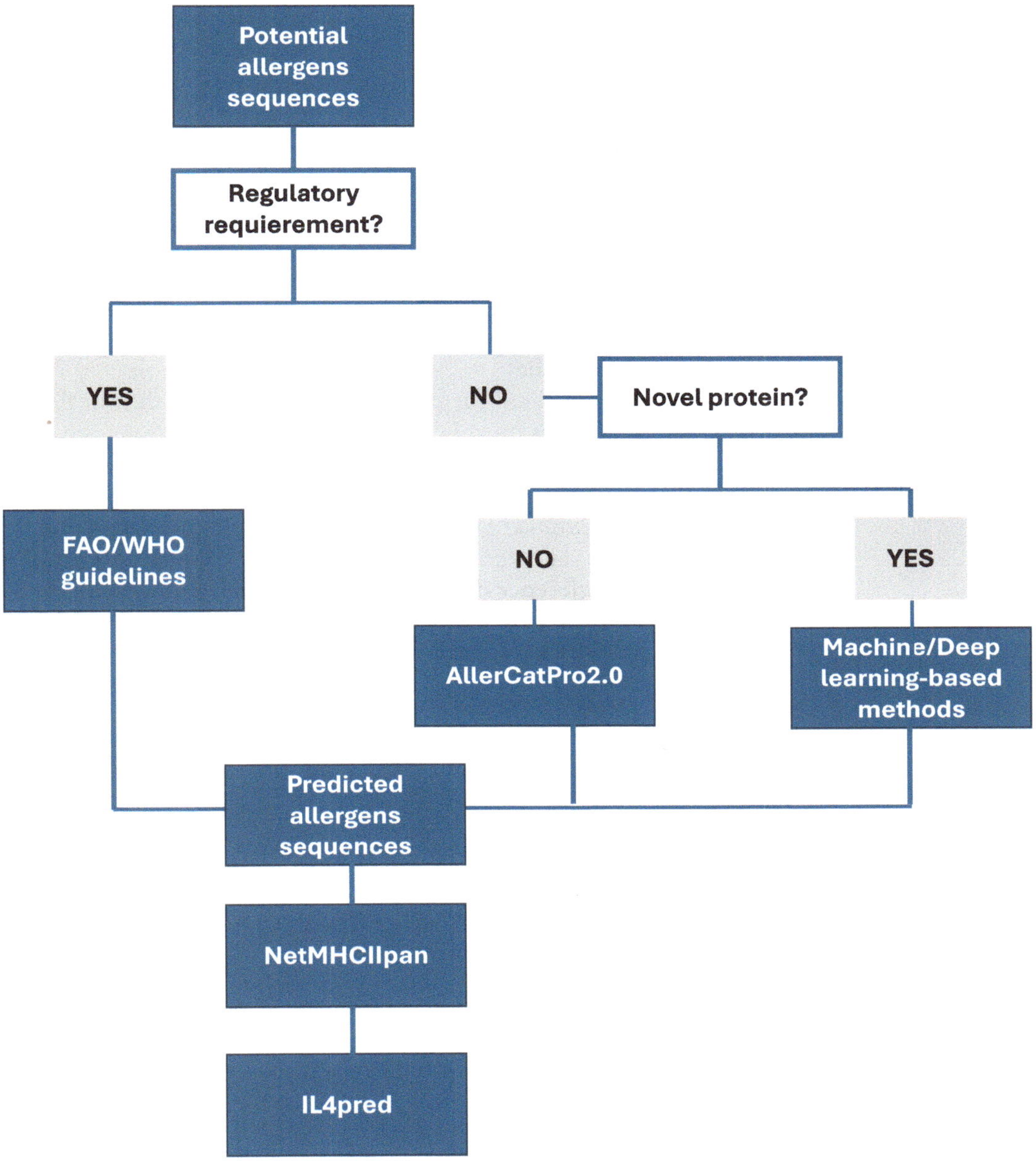

Fig. 1 Decision workflow for allergenicity assessment. The workflow begins with a set of potential allergen protein sequences. If regulatory evaluation is required, the FAO/WHO sequence-based criteria are applied as a primary screening step. In the absence of regulatory constraints, the strategy diverges depending on whether the protein is novel. For proteins with known homologues, structure-informed homology methods (e.g., AllerCatPro 2.0) are applied, whereas novel or engineered proteins are further evaluated using machine- and deep learning–based prediction approaches. Sequences predicted as potential allergens are subsequently subjected to downstream immunological characterization, including T-cell epitope prediction and IL-4 induction potential analysis

method is based solely on homology with known allergens and is summarized on two rules:

- Rule 1: A protein is a possible allergen if that protein has an exact match on a stretch of six or more consecutive identical amino acids
- Rule 2: A protein is a possible allergen if there is any window of 80 amino acids with more than 35% identity with any known allergen (Rule 2).

This method has been shown to have very high sensitivity but low specificity (*see* **Note 1**).

To apply the rules:

Step 1: Download the protein sequences of interest in FASTA files. The sequences must be amino acid ones, not the source genes or transcripts. It can be obtained from Uniprot if accession IDs are available.

Step 2: Download a comprehensive and well-curated database of experimentally validated allergens. Increasing the size and curation quality of the reference dataset improves the reliability and accuracy of subsequent predictive analyses (*see* **Note 2**).

Step 3: Make a blast database using the makeblastdb command in bash.

```
(bash)
makeblastdb -in allergen_database.fasta -dbtype
prot -out db_allergens
```

Step 4: Execute a blastp with your FASTA files as query against the downloaded database. It is important to reduce the word size to 6 or fewer to capture alignments within a window compatible with Rule 1.

```
(bash)
blastp -query saliva_proteins.fasta \
-db db_allergens \
-word_size 3 \
-matrix BLOSUM62 \
-evalue 1000 \
-outfmt "6 qseqid sseqid pident length qlen slen
qstart qend sstart send" \
-out raw_results.tab
```

Step 5: Parse the output with a Python script. Iterate through the different protein queries, searching for exact matches of length six or more, or for 35% identity on matches of 80 aa or more. If the protein length is below 80, search for full-length matches (*see* **Note 3**). Here, there is an example of python function coded to parse the output, the output is shown in Fig. 2.

	Protein	Allergenic	Details
0	tr|A0A0K8RB81|A0A0K8RB81_IXORI	NO	
1	tr|A0A0K8RKR7|A0A0K8RKR7_IXORI	YES	Rule 1: 1 match(s) of exact 6-mer.
2	tr|A0A4D5RL95|A0A4D5RL95_IXOSC	NO	
3	tr|B7P1G7|B7P1G7_IXOSC	NO	
4	tr|B7P208|B7P208_IXOSC	YES	Rule 1: 2 match(s) of exact 6-mer.
5	tr|B7P904|B7P904_IXOSC	NO	
6	tr|B7PBG2|B7PBG2_IXOSC	YES	Rule 1: 1 match(s) of exact 6-mer.
7	tr|B7PH04|B7PH04_IXOSC	NO	
8	tr|B7PJC0|B7PJC0_IXOSC	NO	
9	tr|B7PST4|B7PST4_IXOSC	YES	Rule 1: 4 match(s) of exact 6-mer.
10	tr|B7PU24|B7PU24_IXOSC	NO	
11	tr|B7Q2B8|B7Q2B8_IXOSC	NO	
12	tr|B7Q406|B7Q406_IXOSC	NO	
13	tr|B7Q407|B7Q407_IXOSC	YES	Rule 1: 1 match(s) of exact 6-mer.
14	tr|B7QGE3|B7QGE3_IXOSC	NO	
15	tr|B7QKC1|B7QKC1_IXOSC	NO	
16	tr|B7QM90|B7QM90_IXOSC	NO	
17	tr|B7QM91|B7QM91_IXOSC	NO	
18	tr|B7QM92|B7QM92_IXOSC	NO	
19	tr|B7QMC8|B7QMC8_IXOSC	NO	
20	tr|Q4PMH7|Q4PMH7_IXOSC	NO	

Fig. 2 Output of BLAST-sparse script. The figure shows a screenshot of the script's output, as shown below. It is a tabular output with 3 fields: the protein id, the prediction (allergenic or not) and the details showing the matched rule. This output is the result of applying an in-house set of tick saliva protein FASTA files

```
(python)
def evaluate(file_path):
# charge blast results
# query, subject, % id, length, q_len, s_len, q_
start, q_end, s_start, s_end
cols = ['query', 'subject', 'pident', 'length',
'qlen', 'slen', 'qstart', 'qend', 'sstart', 'send']
df = pd.read_csv(file_path, sep='\t', names=cols)
results = []
#iterate rows
for name, group in df.groupby('query'):
possible_allergen = False
reasons = []
# RULE 1 (6-mer exact match)
r1_matches = group[(group['pident'] == 100) &
(group['length'] >= 6)]
if not r1_matches.empty:
possible_allergen = True
reasons.append(f"Rule 1: {len(r1_matches)} match(s)
of exact 6-mer.")
# RULE 2 (35% id on 80aa window)
r2_matches = group[(group['length'] >= 80) &
(group['pident'] > 35)]
# if protein lenght <80, evaluate complete protein
short_prot_matches = group[(group['qlen'] < 80)
& (group['pident'] > 35) & (group['length'] >=
group['qlen'] * 0.9)]
if not r2_matches.empty or not short_prot_matches.
empty:
```

```
possible_allergen = True
reasons.append("Rule 2: >35% id in 80aa (or
complete protein).")
results.append({
'Protein': name,
'Allergenic': 'YES' if possible_allergen else 'NO',
'Details': " | ".join(reasons)
})
return pd.DataFrame(results)
```

3.2 Structure-Based Prediction

A more complex prediction, based on homology but considering 3D structure, gluten-like motifs, and clinical data, can be performed using the method developed by AllerCatPro2.0. This method is based on homology, following a modified version of FAO/WHO methodology (Rule 1 is only accepted if there are 3 or more 6-mer exact matches to improve the specificity), but also incorporates a rule about similarity in three-dimensional structure data. This improves prediction performance, as shown in various benchmarks [24, 25].

Step 1: Download the fasta archive.

Step 2: Open the web tool AllerCatPro (https://allercatpro.bii.a-star.edu.sg/) and submit the fasta archive. This will run the database alignments and display a table of outputs. Each row corresponds to a query of the fasta. An example of typical output is on Fig. 3.

Step 3: Evaluate the output table, with particular attention to the grey columns under *"Similarity to allergen and resulting predicted evidence for allergenicity."* The *"Result"* column indicates the predicted level of evidence supporting the allergenic potential of the query protein (*strong*, *weak*, or *no evidence*). This classification is derived from the internal decision workflow implemented in AllerCatPro 2.0, which integrates sequence and structural similarity criteria (Fig. 4) (*see* **Note 4**).

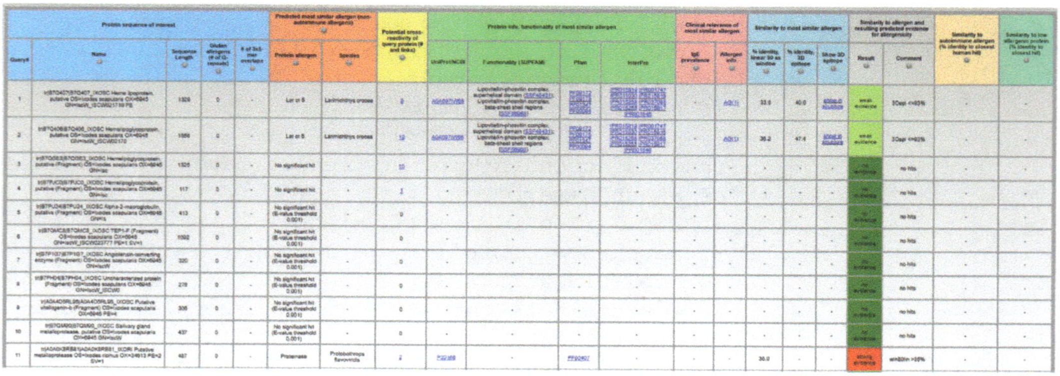

Fig. 3 Output of AllerCatPro2.0. The tool generates a summary table in which each row corresponds to one input protein sequence. For each query, the most similar known allergen, functional annotation, sequence similarity metrics, and predicted evidence of allergenicity are reported. In this example, three proteins are classified as potential allergens (two with weak evidence and one with strong evidence), while the remaining sequences show no evidence of allergenicity

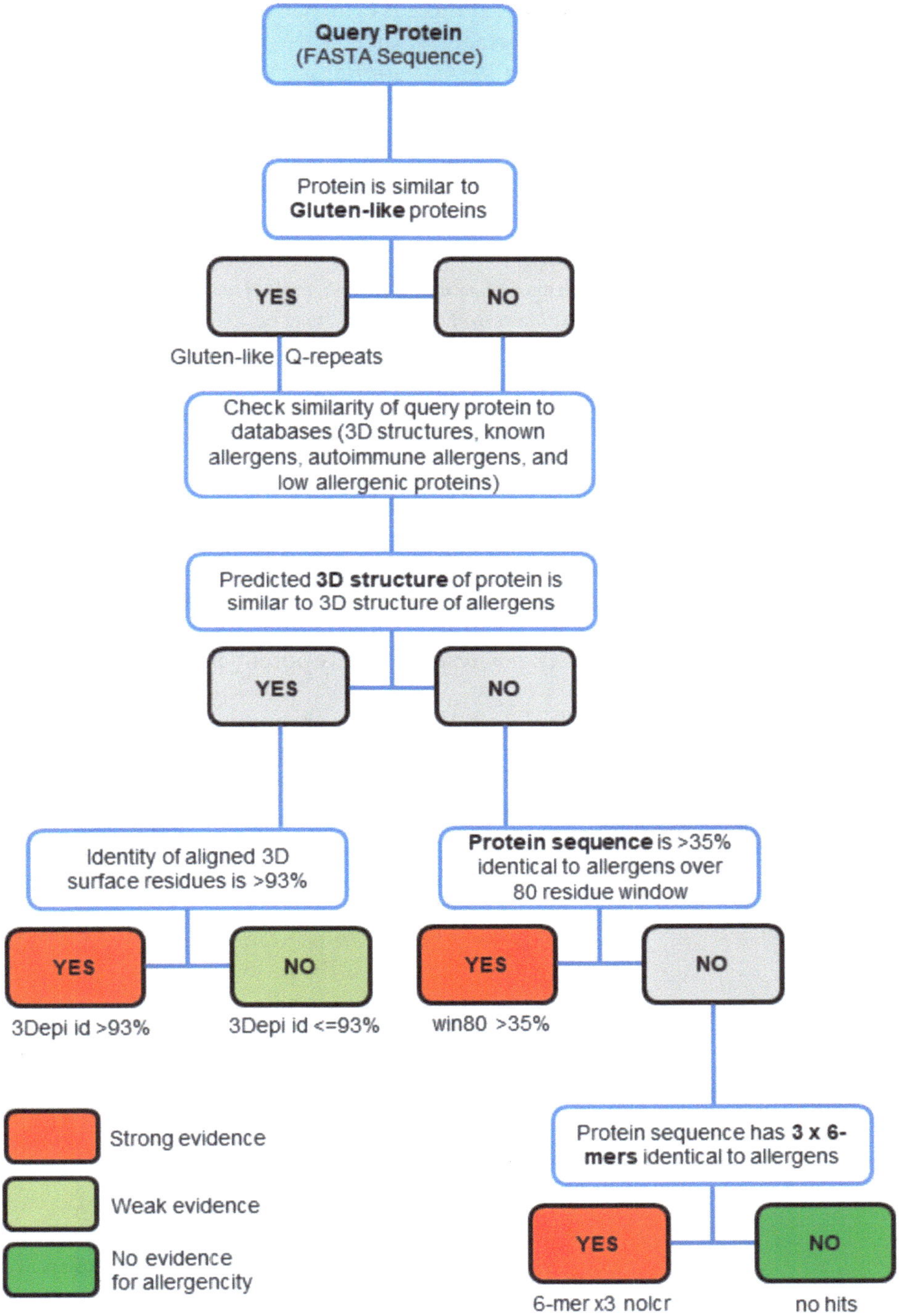

Fig. 4 Workflow of the prediction strategy implemented in AllerCatPro 2.0. The algorithm integrates sequence- and structure-based similarity criteria to classify the allergenic potential of a query protein. The workflow includes screening for gluten-like Q-repeats, comparing against databases of known allergens and low-allergenicity proteins, evaluating predicted 3D structural similarity (surface residue identity >93%), and applying sequence identity thresholds (>35% over an 80-amino acid window or the presence of three identical 6-mer matches). Based on these criteria, proteins are classified as having strong, weak, or no evidence of allergenicity

Step 4: If there is evidence of allergenicity, further review additional information, such as the green columns (*Protein info, functionality of the most similar allergen*) or the pink columns (*Clinical relevance of the most similar allergen*).

Step 5: If there is no evidence of allergenicity, investigate gluten-like allergens, potential cross-reactivity, and similar allergens. These are columns displaying additional information indicating no evidence of sensitization but a potential immune-related hazard (*see* **Note 5**).

Step 6: Look into the *Similarity to low allergenic protein (% identity to closest hit)* column. It can be a proxy of nonallergenic potential.

3.3 Machine/Deep Learning–Based Prediction

Although homology-based prediction, especially the method implemented by AllerCatPro2.0, has shown very good sensitivity and specificity, there are concerns about how well those predictions generalize to novel proteins, modified proteins, and other protein sources that may be far from the databases. Relying on previously discovered allergens has several limitations when evaluating new allergens. If this is the case with our data, it could be better to rely on machine learning/deep learning models. Those models are trained on the same databases but can learn structural, chemical, and other features, and can extrapolate allergenicity potential to non-homologous proteins.

3.3.1 Machine Learning–Based Models

AllerTOP and AlgPred 2.0 are among the most widely used machine learning–based tools for allergenicity prediction. Both provide publicly accessible web servers that allow users to submit protein sequences without requiring programming expertise. For both tools:

Step 1: Open the webtools. AlgPred2.0 (https://webs.iiitd.edu.in/raghava/algpred2/) and AllerTOP (https://www.ddg-pharmfac.net/allertop_test/).

Step 2: Verify that no noncanonical amino acids are present in the FASTA sequences, as the models do not handle them properly. This can be checked, for example, using a Bash command that prints the headers of sequences containing non-canonical amino acids.

```
(bash)
awk '/^>/ {header=$0}!/^>/ && /[^ACDEFGHIKLMNPQRSTVWY]/
{print header}' tu_fasta.fasta
```

Step 3: Copy the protein sequences or upload the FASTA file and submit them to the server to initiate the prediction. When using AlgPred 2.0, it is recommended to select the hybrid model, as this was the configuration evaluated in the original publication. Additionally, a decision threshold can be adjusted according to the desired balance between sensitivity and specificity (*see* **Note 6**).

Step 4: Download the prediction results once the analysis is completed. In the case of AlgPred 3, the output can be downloaded directly as a CSV file, with the predicted class in the last column.

In contrast, AllerTOP does not provide a structured downloadable file; therefore, the results must be manually copied from the web interface and parsed into a tabular format (an example Python script is provided below to facilitate this step).

```
(python)
import csv
input_file = "input.txt"
output_file = "results.csv"
rows = []
with open(input_file) as f:
lines = f.readlines()
header = ""
similar = ""
classification = ""
for line in lines:
if line.startswith(tuple(str(i) + "." for i in
range(1, 1000))):
header = line.split("Results for protein")[1].
strip().rstrip(":")
elif line.startswith("Most similar protein:"):
similar = line.replace("Most similar protein:",
"").strip()
elif line.startswith("Classification based on"):
classification = line.split(":")[1].strip()
rows.append(ss[header, similar, classification])
with open(output_file, "w", newline="") as f:
writer = csv.writer(f)
writer.writerow(["Header", "Most_Similar_Protein",
"Classification"])
writer.writerows(rows)
```

3.3.2 Deep Learning–Based Models

Deep learning–based models can be more powerful than traditional machine learning approaches, especially with the incorporation of protein large language models in recent years. These models encode evolutionary, structural, and physicochemical information learned from millions of protein sequences into numerical embeddings, which enhance predictive performance.

However, they require substantial computational resources and, to date, are not generally available through web-based or non-programming tools. In addition, because they are relatively recent, independent external evaluations are still limited.

Since AllerCatPro 2.0 and conventional machine learning models have already demonstrated strong predictive performance, the cost–benefit ratio of using deep learning approaches is not yet fully clear. Nevertheless, in certain contexts, their implementation may be worthwhile (*see* **Note 7**).

Several deep learning–based predictors have been described. Here, DeepAlgPro is presented as an example; however, as mentioned above, there are no comparative benchmarking studies establishing which model performs best, and most publications report similar results.

Step 1: Open the terminal and clone from GitHub the DeepAlg-Pro repository, this will create a homonime folder.

```
(bash)
git clone https://github.com/chun-he-316/
DeepAlgPro.git
```

Step 2: Move to the DeepAlgPro folder and install the package requirements. You can do it by pip or by creating a new conda (or other) environment.

```
(bash)
pip3 install torch==1.12.1 torchvision==0.13.1
torchaudio==0.12.1 --extra-index-url https://
download.pytorch.org/whl/cu116
(bash)
pip install -r requirements.txt
```

Step 3: If you want to use the pretrained model, you can simply use the following command, being -i the input parameter to choose the input fasta and -o the parameter to choose an output name

```
(bash)
python predict.py -i saliva_proteins.fasta -o
allergen.predict.txt
```

Step 4: In case you have your own data, for example, a specific dataset of plant allergens, it is possible to retrain the model by yourself, taking advantage of the model architecture.

1. Adapt the dataset to the required format. In the case of DeepAlgPro, you will need to change the headers of the sequences to an ID containing "allergen" or "non-allergen."
2. Train the model with training data. Depending on your availability of GPUs, you should adjust the parameters to balance the computation time (*see* **Note 8**).

```
(bash)
python main.py -i data/all.train.fasta --epochs 120
--lr 0.0001 -b 72 --mode train
```

3. Check the validation outputs, choose the best model and change its name to model.pt (this will overwrite the original model weights). The validation outputs are on "model.X.fold.everyepoch.valid.txt".
4. Perform independent testing on your test set.

3.4 Additional Analysis and Characterization

It may be necessary to further characterize allergenic epitopes, as well as the physicochemical properties and structural motifs of the protein. Allergic sensitization is a complex process that involves multiple pathways and the interaction of different immune compartments. In classical type I hypersensitivity, sensitization is largely driven by $CD4^+$ T cell epitopes presented by MHC class II molecules. These epitopes activate $CD4^+$ T cells, and if the cytokine milieu promotes a Th2-biased differentiation, the activated T cells will secrete IL-4, IL-5, and IL-13. IL-4 drives B cell class-switch recombination towards IgE production, leading to the generation of IgE-secreting plasma cells and the establishment of allergic sensitization. Once the potential allergens have been identified by homology, machine, or deep-learning approaches we will evaluate each sequence by systems biology approaches.

3.4.1 Epitope Predictions

Step 1: Open the NetMHCIIpan web server. This tool is specifically designed to predict peptide binding to MHC class II molecules and, therefore, to identify potential $CD4^+$ T cell epitopes. Upload or paste the protein sequence in FASTA format.

Step 2: Set the peptide length to 15 amino acids, which is the most used length for MHC-II epitope prediction. Select the HLA class II alleles of interest. If the HLA genotype of the target population is known, specific alleles (e.g., HLA-DRB1*01:01) can be selected. Otherwise, it is recommended to choose a broader panel of common alleles to improve population coverage.

Step 3: Download and interpret the output table (Fig. 5A). Each row corresponds to one predicted peptide. The most relevant fields for interpretation are:

1. Pos: Starting position of the peptide in the protein sequence.
2. Peptide: The 15-mer evaluated.
3. Core: The nine-amino acid binding core that directly interacts with the MHC-II molecule.
4. Score_EL: Prediction score based on eluted ligand data (internal model score; higher values indicate stronger predicted binding).
5. %Rank_EL: Percentile rank compared to random natural peptides. This is the most important metric for interpretation.
6. BindLevel: Automatic classification based on %Rank_EL.

Predicted binders are classified as:

1. Strong binders (SB): %Rank_EL < 1
2. Weak binders (WB): %Rank_EL between 1 and 5

For downstream analyses, strong binders should be prioritized. If only a limited number of strong binders are detected, weak binders

A

```
# NetMHCIIpan version 4.1

# Input is in FASTA format

# Peptide length 15

# Prediction Mode: EL

# Threshold for Strong binding peptides (%Rank) 1%
# Threshold for Weak binding peptides (%Rank)   5%

# Allele: DRB1_0101
-----------------------------------------------------------------------------------------------------------------
 Pos         MHC             Peptide   Of       Core  Core_Rel        Identity      Score_EL %Rank_EL Exp_Bind  BindLevel
-----------------------------------------------------------------------------------------------------------------
 719    DRB1_0101    QIRIEEVEANALAGG    3  IEEVEANAL     1.000 sp_Q8MUF6_MYSP_     0.894783     0.42       NA   <=SB
 612    DRB1_0101    EMRVNLEQALRAKRA    3  VNLEQALRA     1.000 sp_Q8MUF6_MYSP_     0.875481     0.52       NA   <=SB
   5    DRB1_0101    SAKYMYQSSRAGHGG    3  YMYQSSRAG     0.993 sp_Q8MUF6_MYSP_     0.853754     0.62       NA   <=SB
 225    DRB1_0101    ANHLKGQIAQQLEDT    3  LKGQIAQQL     1.000 sp_Q8MUF6_MYSP_     0.766384     1.03       NA   <=WB
 296    DRB1_0101    KSKYEAELQAHVDEV    3  YEAELQAHV     1.000 sp_Q8MUF6_MYSP_     0.763070     1.05       NA   <=WB
 201    DRB1_0101    RQRLSQENSELIKEV    3  LSQENSELI     1.000 sp_Q8MUF6_MYSP_     0.558525     2.24       NA   <=WB
 217    DRB1_0101    EYKISLDNANHLKGQ    3  ISLDNANHL     0.993 sp_Q8MUF6_MYSP_     0.548913     2.31       NA   <=WB
   7    DRB1_0101    KYMYQSSRAGHGGDI    3  YQSSRAGHG     0.927 sp_Q8MUF6_MYSP_     0.522783     2.50       NA   <=WB
 150    DRB1_0101    AEVFELLAQVETANK    3  FELLAQVET     1.000 sp_Q8MUF6_MYSP_     0.499502     2.69       NA   <=WB
  17    DRB1_0101    HGGDISIEYGTDLGA    4  ISIEYGTDL     1.000 sp_Q8MUF6_MYSP_     0.478600     2.87       NA   <=WB
 510    DRB1_0101    LRKQYQIEIEQLNMR    4  YQIEIEQLN     1.000 sp_Q8MUF6_MYSP_     0.338628     4.28       NA   <=WB
 549    DRB1_0101    ELELSLDAANKANID    3  LSLDAANKA     0.993 sp_Q8MUF6_MYSP_     0.269904     5.25       NA
 539    DRB1_0101    LKKKYQAQITELELS    4  YQAQITELE     1.000 sp_Q8MUF6_MYSP_     0.267339     5.29       NA
 785    DRB1_0101    QIQLLQEMSDKLNEK    3  LLQEMSDKL     0.993 sp_Q8MUF6_MYSP_     0.253425     5.53       NA
```

B

Your job id is 98068

Query Peptides

Peptide ID	Peptide Sequence	SVM Score	Prediction	Hydrophobicity	Hydropathicity	Hydrophilicity	Charge	Mol wt
epitope1	QIRIEEVEANALAGG	0.32	IL4 inducer	-0.09	-0.03	0.27	-2.00	1569.96
Epitope2	EMRVNLEQALRAKRA	0.44	IL4 inducer	-0.42	-0.82	0.70	2.00	1785.31
Epitope3	SAKYMYQSSRAGHGG	0.00	Non IL4 inducer	-0.23	-1.05	-0.02	2.50	1599.95

1/1 50

Fig. 5 Outputs of downstream analysis tools. **A,** Output of NetMHCIIpan 4.1. The tool generates a tabulated output in which each row corresponds to a predicted peptide derived from the input protein sequence. For each peptide, the associated HLA allele, binding core, predicted binding score (Score_EL), percentile rank (%Rank_EL), and binding level classification (strong binder, SB; weak binder, WB) are reported. In this example, several 15-mer peptides are predicted to bind HLA-DRB1*01:01 with high affinity, including strong binders (%Rank $\leq$ 1) and additional weak binders (%Rank $\leq$ 5), indicating potential CD4$^+$ T-cell epitope candidates. **B**, Output of IL4Pred. Each row corresponds to a queried peptide. The table reports the SVM score and predicted classification (IL-4 inducer or non–IL-4 inducer), together with basic physicochemical properties. In this example, two peptides are predicted as IL-4 inducers and one as non–IL-4 inducer

may also be considered. Selected epitopes can then be used for further immunological characterization, such as population coverage analysis or evaluation of Th2-associated responses.

Step 4: Write a new FASTA file containing the sequences of the potential epitopes.

3.4.2 Immune Mediator Induction

As before, one of the signals that induces an IgE isotype change is the production of IL4 (between other interleukins). Some peptides have a higher potential to induce CD4+ naïve cells to produce IL4, making Th2-biased differentiation easier. Using the predicted T-epitopes from NetMHCIIpan, we can determine whether some of them promote IL4 production.

Step 1: Open the IL4Pred web tool; this tool is specially designed to check the potential of a peptide to induce IL-4 production when bonded to MHC-II.

Step 2: Open the peptide mapping tool and paste a FASTA file, with each sequence corresponding to a predicted T epitope.

Step 3: Establish a proper threshold and compute the prediction; 0.3 is set as the default.

Step 4: The output is shown in a table format as shown in Fig. 5B. The potential IL4 inducers are shown on the "Prediction" column.

3.4.3 Additional Steps

Depending on the biological context of the allergen (e.g., route of exposure) you can compute additional predictions that could help understand the potential hazard of the allergen. For example, for food allergens, processed by gastrointestinal absorption, it is common to consider the hydrolysis by gastrointestinal enzymes.

Step 1: Open the web tool PeptideCutter (https://web.expasy.org/peptide_cutter/).

Step 2: Paste the sequence of the protein we are evaluating and select the enzymes we want to consider, for example, pepsin if we are working with food allergens.

Step 3: Perform the analysis and check the output. The output will be a table with all cleavage sites on your protein

Step 4: Check if these cleavage sites interrupt the T-cell epitopes. If some of the epitopes are enzyme-resistant, it could be a signal of allergic hazard.

4 Notes

1. The FAO/WHO rules are the official ones and are designed to be very sensitive. This means that there will probably be false positives, as a caution principle. If you want to be more restrictive, you can replace Rule 1 with a modified one: three exact matches of six or more amino acids. This is implemented in new methods as AllerCatPro2.0 because it has been proven to improve specificity without affecting sensitivity.
2. To enhance the reliability and precision of predictive analyses, it is recommended to download and utilize comprehensive

databases of experimentally validated allergens. Some of the most highly curated sources are mentioned in Sect. 2. Sometimes databases can be mixed.

3. This is not defined in the FAO/WHO rules, but it is necessary to add this exception in order not to obtain errors in the code.
4. As a result of the decision workflow, a protein that fulfills the FAO/WHO sequence-based criteria and exhibits high three-dimensional structural similarity to known allergens will be classified as either strong or weak evidence depending on the percentage of 3D similarity, regardless of the overall sequence homology. AllerCatPro 2.0 provides the underlying similarity metrics and suggests a preliminary categorization; however, it ultimately allows researchers to define their own decision thresholds and interpretation criteria during downstream analysis. Importantly, if a protein meets the FAO/WHO requirements for strong evidence at the sequence level, the absence of sufficient structural similarity may not necessarily invalidate its classification. In such cases, sequence-based evidence can be considered sufficient to support a strong allergenicity indication, depending on the analytical framework adopted.
5. AllerCatPro 2.0 and other homology-based prediction methods are primarily designed to detect similarity to known IgE-mediated allergens, which represent the most common and well-characterized form of allergic sensitization. However, other types of hypersensitivity reactions and immunological risk factors may also be relevant depending on the biological context. These include, for example, the presence of repetitive gluten-like motifs associated with T-cell–mediated responses (e.g., celiac disease) or sequence/structural similarity to autoantigens that could suggest a theoretical risk of molecular mimicry. AllerCatPro 2.0 reports these additional features to provide complementary immunological information. Their relevance should be determined by the researcher according to the specific scope of the study, as they do not constitute direct evidence of IgE-mediated allergenicity but may indicate broader immune-related considerations.
6. The performance of AlgPred2 strongly depends on the selected threshold value, which directly affects the balance between sensitivity and specificity. Lower thresholds increase sensitivity (reducing false negatives) but may lead to higher false positive rates, whereas higher thresholds increase specificity at the expense of sensitivity. The threshold should therefore be selected according to the study objective (e.g., screening vs. risk assessment).
7. AllerCatPro 2.0 and classical machine learning methods typically achieve high performance, with sensitivity and specificity

around 0.85 across different benchmark studies. Therefore, in most cases, these approaches are recommended as a first option, as they offer good accuracy and are less computationally intensive. However, in some situations, such as evaluating de novo proteins, engineered sequences, or proteins with low homology to known allergens, it may be useful to apply deep learning models. These models do not only rely on direct sequence similarity but also rely on learned features such as physicochemical properties, motif combinations, and other hidden patterns in protein sequences. Even so, deep learning approaches require more computational resources and have fewer independent validation studies. For this reason, their use should be justified depending on the specific research question and the type of sequences being analyzed.

8. If no GPU is available and the model must be trained on a CPU, it is necessary to adjust the training parameters to reduce computational time and memory usage. Training deep learning models on CPU can be significantly slower, especially with large datasets.

   To adapt the training process to CPU-based execution, consider the following adjustments:

   (a) Reduce the batch size (-b): Lowering the batch size decreases memory consumption and may prevent crashes due to limited RAM.

   (b) Reduce the number of epochs (–epochs): Instead of 120 epochs, start with 30–60 epochs and monitor validation performance.

   (c) Augment the learning rate –lr to 0.01, this will allow the model to learn faster.

   Consider that this can significantly reduce the accuracy of the model.

Competing Interests

The authors have no conflicts of interest to declare that are relevant to the content of this chapter.

References

1. Pomés A (2002) Intrinsic properties of allergens and environmental exposure as determinants of allergenicity. Allergy 57(8):673–679. https://doi.org/10.1034/j.1398-9995.2002.02160.x
2. Costa J, Villa C, Verhoeckx K et al (2022) Are physicochemical properties shaping the allergenic potency of animal allergens. Clin Rev Allergy Immunol 62(1):1–36. https://doi.org/10.1007/s12016-020-08826-1
3. Renz H (1995) The central rôle of T-cells in allergic sensitization and IgE regulation. Exp Dermatol 4(4 Pt 1):173–182.

https://doi.org/10.1111/j.1600-0625.1995.tb00242.x
4. Paul WE, Zhu J (2010) How are TH2-type immune responses initiated and amplified. Nat Rev Immunol 10(4):225–235. https://doi.org/10.1038/nri2735
5. Pomés A, Davies JM, Gadermaier G et al (2018) WHO/IUIS allergen nomenclature: providing a common language. Mol Immunol 100:3–13. https://doi.org/10.1016/j.molimm.2018.03.003
6. McGowan EC, Saini S (2013) Update on the performance and application of basophil activation tests. Curr Allergy Asthma Rep 13(1):101–109. https://doi.org/10.1007/s11882-012-0324-x
7. Wachholz PA, Dearman RJ, Kimber I (2005) Detection of allergen-specific IgE antibody responses. J Immunotoxicol 1(3):189–199. https://doi.org/10.1080/15476910490919140
8. Foo ACY, Mueller GA (2021) Abundance and stability as common properties of allergens. Front Allergy 2:769728. https://doi.org/10.3389/falgy.2021.769728
9. Huby RDJ, Dearman RJ, Kimber I (2000) Why are some proteins allergens. Toxicol Sci 55(2):235–246. https://doi.org/10.1093/toxsci/55.2.235
10. Hileman RE, Silvanovich A, Goodman RE et al (2002) Bioinformatic methods for allergenicity assessment using a comprehensive allergen database. Int Arch Allergy Immunol 128(4):280–291. https://doi.org/10.1159/000063861
11. Ladics GS, Cressman RF, Herouet-Guicheney C et al (2011) Bioinformatics and the allergy assessment of agricultural biotechnology products: industry practices and recommendations. Regul Toxicol Pharmacol 60(1):46–53. https://doi.org/10.1016/j.yrtph.2011.02.004
12. Zhou F, He S, Zhang Y et al (2022) Prediction and characterization of the T cell epitopes for the major soybean protein allergens using bioinformatics approaches. Proteins 90(2):418–434. https://doi.org/10.1002/prot.26233
13. Camacho C, Coulouris G, Avagyan V et al (2009) BLAST+: architecture and applications. BMC Bioinformatics 10:421. https://doi.org/10.1186/1471-2105-10-421
14. Nguyen MN, Krutz NL, Limviphuvadh V et al (2022) AllerCatPro 2.0: a web server for predicting protein allergenicity potential. Nucleic Acids Res 50(W1):W36–W43. https://doi.org/10.1093/nar/gkac446
15. Sharma N, Patiyal S, Dhall A et al (2021) AlgPred 2.0: an improved method for predicting allergenic proteins and mapping of IgE epitopes. Brief Bioinform 22(4):bbaa294 https://doi.org/10.1093/bib/bbaa294
16. Dimitrov I, Bangov I, Flower DR et al (2014) AllerTOP v.2—a server for in silico prediction of allergens. J Mol Model 20(6):2278. https://doi.org/10.1007/s00894-014-2278-5
17. He C, Ye X, Yang Y et al (2023) DeepAlgPro: an interpretable deep neural network model for predicting allergenic proteins. Brief Bioinform 24(4):bbad246 https://doi.org/10.1093/bib/bbad246
18. Goodman RE, Ebisawa M, Ferreira F et al (2016) AllergenOnline: a peer-reviewed, curated allergen database to assess novel food proteins for potential cross-reactivity. Mol Nutr Food Res 60(5):1183–1198. https://doi.org/10.1002/mnfr.201500769
19. Negi SS, Schein CH, Braun W (2023) The updated Structural Database of Allergenic Proteins (SDAP 2.0) provides 3D models for allergens and incorporated bioinformatics tools. J Allergy Clin Immunol Glob 2(4):100162. https://doi.org/10.1016/j.jacig.2023.100162
20. Radauer C, Nandy A, Ferreira F et al (2014) Update of the WHO/IUIS allergen nomenclature database based on analysis of allergen sequences. Allergy 69(4):413–419. https://doi.org/10.1111/all.12348
21. van Ree R, Sapiter Ballerda D, Berin MC et al (2021) The COMPARE database: a public resource for allergen identification, adapted for continuous improvement. Front Allergy 2:700533. https://doi.org/10.3389/falgy.2021.700533
22. Fiers MWEJ KleterGA, Nijland H et al (2004) Allermatch™: a webtool for the prediction of potential allergenicity according to current FAO/WHO Codex alimentarius guidelines. BMC Bioinformatics 5:133. https://doi.org/10.1186/1471-2105-5-133
23. Codex Alimentarius Commission (2003) Principles for the risk analysis of foods derived from modern biotechnology. CAC/GL 44-2003. FAO/WHO, Rome. Available at: https://www.fao.org. Accessed 24 Feb 2026
24. Wang J, Yu Y, Zhao Y et al (2013) Evaluation and integration of existing methods for computational prediction of allergens. BMC Bioinformatics 14(Suppl 4):S1. https://doi.org/10.1186/1471-2105-14-S4-S1
25. Maurer-Stroh S, Krutz NL, Kern PS et al (2019) AllerCatPro—prediction of protein allergenicity potential from the protein sequence. Bioinformatics 35(17):3020–3027. https://doi.org/10.1093/bioinformatics/btz029

Part VI

Translational and Participatory Models

Chapter 19

Participatory Citizen Science to Evaluate Risks and Interventions to Allergic Reactions to Arthropod Bites

José de la Fuente and Rita Vaz-Rodrigues

Abstract

Participatory citizen science aligns with the principles of Open Science, promoting transparency, accessibility, and public engagement in scientific research. Collecting relevant information from the population is important to evaluate epidemiological data, geographical distribution, incidence, and risks associated with allergic reactions to arthropod bites. In this chapter, the methods for implementation of participatory citizen science are described to evaluate risks and interventions to allergic reactions to arthropod bites.

Keywords Citizen science, Participatory science

1 Introduction

Participatory citizen science refers to the involvement of citizens in scientific research through public participation in different activities such as community-based research, civic science, and crowd-sourced data collection. According to the European Citizen Science Association (ECSA; https://www.ecsa.ngo), Citizen Science is defined as "an 'umbrella' term that describes a variety of ways in which the public participates in science, with two main characteristics in common: (a) citizens are actively involved in research, in partnership or collaboration with scientists or professionals; and (b) there is a genuine outcome, such as new scientific knowledge, conservation action or policy change."

The wide range of symptoms and case-to-case differences, together with the limited awareness among healthcare professionals and the general population, together with the limited diagnostic capacities and lack of standardization in diagnostics, affect the diagnosis and treatment of allergic reactions to arthropod bites [1].

In this context, participatory citizen science contributes to providing information on epidemiology, geographical distribution,

José de la Fuente and Alejandro Cabezas-Cruz (eds.), *Allergic Disease Models*, Methods in Molecular Biology, vol. 3092,
https://doi.org/10.1007/978-1-0716-5635-8_19,

incidence, and risks associated with allergic reactions to arthropod bites. This information contributes to improving awareness about risks and preventive measures for these diseases in both the general population and the healthcare system [2–5].

2 Materials

2.1 Ethical Approval

The use of samples and individual data requires approval by Ethics Committees of the scientific institutions and local authorities, as regulated in each country and region. The statement should be included in publications, for example, in reference [5], "Written informed consent was obtained from all participants in this study. The use of samples and individual data was approved by the Ethical and Scientific Committees (University Hospital of Ciudad Real C-352 and SESCAM C-73)."

2.2 Survey Platforms

Surveys for participatory citizen science can be implemented by directly contacting individuals [5], using a web-based questionnaire [4], or combining both approaches.

3 Methods

3.1 Questionnaire

As described in reference [4], the questionnaire can present:

1. Demographic questions such as age, gender, and country of residence.
2. Data about geographical bite location (country and region).
3. Information on the development of local reactions to the bite, such as redness, pruritus, inflammation and hives, or systemic reactions after the ingestion of mammalian meat.
4. Systemic AGS-related symptomatology including cutaneous (pruritus, edema and urticaria), cardiorespiratory (anaphylaxis, dyspnea, and cardiovascular distress) and gastrointestinal (diarrhea, abdominal pain, vomiting, and reflux) symptoms.
5. Questions to determine if the respondents visited a health center after the bite and if a proper diagnosis was obtained later.
6. An open-ended question to allow respondents to provide additional information if needed.
7. A question regarding participants' consent for the use and publication of this anonymous information.

3.2 Survey

1. Sex *
   Check only one option.
   ☐ Women ☐ Men ☐ Other
2. Age * _____y/o
3. Country of residence * _________________
4. Have you been bitten by the arthropod(s) on which the study is focused? *
   Check only one option.
   ☐ Yes. If Yes, one or multiple bites? ONE___ MULTIPLE___
   ☐ No. If No, *go to question 16*
5. Country where you have been bitten *_____________________
6. Approximately when was the last bite recorded? DATE_______
7. Region of the country where you have been bitten ___________________
8. Have you developed allergies on the bitten site? *
   Check all that is applicable.
   ☐ No reaction ☐ Redness
   ☐ Inflammation ☐ Itch
   ☐ Do not remember
9. After being bitten by the tick, do you have any of the following reactions when you
   eat red meat (1–6 hrs after consumption)? *
   Check all that are applicable.
   ☐ No reaction
   ☐ Respiratory distress ☐ Itch
   ☐ Gastrointestinal symptoms ☐ Diarrhea
   ☐ Abdominal pain ☐ Vomiting
   ☐ Reflux
   ☐ Urinary incontinence ☐ Cardiovascular disease
   ☐ Other:
10. Did you require medical attention, and have you been to the health center? *
    Check only one option.
    ☐ Yes
    ☐ No. If No, *go to question 16*
11. At the health center, you were diagnosed with? ___________________
12. Did you recover from the experienced symptomatology? YES_____ NO________
13. If yes, approximately how long did it take to recover? _______ days
14. Were you diagnosed with human vector-borne diseases? YES____ NO______

15. Please disclose if you have any extra information you want to share.
16. Do you consent to the use of this anonymous information for research and scientific publications?

Check only one option.
☐ Yes ☐ No
* Indicates a required question to be included in the analysis (*see* **Notes 1–3**).

3.3 Spatial and Statistical Analysis

1. The Spatial distribution of the respondents is obtained using the Nomenclature of Territorial Units for Statistics (NUTS) code number 2 for the countries involved, referring to the administrative divisions of these countries.
2. From the final sample used to show demographic information, territories with cases without responses are removed from mapping.
3. Relevant variables such as number of bites, local cutaneous reactions, and systemic symptoms after mammalian meat consumption are registered at the NUTS2 level. Data on local skin reactions to tick bites are overlaid on the percentage of health center visits after exposure to arthropods. Data on the development of systemic symptomatology after consumption of mammalian meat after a bite are spatially overlapped with the percentage of multiple bites registered.
4. The obtained results are depicted using the open-source software QGIS Geographic Information System (version 3.34, *Prizren*, QGIS Association).
5. A generalized linear model (GLM) with a binomial regression and logit link function is used to analyze the most determinant factors influencing the presence of local skin reactions after arthropod bite and the development of systemic symptomatology after the consumption of mammalian meat post-bite.
6. Local skin reactions and systemic symptoms are set as binomial response variables in two different models using the *glm* function from the *lme4* package [6]. Predictors (fixed factors) included in the models are age, gender, number of tick bites, and the geographical location of the bites.
7. The relationship between the outcome variable (presence/absence of local or systemic reactions) and the predictors is expressed as: outcome variable, age + gender + tick bites + location, to investigate predictors that may be associated with the development of the most common onsite post-bite signs such as redness, pruritus, inflammation, and hives, and without including systemic symptomatology following bite for analysis.

8. The development of Alpha-Gal syndrome (AGS)-related systemic symptomatology is IgE-mediated and does not occur immediately after a tick bite, which only causes sensitization, but usually between 2 and 6 h after the consumption of mammalian meat or derived bioproducts. Accordingly, in the questionnaire, respondents were asked about previous history of tick bites and development of allergies at the bite site and/or after tick bite and 1–6 h after consumption of red meat. Another GLM with a logit link was fitted to study the effects of age, gender, localized skin reactions, and number of bites on visits to healthcare centers after tick exposure. The fitted models were evaluated based on predictive capacity, normality, and the presence or absence of residual patterns in data variation. The obtained plots were created using the *ggplot2* package [7].
9. All analyses are performed using the R software, version 4.4.0 [8]. Statistical significance is considered using a *p*-value lower than 0.05 with a confidence level (CL) of 95%.

4 Notes

1. For the survey conducted by directly contacting individuals, after consent is obtained, a signature can be requested for internal records ONLY.
2. The survey by web-based questionnaire can be run by Google Forms to collect anonymous single response to a web-based questionnaire consisting of open-ended, multiple-choice, and closed questions.
3. The survey link can be shared on public communication platforms such as WhatsApp, Facebook, Twitter, and Instagram, as well as on organization websites (e.g., AGI Alpha-gal Information; https://alphagalinformation.org) to improve data collection.

Competing Interests

The authors have no conflicts of interest to declare that are relevant to the content of this chapter.

References

1. de la Fuente J (2024) The Alpha-Gal syndrome is underdiagnosed. Actas Dermosifiliogr 115:T116–T117
2. Vaz-Rodrigues R, Feo Brito F, Guzmán Rodríguez R, Mazuecos L, de la Fuente J (2023) Allergic reactions associated with medically relevant arthropods. Ann Med 55:2286531
3. de la Fuente J, Estrada-Peña A, Gortázar C, Vaz-Rodrigues R, Sánchez I, Carrión Tudela J (2023) Citizen science on lyme borreliosis in Spain reveals disease-associated risk factors and control interventions. Vector Borne Zoonotic Dis 23(9):441–446
4. Vaz-Rodrigues R, Rafael M, Carniato D, de la Fuente J (2025) Public perspectives on tick bite exposure, healthcare visits and associated allergies in Iberia. Ann Med 57(1): 2499028
5. Vaz-Rodrigues R, Odyek Abila P, Kasaija P, Kibuuka M PD, de la Fuente J (2025) Allergic reactions associated with tsetse fly bites. Ann Med 57(1):2514096
6. Bates D, Mächler M, Bolker B, Walker S (2015) Fitting linear mixed-effects models using lme4. J Stat Soft 67:1–48
7. Wickham H (2016) ggplot2: elegant graphics for data analysis. Available from: https://ggplot2.tidyverse.org
8. R Core Team (2023) R: the R project for statistical computing. Available from: https://www.r-project.org/

Check for updates

Chapter 20

ImmunoCAP-Based Detection of Alpha-Gal Syndrome: A Reproducible Diagnostic Protocol for Human Patients

Kinga Lis

Abstract

Alpha-GAL syndrome (AGS) is a type of allergy to mammalian meat and other foods or pharmaceuticals made from mammalian substrates. It is an IgE-mediated reaction caused by sensitization to the polysaccharide galactose-α-1,3-galactose (alpha-GAL) found in tick saliva after a previous tick bite. Alpha-GAL is present on proteins of mammals except primates. Sensitization to alpha-GAL in tick saliva predisposes to cross-reactivity following exposure (oral or non-oral) to glycated galactose-α -1,3-galactose proteins, regardless of their origin. Symptoms of varying severity (from dyspepsia to anaphylaxis) appear 3–6 h after exposure (delayed hypersensitivity) and are usually in the presence of a cofactor. Due to the unusual mechanism and source of allergy, the diagnosis of AGHS is difficult and long-lasting. It is estimated that it takes an average of 7 years from the appearance of the first symptoms to the diagnosis of AGS. Diagnostic procedures include a review of the patient's clinical history, a physical examination, and additional tests, such as alpha-GAL-specific immunoglobulin E (sIgE) or, optionally, a basophil activation test (BAT) with alpha-GAL. Alpha-GAL sIgE levels can be measured using screening methods (ImmunoCAP $ISAC_{E112i}$ or ALEX3) and the quantitative ImmunoCAP method (ThermoFisher Scientific), which is the reference method. Adherence to appropriate diagnostic procedures and careful analysis of clinical data and additional test results is an effective tool for diagnosing AGS and implementing appropriate therapeutic strategies.

Keywords Alpha-GAL, Alpha-GAL syndrome, Diagnostics, sIgE, ImmunoCAP

1 Introduction: Alpha-GAL Syndrome (AGS)—Basic Information

1.1 AGS—Definition

Alpha-GAL Syndrome (AGS) is a type of allergy to mammalian meat (e.g., beef, pork, or lamb) and other mammalian food or pharmaceutical products (e.g., milk and dairy food products, gelatin-rich food or medicines, therapeutic monoclonal antibodies, vaccines) that develops as a result of a tick bite. The reaction is caused by an IgE-mediated sensitization to the polysaccharide galactose-α -1,3-galactose (alpha-GAL) found in tick saliva. This polysaccharide is also expressed on mammalian proteins (except primates).

José de la Fuente and Alejandro Cabezas-Cruz (eds.), *Allergic Disease Models*, Methods in Molecular Biology, vol. 3092,
https://doi.org/10.1007/978-1-0716-5635-8_20,

Sensitization to alpha-GAL from tick saliva may result in a cross-reaction to galactose-α-1,3-galactose glycated proteins, regardless of their origin [1–5].

The polysaccharide galactose-α-1,3-galactose is synthesized by the glycosylation enzyme α-1-3 galactosyltransferase (α-1-3GT), which is inactive in humans and most primates. In this group of mammals, this enzyme is deactivated at the molecular level by inactivation of the gene encoding the α-1-3GT enzyme, which translates into the absence of alpha-GAL on the surface of their proteins [6–8]. The alpha-GAL epitope is highly immunogenic in humans. It has been shown that sensitization with the alpha-GAL epitope may result in the synthesis of immunoglobulins of classes A, M, G, and E (IgA, IgM, IgG, and IgE) [3, 7, 9–12]. It has also been shown that all humans have natural antibodies (IgM/IgG) against the "B-like" blood group oligosaccharide found on non-primate red blood cells, which is galactose-alpha-1,3-galactose (alpha-GAL) [13–15]. Alpha-GAL is one of the main xenoantigens that prevents organ transplantation from various mammals (e.g., pigs) to humans [14, 16, 17]. Particularly strong alpha-GAL glycation is demonstrated by proteins of red meat (mainly beef) and offal (kidney, liver), as well as thyroid gland proteins [4, 18, 19].

1.2 AGS—History

The first references to an association with alpha-gal allergy and red meat allergy appeared around 2006 and concerned unusual allergic reactions in people treated with Cetuximab [20, 21]. Cetuximab is a recombinant human-murine chimeric IgG1 monoclonal antibody to epidermal growth factor receptor (EGFR), synthesized and produced by the SP2/0 mouse cell line, which expresses the gene encoding α -1,3-galactosyltransferase [22]. Cetuximab blocks the EGFR receptor. This antibody is used to treat inoperable, progressive, or recurrent EGFR-positive colorectal cancer and head and neck cancer [23]. The association between exposure to tick saliva and the occurrence of hypersensitivity reactions to Cetuximab was inferred based on the observation that the area with higher rates of hypersensitivity reactions to cetuximab (primarily the southeast United States, particularly Tennessee, Arkansas, and North Carolina) overlaps with the endemic area of Rocky Mountain spotted fever, which is vectored by *Amblyomma americanum* (*A. americanum*) ticks. It was also noted that patients who experienced hypersensitivity reactions to cetuximab were more likely to have been previously bitten by ticks than those who did not [20, 24, 25]. Furthermore, individuals hypersensitive to cetuximab were also more likely to report various types of hypersensitivity reactions associated with the consumption of red meat or dairy products [25, 26].

In patients with hypersensitivity to Cetuximab, the presence of Cetuximab-specific antibodies (IgE class) has been demonstrated, and these antibodies have been shown to bind alpha-GAL to bovine thyroglobulin [20]. Moreover, it was noticed that the anti-EGFR monoclonal antibody synthesized by a Chinese hamster ovary (CHO) cell line, with an inactive α-1,3-galactosyltransferase gene, did not react with the sera of patients exhibiting cetuximab-related anaphylaxis [20]. All these observations, research results, and analysis of reported clinical cases confirmed that alpha-GAL plays a key role as a target epitope in anaphylaxis to Cetuximab and other carriers of this carbohydrate determinant in the mechanism of cross-allergy with primary sensitization to alpha-GAL from tick saliva following a previous bite by this arachnid [1, 20, 26].

1.3 AGS—Pathomechanism

A hypersensitivity reaction to alpha-GAL from foods or medications is preceded by a tick bite, which is the primary cause of alpha-GAL Syndrome (Fig. 1).

Tick saliva contains proteins or lipids that bind alpha-GAL and various factors with immunomodulatory properties. A tick bite involves the introduction of this arachnid's saliva into the skin. The introduced foreign antigens are recognized by immunocompetent cells of the bitten organism's immune system, resulting in a limited inflammatory response, including the recruitment of basophils and antigen-presenting cells capable of secreting interleukin-4 (IL-4). This leads to the creation of a cytokine environment that favors a type 2 (Th2) T-helper lymphocyte-dependent response, which recognizes antigens from tick saliva. Th2 cells then interact with B lymphocytes and induce the synthesis of alpha-GAL-specific antibodies, with a final class switching of the produced immunoglobulin toward the synthesis of immunoglobulin E (IgE). The alpha-GAL-specific IgE produced at this stage is bound to the surface of basophils and mast cells via high-affinity IgE receptors (FcERI) [17, 27, 28].

The clinical reaction of AGS develops in initially sensitized individuals approximately 3–6 h after consuming products containing alpha-GAL. Due to the variability of the clinical picture and the delayed nature of the reaction, various mechanisms and causes of this phenomenon are considered. Environmental factors (concentration

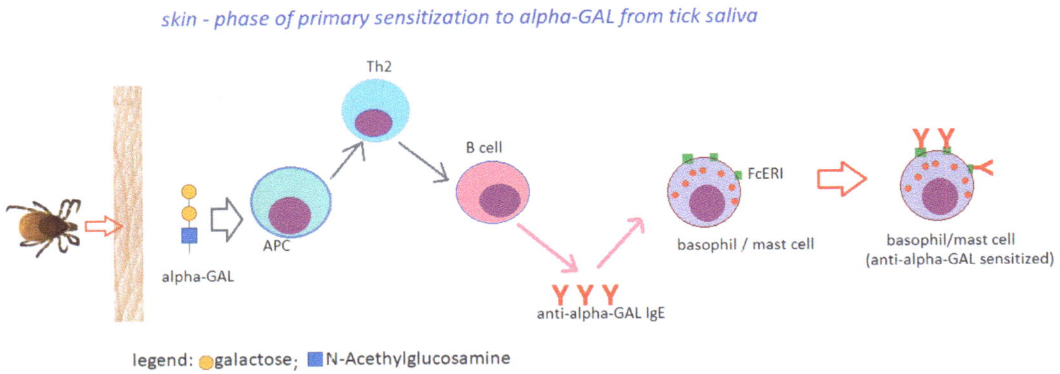

Fig. 1 Mechanism of primary sensitization to alpha-GAL from tick saliva

and method of alpha-GAL exposure), the efficiency of the patient's immune system, and the kinetics of allergen digestion and absorption are considered, especially in relation to animal fats as a carrier of this polysaccharide [17, 27, 28]. The alpha-GAL oligosaccharide occurs in both glycoprotein and glycolipid forms [29]. The glycolipid fraction of alpha-GAL likely accounts for the delayed clinical manifestation of AGS. The peak of triglyceride digestion products in the blood after a meal occurs 3–4 h after a meal [30, 31], which coincides with the time of clinical manifestation of AGS in allergic individuals [17, 27, 28].

1.4 AGS—Clinical Symptoms

A hypersensitivity reaction to alpha-GAL is classified as a late or delayed reaction. Symptoms usually appear 3–6 h after consuming mammalian meat or products of mammalian origin (e.g., dairy products, gelatin, and offal). The spectrum of clinical manifestations of AGS is very diverse and may include gastrointestinal symptoms, skin symptoms, respiratory and circulatory system symptoms, and systemic reactions (*see* **Note 1**). Heat treatment of foods does not affect the likelihood, timing, or severity of a hypersensitivity reaction. It has been noted that the consumption of fatty foods may accelerate the onset of a hypersensitivity reaction to alpha-GAL, while the simultaneous action of a cofactor (mainly ethyl alcohol, nonsteroidal anti-inflammatory drugs [NSAIDs], or physical activity) may accelerate or trigger the reaction. It is also significant that due to the heterogeneity of symptoms, AGS is often misdiagnosed as food indigestion, indigestion, irritable bowel syndrome, or alcohol allergy [25, 32–41].

The occurrence of Alpha-GAL Syndrome is also more frequent in people who are particularly exposed to ticks due to their place of residence (rural areas, proximity to parks or forest areas), lifestyle (hiking), or occupation (foresters, animal breeders [42, 43]. Frequent or repeated exposure to ticks lowers the sensitivity threshold to alpha-GAL from food and/or increases the intensity of symptoms [37].

1.5 AGS—Epidemiology

The average prevalence of alpha-GAL allergy in Europe is estimated to range from 5.5% to 24.5%, while in the United States, the percentage of individuals with sIgE antibodies to alpha-GAL is approximately 2% of the population. This rate is higher in rural areas and among forest workers. Nearly half of individuals with a positive sIgE-alpha-GAL result also have another atopic disease [44–47].

2 Materials

The ImmunoCAP (ThermoFisher Scientific, Waltham, Massachusetts, USA) is a certified, automated in vitro diagnostic (CE IVD) system for the quantitative measurement of allergen-specific IgE concentrations in human serum or plasma. ImmunoCAP is the reference method for measuring specific IgE in allergy diagnosis. The ImmunoCAP system utilizes a fluoroimmunoenzymatic assay (FEIA). The ImmunoCAP system is a fluoroimmunoenzymatic (FEIA) technique. It is compatible with PHADIA system immunochemical platforms (ThermoFisher Scientific, Waltham, Massachusetts, USA) [48, 49].

3 Methods

3.1 AGS Diagnostics—General Guidelines

Alpha-GAL Syndrome is an unusual form of food allergy with a wide range of symptoms. This often results in misdiagnosis as irritable bowel syndrome, meat allergy, milk allergy, food intolerance, or anaphylaxis of unknown cause. It is estimated that the average time to accurate diagnosis in a patient with AGS is up to 7 years [50].

Maintaining proper diagnostic procedures and careful analysis of clinical data and additional test results is an effective tool for making the correct diagnosis and implementing appropriate therapeutic procedures (*see* **Note 2**).

3.2 AGS Diagnostics—Laboratory Methods

Currently, sIgE-alpha-GAL (allergen code: o215) can only be quantitatively measured using the ImmunoCAP method (Thermo Fisher Scientific, Waltham, Massachusetts, USA) on the Phadia System (Thermo Fisher Scientific, Waltham, Massachusetts, USA). The ImmunoCAP technique is the reference method.

Screening for sIgE-alpha-GAL is also available using a semi-quantitative technique on the ImmunoCAP $ISAC_{E112i}$ (ThermoFisher Scientific, Waltham, Massachusetts, USA) or ALEX3 (ALLERGY XPLORER, MacroArray Diagnostics GmbH (MADx), Vienna, Austria) multiplex molecular allergy diagnostic platforms. A positive result obtained by any screening method always requires confirmation by a quantitative technique (ImmunoCAP, allergen code: o215).

3.3 sIgE-anti-alpha-GAL—ImmunoCAP

During the ImmunoCAP analytical procedure (Fig. 2), allergen-specific IgE from the patient's serum reacts with the corresponding allergens, which are covalently bound to the ImmunoCAP. The antigenic substrate (source of galactose-α-1,3-galactose, alpha-GAL) in the sIgE-alpha-GAL assay is bovine thyroglobulin (allergen code: o215) [48]. After washing unbound patient serum components from the reaction system, an enzyme-labeled anti-human IgE conjugate (a mouse monoclonal anti-human IgE antibody labeled with β-galactosidase) is added to detect specific IgE bound to allergens coated on the ImmunoCAP. After incubation, excess unbound conjugate (enzyme-labeled anti-IgE) is washed from the reaction medium. Next, a developing reagent (a chemiluminophore compatible with the labeling enzyme; 4-methylumbelliferyl-β-D-galactoside) is added to detect complexes

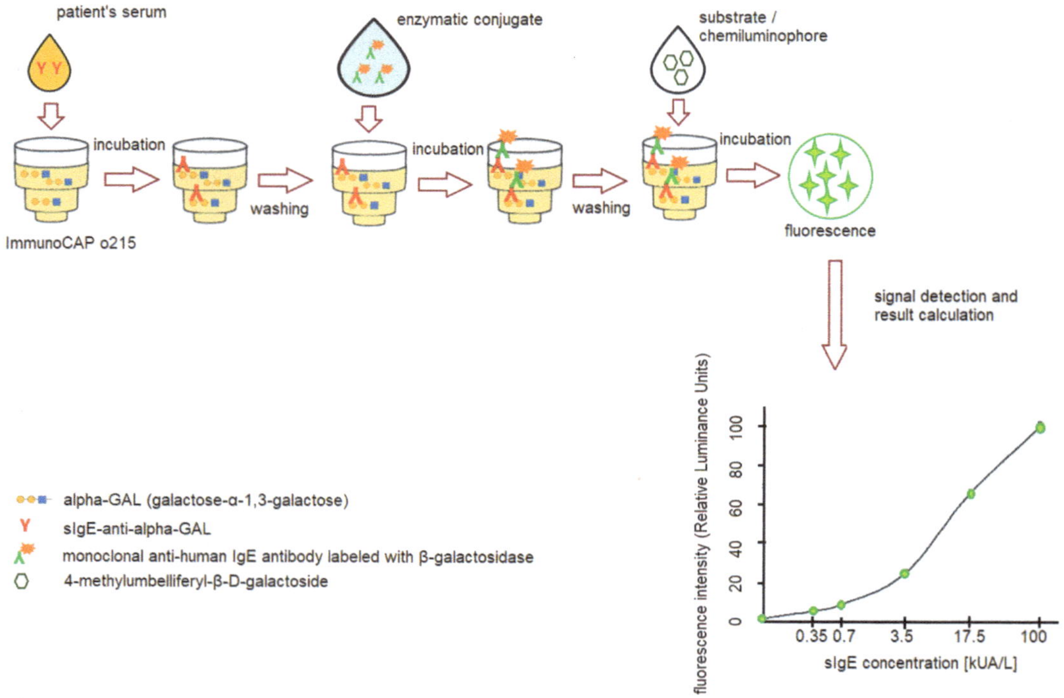

Fig. 2 Scheme of sIgE-anti-alpha-GAL measurement using the ImmunoCAP method

formed on the ImmunoCAP. The enzymatic reaction is stopped by adding 4% sodium carbonate to the reaction medium. The automatic detection of the chemiluminescent signal is converted by the analyzer software to the concentration of specific IgE in the test sample. The sIgE concentration is expressed in kUA/L. The chemiluminescence intensity is proportional to the sIgE concentration in the test sample. The method is calibrated in the range 0–100 kUA/L. Reference material: IgE calibrators are traceable in an unbroken calibration chain according to the 2nd International Reference Regulation (IRP) 75/502 for human immunoglobulin E in blood serum of the World Health Organization (WHO). The minimum limit of quantification for sIgE antibodies is 0.1 kUA/L. Cross-reactivity with other human immunoglobulins (IgA, IgD, IgM, IgG) does not occur within the physiological concentration range [49, 51].

3.4 sIgE-anti-alpha-GAL—ISAC$_{E112i}$

The ISAC$_{E112i}$ (Thermo Fisher Scientific, Waltham, Massachusetts, USA) test is an immunoenzymatic test based on biochip (microarray) technology. The ISAC$_{E112i}$ biochip is a glass slide on which specific biological molecules are precisely deposited at specific locations. The source of the alpha-GAL allergen in the ISAC$_{E112i}$ test is bovine thyroglobulin (as in ImmunoCAP). The ISAC$_{E112i}$ test is a multiplex platform for the transdermal detection of sIgE for 112 individual allergen components from 48 different allergen sources (plant and animal). It is a solid-phase enzyme immunoassay with fluorescence detection (Fig. 3). Specific IgE from the patient's serum is captured by allergens placed on a slide (test matrix). In the next step (after washing unbound reaction components), allergen/sIgE immune complexes are detected using fluorochrome-labeled anti-human IgE antibodies. The fluorescence intensity is measured using a microarray scanner [e.g., CapitalBio LuxScan 10k (Capitalbio, San Diego, California, USA)] and converted to a numerical value using the Phadia Microarray Image (MIA) program. sIgE results are reported in empirical ISAC Standardized Units (ISU-E). sIgE values are assigned to four categories: undetectable sIgE (<0.3 ISU-E); low sIgE (0.3–0.9 ISU-E); moderate or high sIgE (1–14.9 ISU-E); very high sIgE (>15 ISU-E). Please note that the $ISAC_{E112i}$ test is a screening test, and positive sIgE results measured with this method should be confirmed by a quantitative method (preferably ImmunoCAP) [51–55].

3.5 AGS Diagnostics—sIgE-alpha-GAL Clinical Interpretation

The analytical sensitivity threshold of the ImmunoCAP sIgE test is 0.1 kUA/L. All results above 0.1 kUA/L are considered positive (sIgE for the given allergen is present). However, although in healthy individuals the 95th percentile of ImmunoCAP specific IgE concentration is usually below 0.1 kUA/L, in clinical practice it is commonly accepted that the cut-off point for a clinically significant positive result is 0.35 kUA/L. When interpreting sIgE results, it

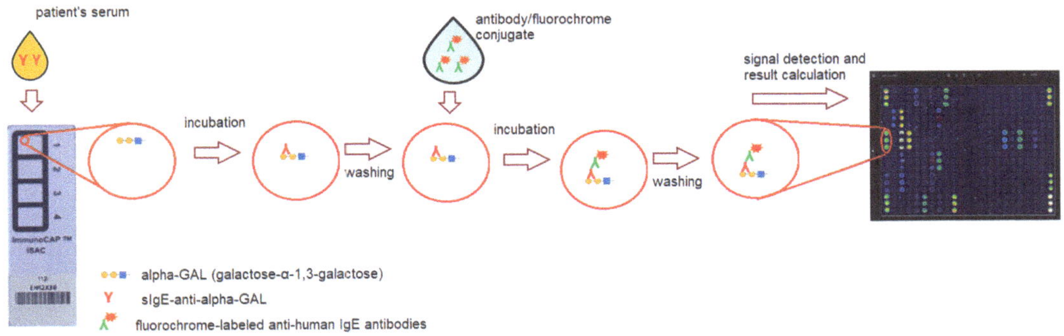

Fig. 3 Scheme of sIgE-anti-alpha-GAL measurement using the $ISAC_{E112i}$ method

Table 1
The most common clinical symptoms of Alpha-GAL Syndrome [5, 35]

Body area	Type/kind of symptoms
Digestive system	epigastric pain (generalized or localized), nausea, diarrhea, vomiting
Skin and/or mucous membranes	generalized or localized itching and/or hives, edema
Circulatory system	chest tightness, heart palpitations
Respiratory system	shortness of breath, cough
Whole body	anaphylactic shock

Table 2
Important guidelines for the diagnosis of Alpha-GAL Syndrome [1, 58–63]

Diagnostic steps	Objectives	Important details
Patient's clinical history	Collecting detailed information about symptoms and their triggers Preliminary assessment of the likelihood of allergy (family history of allergy, patient's atopic history) Preliminary identification of suspect foods Planning and selecting appropriate additional tests	Pay particular attention to: – Time from meal ingestion to the onset of the reaction – Type of symptoms (general/local) – Frequency of symptoms and their cyclical nature – Repeated accompanying factors (e.g., exposure to cofactors—alcohol, physical activity, NSAIDs) – Other unusual suggestions from the patient (e.g., always after drinking, always or often at events, etc.) – In adults—new or first-ever allergy Ask about a history of tick bites or (if the patient doesn't remember or doesn't want to be bitten) determine the likelihood of tick exposure: – Determine the patient's place of residence (rural, urban, near a park, garden or meadow) – Determine the type of work or hobby (forester, hunter) – Determine how the patient spends their free time (does he or she enjoy outdoor activities) Ask about meat—does the patient eat meat at all? – If not: ask about dairy products—if the patient does not eat meat and does not consume dairy products, AGS can basically be ruled out; if he does not eat meat but consumes dairy products, AGS is possible but very unlikely – If so: ask what type (poultry, beef, pork, or other mammals, offal), how often, in what culinary forms (raw/boiled/roasted/fried), and ask if the patient consumes meat with various cofactors (mainly alcohol)
Physical examination	Assessment of current health status (at the time of examination) Assessing the likelihood of allergies or other (non-allergic) diseases	In AGS, no characteristic symptoms are present during the current physical examination (unless the examination is performed during anaphylaxis). Therefore, pay attention to the symptoms occurring during the allergic reaction, as reported by the patient (refer to the clinical history) If the patient has been bitten by a tick in the past, note whether there are persistent local reactions at the bite site, such as itching for two or more weeks and skin lesions resembling hives
Additional tests	Confirming or ruling out an allergic cause of the disease	Select tests (sIgE) based on the patient's clinical history and the suspected allergy profile

Diagnostic steps	Objectives	Important details
	Identifying the causative allergen	Order a total IgE (tIgE) using a quantitative method (ImmunoCAP, Thermo Fisher)—the result will be useful for interpreting the sIgE results (sIgE/tIgE ratio) If you suspect AGS order: – A specific IgE for alpha-GAL (sIgE-alpha-GAL) (ImmunoCAP, ThermoFisher singleplex) – Optionally—to differentiation from other meat allergies, sIgE (ImmunoCAP, ThermoFisher singleplex) for: Meat allergen extracts (according to the patient's clinical history) and main allergenic components of meat (Bos d 6, Sus d 1)—differentiation of primary meat allergies (beef, pork) Serum albumin allergy marker component (Fel d 2, Sus d 2)—differentiation of cross-allergy to serum albumin (pork-cat syndrome) Cow's milk allergen extract and main milk allergen components (Bos d 8, Bos d 4, Bos d 5, Bos d 6)—differentiation of milk, beef and milk allergies If you have not considered AGS (sIgE-alpha-GAL has not been measured), it's worth considering the possibility of AGS if: – sIgE to mammalian meat allergens is present (especially if it is not accompanied by sIgE to poultry or these are the only positive sIgE results for food allergens in multiplex tests (e.g., ALEX, MADx)—order the sIgE-alpha-GAL (ImmunoCAP, ThermoFisher singleplex) to confirm or rule out AGS – sIgE for alpha-GAL in the multiplex ISAC test (ThermoFisher) is positive—confirm this result with the sIgE-alpha-GAL (ImmunoCAP, ThermoFisher singleplex) Other tests: – Skin Prick Tests (SPT): May be considered with pork kidneys, pork, beef, or gelatin colloid It should be taken into account that the sensitivity of these tests in AGS is low – Oral challenge (OC): May be considered—with boiled pork kidneys or beef with minimum 2-h observation, dose escalation on the second day, and optionally in combination with physical exercise (treadmill activity) It should be considered that the interpretation of the OC result in AGS is difficult due to the delayed nature of the hypersensitivity reaction – Basophil Activation Test (BAT): May be considered, using kits certified for in vitro diagnostics (CE IVD) and a standardized allergen (alpha-GAL) solution (CE IVD) Good for distinguishing patients with clinical AGS from those with asymptomatic alpha-GAL sensitization (sIgE-alpha-GAL positive, no clinical symptoms)

should also be taken into account that sIgE may be present in the absence of allergy symptoms and that sIgE concentrations in the range of 0.1–0.35 kUA/L may be clinically significant in some individuals, depending on age and the nature of clinical symptoms and in relation to some allergens (especially food allergens) [56, 57].

Mabelane et al [58] estimated that if the serum sIgE-alpha-GAL concentration is 5.5 kUA/L (or higher) or the sIgE-alpha-GAL/tIgE ratio is 2.1% (or higher), the probability of clinical symptoms of meat allergy in AGS is 95%. However, one should always consider the individual patient's response, the variable nature of the allergic reaction characteristic of the alpha-GAL syndrome, and the fact that the presence of cofactors can lower the patient's reactivity threshold [1, 58–64].

4 Notes

1. The most common clinical symptoms of Alpha-GAL Syndrome [5, 35] (see Table 1).
2. Important guidelines for the diagnosis of Alpha-GAL Syndrome [1, 58–63] (see Table 2).

Competing Interests

The author has no conflicts of interest to declare that are relevant to the content of this chapter.

References

1. Platts-Mills TAE, Li RC, Keshavarz B et al (2020) Diagnosis and management of patients with the α-gal syndrome. J Allergy Clin Immunol Pract 8(1):15–23. https://doi.org/10.1016/j.jaip.2019.09.017
2. Avila SA, Wojno T (2022) Alpha-gal syndrome: a new etiology for periorbital edema. Ophthalmic Plast Reconstr Surg 38(4):e99–e100. https://doi.org/10.1097/IOP.0000000000002135
3. Perusko M, Grundström J, Eldh M et al (2024) The α-gal epitope—the cause of a global allergic disease. Front Immunol 15:1335911. https://doi.org/10.3389/fimmu.2024.1335911
4. Macdougall JD, Thomas KO, Iweala OI (2022) The meat of the matter: understanding and managing alpha-gal syndrome. Immunotargets Ther 11:37–54. https://doi.org/10.2147/ITT.S276872
5. Propst SBH, Thompson DK (2025) Alpha-gal syndrome and the gastrointestinal reaction: a narrative review. Front Allergy 6:1535103. https://doi.org/10.3389/falgy.2025.1535103
6. Galili U, Shohet SB, Kobrin E et al (1988) Man, apes, and old world monkeys differ from other mammals in the expression of alpha-galactosyl epitopes on nucleated cells. J Biol Chem 263(33):17755–17762. https://doi.org/10.1016/S0021-9258(19)77900-9
7. Macher BA, Galili U (2008) The Galalpha1,3Galbeta1,4GlcNAc-R (alpha-Gal) epitope: a carbohydrate of unique evolution and clinical relevance. Biochim Biophys Acta 1780(2):75–88. https://doi.org/10.1016/j.bbagen.2007.11.003
8. Hinterholzer A, Moises J, Regl C et al (2022) Unambiguous identification of α-Gal epitopes in intact monoclonal antibodies by NMR

spectroscopy. MAbs 14(1):2132977. https://doi.org/10.1080/19420862.2022.2132977
9. Hamadeh RM, Galili U, Zhou P, Griffiss JM (1995) Anti-alpha-galactosyl immunoglobulin A (IgA), IgG, and IgM in human secretions. Clin Diagn Lab Immunol 2(2):125–131. https://doi.org/10.1128/cdli.2.2.125-131.1995
10. Apostolovic D, Rodrigues R, Thomas P et al. (2018) Immunoprofile of α-Gal- and B-antigen-specific responses differentiates red meat-allergic patients from healthy individuals. Allergy 73(7):1525–1531. https://doi.org/10.1111/all.13400
11. Martín-Lázaro J, Núñez-Orjales R, González-Guzmán LA et al (2020) Galactose-α-1,3-galactose (alpha-gal) allergy: first pediatric case in a series of patients in Spain. Allergol Immunopathol (Madr) 48(3):251–258. https://doi.org/10.1016/j.aller.2019.07.004
12. Hilger C, Fischer J, Wölbing F, Biedermann T (2019) Role and mechanism of galactose-alpha-1,3-galactose in the elicitation of delayed anaphylactic reactions to red meat. Curr Allergy Asthma Rep 19(1):3. https://doi.org/10.1007/s11882-019-0835-9
13. Galili U, Rachmilewitz EA, Peleg A, Flechner I (1984) A unique natural human IgG antibody with anti-alpha-galactosyl specificity. J Exp Med 160(5):1519–1531. https://doi.org/10.1084/jem.160.5.1519
14. Milland J, Sandrin MS (2006) ABO blood group and related antigens, natural antibodies and transplantation. Tissue Antigens 68(6):459–466. https://doi.org/10.1111/j.1399-0039.2006.00721.x
15. Koike C, Uddin M, Wildman DE et al (2007) Functionally important glycosyltransferase gain and loss during catarrhine primate emergence. Proc Natl Acad Sci USA 104(2):559–564. https://doi.org/10.1073/pnas.0610012104
16. Daugs A, Lehmann N, Eroglu D et al (2018) In vitro detection system to evaluate the immunogenic potential of xenografts. Tissue Eng C Methods 24(5):280–288. https://doi.org/10.1089/ten.TEC.2017.0532
17. Román-Carrasco P, Hemmer W, Cabezas-Cruz A et al (2021) The α-gal syndrome and potential mechanisms. Front Allergy 2:783279. https://doi.org/10.3389/falgy.2021.783279
18. Krstić Ristivojević M, Grundström J, Apostolović D et al (2020) Alpha-gal on the protein surface hampers transcytosis through the Caco-2 monolayer. Int J Mol Sci 21(16):5742. https://doi.org/10.3390/ijms21165742
19. Carson AS, Gardner A, Iweala OI (2022) Where's the beef? understanding allergic responses to red meat in alpha-gal syndrome. J Immunol 208(2):267–277. https://doi.org/10.4049/jimmunol.2100712
20. Chung CH, Mirakhur B, Chan E et al (2008) Cetuximab-induced anaphylaxis and IgE specific for galactose-alpha-1,3-galactose. N Engl J Med 358(11):1109–1117. https://doi.org/10.1056/NEJMoa074943
21. Lammerts van Bueren JJ, Rispens T, Verploegen S et al (2011) Anti-galactose-α-1,3-galactose IgE from allergic patients does not bind α-galactosylated glycans on intact therapeutic antibody Fc domains. Nat Biotechnol 29(7):574–576. https://doi.org/10.1038/nbt.1912
22. Wasif Saif M, Diasio RB (2009) Chapter 68—Colon cancer. In: Waldman SA, Terzic A, Egan LJ et al (eds) Pharmacology and therapeutics, W.B. Saunders, pp 959–968. https://doi.org/10.1016/B978-1-4160-3291-5.50072-X
23. Chinuki Y, Morita E (2019) Alpha-Gal-containing biologics and anaphylaxis. Allergol Int 68(3):296–300. https://doi.org/10.1016/j.alit.2019.04.001
24. O'Neil BH, Allen R, Spigel DR et al (2007) High incidence of cetuximab-related infusion reactions in Tennessee and North Carolina and the association with atopic history. J Clin Oncol 25(24):3644–3648. https://doi.org/10.1200/JCO.2007.11.7812
25. Commins SP, James HR, Kelly LA et al (2011) The relevance of tick bites to the production of IgE antibodies to the mammalian oligosaccharide galactose-α-1,3-galactose. J Allergy Clin Immunol 127(5):1286–1293. https://doi.org/10.1016/j.jaci.2011.02.019
26. Steinke JW, Platts-Mills TA, Commins SP (2015) The alpha-gal story: lessons learned from connecting the dots. J Allergy Clin Immunol 135(3):589–596. https://doi.org/10.1016/j.jaci.2014.12.1947
27. Wilson JM, Erickson L, Levin M et al (2024) Tick bites, IgE to galactose-alpha-1,3-galactose and urticarial or anaphylactic reactions to mammalian meat: The alpha-gal syndrome. Allergy 79(6):1440–1454. https://doi.org/10.1111/all.16003
28. Platts-Mills TAE, Gangwar RS, Workman L, Wilson JM (2025) The immunology of alpha-gal syndrome: history, tick bites, IgE, and delayed anaphylaxis to mammalian meat. Immunol Rev 332(1):e70035. https://doi.org/10.1111/imr.70035
29. Galili U, Basbaum CB, Shohet SB et al (1987) Identification of the Gal alpha 1-3Gal glycosphingolipids of erythrocytes using the mouse monoclonal antibody Gal-13. J Biol Chem 262:4683–4688. https://doi.org/10.1016/S0021-9258(18)61249-9
30. Lambert JE, Parks EJ (2012) Postprandial metabolism of meal triglyceride in humans. Biochim Biophys Acta 1821:721–726. https://doi.org/10.1016/j.bbalip.2012.01.006

31. Engin AB, Engin A (2024) The checkpoints of intestinal fat absorption in obesity. Adv Exp Med Biol 1460:73–95. https://doi.org/10.1007/978-3-031-63657-8_3
32. Gonzalez-Quintela A, Vidal C, Gude F (2004) Alcohol, IgE and allergy. Addiction Biol 9:195–204. https://doi.org/10.1080/13556210412331292235
33. Kennedy JL, Stallings AP, Platts-Mills TA et al (2013) Galactose-α-1,3-galactose and delayed anaphylaxis, angioedema, and urticaria in children. Pediatrics 131(5):e1545. https://doi.org/10.1542/peds.2012-2585
34. Apostolovic D, Tran TA, Hamsten C et al (2014) Immunoproteomics of processed beef proteins reveal novel galactoseα-1,3-galactose-containing allergens. Allergy 69(10):1308–1315. https://doi.org/10.1111/all.12462
35. Fischer J, Yazdi AS, Biedermann T (2016) Clinical spectrum of α-Gal syndrome: from immediate-type to delayed immediate-type reactions to mammalian innards and meat. Allergo J Int 25:55–62. https://doi.org/10.1007/s40629-016-0099-z
36. Wilson JM, Schuyler AJ, Workman L et al (2019) Investigation into the α-gal syndrome: characteristics of 261 children and adults reporting red meat allergy. J Allergy Clin Immunol Pract 7(7):2348–2358. https://doi.org/10.1016/j.jaip.2019.03.031
37. Croglio MP, Commins SP, McGill SK (2021) Isolated gastrointestinal alpha-gal meat allergy is a cause for gastrointestinal distress without anaphylaxis. Gastroenterol 160(6):2178–2180. https://doi.org/10.1053/j.gastro.2021.01.218
38. Choudhary SK, Karim S, Iweala OI et al (2021) Tick salivary gland extract induces alpha-gal syndrome in alpha-gal deficient mice. Immun Inflamm Dis 9(3):984–990. https://doi.org/10.1002/iid3.457
39. Richards NE, Richards RD (2021) Alpha-gal allergy as a cause of intestinal symptoms in a gastroenterology community practice. South Med J 114(3):169–173. https://doi.org/10.14423/SMJ.0000000000001223
40. Busing JD, Stone CAJ, Nicholson MR (2023) Clinical presentation of alpha-gal syndrome in pediatric gastroenterology and response to mammalian dietary elimination. Am J Gastroenterol 118(7):1293–1296. 10.14309/ajg.0000000000002268
41. McGill SK, Levin ME, Shaheen NJ et al (2024) Gastrointestinal-isolated distress is common in alpha-gal allergic patients on mammalian meat challenge. Am J Gastroenterol 58(1):80–84. https://doi.org/10.3389/fped.2021.801753
42. Shishido AA, Wormser GP (2025) A review of alpha-gal syndrome for the infectious diseases practitioner. Open Forum Infect Dis 12(8):ofaf430. https://doi.org/10.1093/ofid/ofaf430
43. Leth-Møller KB, Hansen FR, van Hage M et al. (2025) Living close to the forest is a risk factor for galactose-α-1,3-galactose sensitization. J Allergy Clin Immunol Glob 4(3):100500. https://doi.org/10.1016/j.jacig.2025.100500
44. Commins SP (2016) Invited commentary: alpha-gal allergy: tip of the iceberg to a pivotal immune response. Curr Allergy Asthma Rep 16(9):61. https://doi.org/10.1007/s11882-016-0641-6
45. Villalta D, Pantarotto L, Da Re M et al (2016) High prevalence of sIgE to Galactose-α-1,3-galactose in rural pre-Alps area: a cross-sectional study. Clin Exp Allergy 46(2):377–380. https://doi.org/10.1111/cea.12655
46. Platts-Mills TA, Hilger C, Jappe U et al (2021) Carbohydrate epitopes currently recognized as targets for IgE antibodies. Allergy 76(8):2383–2394. https://doi.org/10.1111/all.14802
47. Benders-Guedj M, Köberle M, Hofmann H et al. (2023) High-risk groups for alpha-gal sensitization. Allergol Select 7:140–148. https://doi.org/10.5414/ALX02424E
48. https://www.thermofisher.com/phadia/wo/en/resources/allergen-encyclopedia/f27/o215.html. Accessed 25 Oct 2025
49. https://dfu.phadia.com/Data/Pdf/5be441b889c2320848d6fb28.pdf. Accessed 25 Oct 2025
50. Thompson CC, Saracco B, Pruthi A, Cerceo E (2025) Alpha-gal syndrome: often hidden, under-recognized, and in need of attention—a rapid review. Int J Gen Med 18:3477–3488. https://doi.org/10.2147/IJGM.S519844
51. Velázquez-Soto, H., & C. Jimenez Martinez, M. (2023). Fundamentals of Molecular Allergy: From Bench to Bedside. In Allergic Disease - New Developments in Diagnosis and Therapy. IntechOpen. https://doi.org/10.5772/intechopen.107512
52. Westwood M, Ramaekers B, Lang S et al (2016) ImmunoCAP® ISAC and Microtest for multiplet allergen testing in people with difficult to manage allergic disease: a systematic review and cost analysis. Health Technol Assess (Winchester, England) 20(67):1–178. https://doi.org/10.3310/hta20670
53. https://corporate.thermofisher.com/content/dam/phadia/library/en/ImmunoCAP%20ISAC%20Assay%20product%20overview.pdf. Accessed 25 Oct 2025

54. Melioli G, Spenser C, Reggiardo G et al (2014) Allergenius, an expert system for the interpretation of allergen microarray results. World Allergy Organ J 7(1):15–23. https://doi.org/10.1186/1939-4551-7-15
55. Lis K, Bartuzi Z (2023) Selected technical aspects of molecular allergy diagnostics. Curr Issues Mol Biol 45(7):5481–5493. https://doi.org/10.3390/cimb45070347
56. Graham F, Caubet JC, Ramadan S et al (2020) Specific IgE decision point cutoffs in children with IgE-mediated wheat allergy and a review of the literature. Int Arch Allergy Immunol 181(4):296–300. https://doi.org/10.1159/000505728
57. Thorpe M, Movérare R, Fischer C et al (2023) History and utility of specific IgE cutoff levels: what is the relevance for allergy diagnosis?. J Allergy Clin Immunol Pract 11(10):3021–3029. https://doi.org/10.1016/j.jaip.2023.05.022
58. Mabelane T, Basera W, Botha M et al (2018) Predictive values of alpha-gal IgE levels and alpha-gal IgE: total IgE ratio and oral food challenge-proven meat allergy in a population with a high prevalence of reported red meat allergy. Pediatr Allergy Immunol 29(8):841–849. https://doi.org/10.1111/pai.12969
59. Weins AB, Eberlein B, Biedermann T (2019) Diagnostics of alpha-gal syndrome: current standards, pitfalls and perspectives. Hautarzt 70(1):36–43. https://doi.org/10.1007/s00105-018-4288-1
60. Fischer J, Huynh HN, Hebsaker J et al (2020) Prevalence and impact of type I sensitization to alpha-gal in patients consulting an allergy unit. Int Arch Allergy Immunol 181(2):119–127. https://doi.org/10.1159/000503966
61. Commins SP (2020) Diagnosis and management of alpha-gal syndrome: lessons from 2,500 patients. Expert Rev Clin Immunol 16(7):667–677. https://doi.org/10.1080/1744666X.2020.1782745
62. Germán-Sánchez A, Alonso-Llamazares A, García-González F et al (2023) Diagnostic validity of specific immunoglobulin E levels to alpha-gal in alpha-gal syndrome: a cross-sectional analysis. Allergy Asthma Clin Immunol 19(1):102. https://doi.org/10.1186/s13223-023-00856-6
63. Darsow U, Gelincik A, Jappe U et al (2024) Algorithms in allergy: an algorithm for alpha-gal syndrome diagnosis and treatment, 2024 update. Allergy 79(11):3169–3172. https://doi.org/10.1111/all.16291
64. van Nunen S (2025) Quantifying the allergenic potency of mammalian meat sources, an important step in providing a framework for improved management of mammalian meat allergy after tick bite (alpha-gal syndrome/AGS). Allergy 80(1):9–11. https://doi.org/10.1111/all.16381

Chapter 21

Protocols for Diagnosis and Management of Arthropod Bite Allergies in Humans

Francisco Feo Brito, José Miguel Urra Ardanaz, Ana María Burgos Montero, Raúl Guzmán Rodríguez and José de la Fuente

Abstract

Allergic diseases have increased significantly in recent decades, with a higher incidence of severe manifestations such as anaphylaxis. This increase is attributed mainly to environmental factors associated with the Western lifestyle that alter the epithelial barrier, promoting sensitization to allergens, such as those from arthropods. Hymenoptera (bees and wasps) are the most implicated arthropods. The sting introduces a venom composed of highly sensitizing proteins that are capable of triggering reactions ranging from local and mild to potentially fatal systemic reactions. The incidence is higher in rural areas, adults, and exposed professionals such as beekeepers. On the other hand, tick bites can sensitize individuals to alpha-gal, causing allergic reactions 2–6 h later. This delay in the response leads to an underdiagnosis of this allergic disease. The diagnosis is based on the clinical history and complementary tests such as skin tests, specific IgE tests, basophil activation tests, and tryptase determination, although false negatives may occur, especially after a recent reaction or in patients with mastocytosis. Treatment includes preventive measures, adrenaline for systemic reactions, and specific immunotherapy, which are effective at reducing the risk of future severe reactions and improving quality of life. Novel oral immunotherapy food could be an effective alternative in the treatment of alpha-gal allergy.

Keywords Arthropod allergy, Hymenopter, Tick, Bite, Anaphylaxis, Oral immunotherapy food

1 Introduction

Allergic diseases have increased significantly in recent decades, not only in terms of incidence but also in terms of the clinical presentation, with a higher incidence of systemic manifestations such as anaphylaxis. Genetic, infectious, and environmental factors are related to this increase, although genetic factors are unlikely to be responsible for such notable changes in such a short period of time. Therefore, environmental and infectious factors are the most directly involved [1].

José de la Fuente and Alejandro Cabezas-Cruz (eds.), *Allergic Disease Models*, Methods in Molecular Biology, vol. 3092, https://doi.org/10.1007/978-1-0716-5635-8_21,

Immune-mediated diseases are characterized by defective epithelial barriers, which correspond to the onset of the process. Environmental exposures (exposomes), including global warming; changes in and loss of biodiversity; air pollution; the presence of pathogens, allergens, detergents, surfactants, microplastics, nanoparticles, enzymes, and emulsifiers in processed foods; and dietary habits, are responsible for altering the skin and mucosal barrier. Exposure to barrier-damaging agents causes epithelial cell injury and barrier damage, colonization by opportunistic pathogens, a loss of commensal bacteria, decreased microbiota diversity, bacterial translocation, allergic sensitization, and inflammation in the peri-epithelial area. In summary, a combination of environmental and infectious factors associated with the so-called "Western lifestyle" appears to be responsible for this increase in allergies [2, 3].

In parallel to this increase in allergic diseases, a higher prevalence of insect allergies has been observed; insects constitute 80% of the animal kingdom and are highly relevant in nature, from parasite predation to honey production and fruit tree pollination. Among insects, arthropods are small invertebrate animals characterized by an external skeleton and jointed legs. From an allergology perspective, allergic reactions caused by Hymenoptera and ticks to glycan α-Gal are significant. Among the hymenopteran species involved in immediate hypersensitivity responses, wasps (*Vespula orientalis* and *Polistes dominulus*) and bees (*Apis mellifera*) are noteworthy (Fig. 1). The insect introduces venom into the patient through a sting with a lancet of spicules, which, in addition to advancing, prevents its withdrawal until venom inoculation is complete (Fig. 1). The venom consists of a mixture of proteins, enzymes, and peptides with a high sensitizing capacity. Exposure to these components can induce sensitization in predisposed individuals, and subsequent contact with Hymenoptera can trigger an allergic response [4].

Adverse reactions to Hymenoptera have been known for more than 5000 years, with the death of the first Egyptian pharaoh (Menes) occurring after a wasp sting. Pliny the Elder, in his vast work on nature, also highlighted the health risk posed by Hymenoptera. As mentioned, the prevalence of Hymenoptera sting allergies has been on the rise in recent years, affecting up to 7.5% of adults and 3.4% of children. Reactions can be mild and restricted to the skin or moderate to severe, with a risk of life-threatening anaphylaxis. Their prevalence is strongly influenced not only by age but also by habitat. In addition, the probability is increased by the number of stings a person receives. Thus, the number of people affected is three times greater in rural areas than in urban areas, more common in adults than in young people, and higher in males than in females. In terms of occupational risk, up to 30% of beekeepers are allergic to bee venom. Mild symptoms generally manifest as generalized skin symptoms, including redness, hives, and angioedema. Dizziness, dyspnea, and

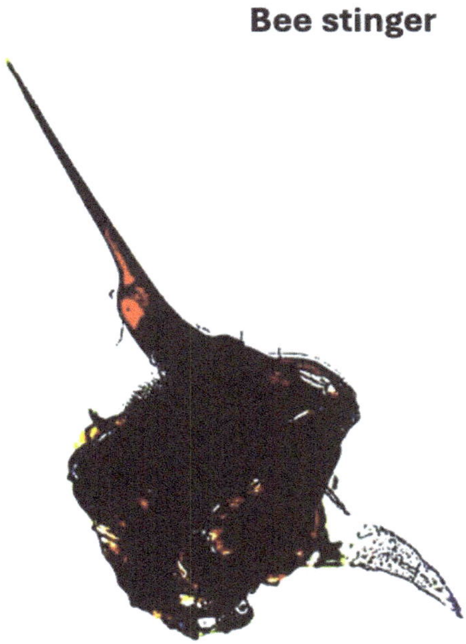

Fig. 1 Bee. Image of a bee in spring with a microscopic photograph of the female stinger, the egg-laying organ used for defense

nausea are examples of moderate reactions, whereas shock and loss of consciousness, or even cardiac or respiratory arrest, correspond to severe reactions that can have fatal consequences. The severity of these reactions negatively impacts patients' quality of life, causing changes in their habits and stressful situations, given the possible exposure to Hymenoptera associated with outdoor activities [5, 6]. The incidence of mortality from stings ranges from 0.09 to 0.45 deaths per million, but some cases of sudden death in agricultural environments or due to a failure to identify the insect could suggest that this figure is underestimated [6–8].

Hymenoptera allergy is an excellent model for immediate hypersensitivity reactions. On the one hand, the identification of the allergen, its consequences upon contact with the sensitized patient through mast cell and basophil activation, the different intensities of each clinical case, accurate diagnosis through a detailed clinical history, and in vivo and laboratory tests are important. Additionally, the relevance of preventive measures, as well as emergency treatment and the high degree of efficacy of specific immunotherapy, complete the prototype of an allergic process [9, 10].

Ticks are among the arthropods that induce allergic responses through their bites. These arachnids rank second after mosquitoes as vectors involved in the transmission of diseases. Allergy to tick bites is known as α-Gal syndrome, a recently diagnosed disease mediated by anti-α-Gal IgE antibodies. Human evolution has led to the loss of the ability to synthesize this glycan and, consequently, to the development of a response to α-Gal that manifests itself in allergic reactions to the consumption of red meat from mammals and some derived products.

Alpha-Gal is a molecule that is present in the saliva of ticks (e.g., lone star tick, *Amblyomma americanum*) and potentially other arthropods, usually after it has fed on mammalian blood [11]. Tick bites are the primary risk factor for allergic sensitization, which can affect between 5% and 8% of the population. However, α-Gal syndrome is underdiagnosed because of its atypical clinical presentation, with delayed symptoms that in many cases are not associated with food intake [11, 12]. However, the clinical manifestation of severe reactions that could be fatal requires greater attention to determine the large number of hidden diagnoses. Given that up to 30% of anaphylaxis cases are diagnosed as idiopathic, with no known allergen, one of the objectives should be to exclude alpha-gal allergy with a simple diagnostic kit. Similar to other cases, such as when anaphylaxis due to hepatic hydatid disease was ruled out by IgE specific to *Echinococcus* [13], the levels of specific IgE antibodies against alpha-gal should be determined in all patients with idiopathic anaphylaxis.

2 Allergy to Hymenoptera Sting

2.1 Clinical Manifestations

Hymenoptera stings can occur in people of any age. Humans are usually stung by accidental contact with a single insect or by being near a nest (multiple stings from several insects). Certain professions and hobbies can also increase the risk of exposure, such as beekeeping, professions related to agriculture and construction, and outdoor activities. The place of residence is also important. More insects are present in areas with a more temperate climate, and they remain active for more months of the year.

2.2 Local Reactions

Most Hymenoptera stings (aphid and vespid) produce a local reaction within minutes, which can last for hours. Clinically, they involve immediate pain at the site of venom injection, followed by erythema, edema, local heat, and itching. In most cases, they are irrelevant, resolve within a few days and do not require any medical attention, except in the exceptional case of stings inside the oral cavity that can cause airway obstruction. These reactions are temporary and normal consequences of the vasoactive and inflammatory actions of some components of the venom [14].

On other occasions, an extensive local reaction occurs, where the erythema and edema that occur are more than 10 cm in diameter, are more painful and may persist longer. These reactions usually worsen within 48 h and resolve within 7–10 days. They are sometimes accompanied by nonspecific symptoms (headache, chills, dizziness, or fever). In addition to the local action of the venom, these reactions may involve a mechanism mediated by immunoglobulin E in people who are allergic to the insect that caused the sting.

2.3 Systemic Reactions

Systemic reactions include symptoms such as generalized urticaria, angioedema, facial swelling, stridor, difficulty breathing, wheezing secondary to bronchospasm, abdominal pain, nausea, vomiting, and flushing. Sometimes the symptoms are so severe that they can lead to death.

Different criteria are available for defining systemic allergic reactions in an attempt to classify the severity of the clinical presentation. The European Academy of Allergy and Clinical Immunology (EAACI) recently proposed a simplification of the criteria for the severity of acute allergic reactions, dividing them into local (grade 1) and systemic (grades 2 and 3) [15].

The most common systemic symptoms are skin symptoms, mainly urticaria, and respiratory symptoms such as dyspnea, wheezing, and upper airway obstruction. Symptoms and signs of hypotension may occur in more than 60% of adults; in half of the patients, these symptoms and signs are accompanied by loss of consciousness. Cardiac involvement during anaphylaxis can cause bradycardia, arrhythmias, and acute coronary syndromes, which may be secondary to decreased venous return caused by histamine-induced vasodilation and blood vessel permeabilization and/or decreased cardiac output due to hypoxia and respiratory acidosis. Gastrointestinal symptoms (abdominal pain, nausea, vomiting, and diarrhea) and uterine cramps with possible spontaneous abortion may occur, as well as neurological symptoms such as seizures.

Biphasic anaphylaxis is a clinical condition characterized by the recurrence of systemic symptoms within 4–12 h (exceptionally 72 h) after resolution (without reexposure) and has been reported in 0.4%–14.7% of patients.

2.4 Concomitant Diseases

Patients who are allergic to Hymenoptera stings are at greater risk of severe systemic reactions when they have underlying diseases such as cardiovascular disease. The components of the venom can induce the release of serotonin and adrenaline, which increase platelet aggregation, thus increasing the likelihood of thrombosis as a result of increased Factor V levels and the release of a thromboplastin-like substance from the vascular wall. These molecules and other substances released by mast cells could have negative inotropic and chronotropic effects. During an anaphylactic reaction, de novo synthesis of LTC4 and PGD2 at the cardiac level can cause vasoconstriction. Similarly, in some patients with coronary artery disease, the stimulation of H1 receptors can cause vasoconstriction of large-caliber coronary arteries, unlike in patients with a healthy myocardium. The activation of metalloproteinases also facilitates the degradation of the connective tissue of atheromatous plaques, thereby increasing the risk of ischemia [15].

In the presence of systemic reactions, baseline serum tryptase levels should always be determined, as adults affected by mast cell diseases or elevated baseline tryptase levels have a significantly higher risk of severe reactions to Hymenoptera stings.

2.5 Toxic Reactions and Other Reactions Following Hymenoptera Stings

Systemic toxicity reactions are caused by the action of venom components with enzymatic activity and organ-specific toxicity and usually occur after multiple simultaneous stings (from several dozen to several hundred). Toxic effects occur within hours or days and include rhabdomyolysis, intravascular hemolysis, coagulation disorders, liver damage, and acute renal failure. Fatal cases are rare.

Other atypical systemic reactions are rare and are caused by a toxic or non-IgE-mediated immunological mechanism, in some cases by autoimmunity. These reactions can occur within hours or days after a single sting and include serum sickness-like manifestations, central nervous system manifestations (acute encephalopathy, Guillain–Barré syndrome, myasthenia, and peripheral neuritis), hematological reactions (thrombocytopenic purpura, Henoch–Schönlein purpura, hemolysis, and coagulation disorders), muscular reactions (rhabdomyolysis), renal reactions (acute renal failure due to interstitial nephritis or tubular damage, and nephrotic syndrome) and respiratory reactions (alveolar hemorrhage) [14].

3 Allergy to Tick Bites

3.1 Ticks

Ticks are obligate hematophagous ectoparasites responsible for transmitting a wide range of infectious agents. They attach themselves to tall grass, bushes, or weeds, from which they attach themselves to animals or people passing by to feed on their blood. Therefore, risk factors for bites include living or performing activities in rural or wooded areas where these parasites are abundant. In addition, tick bites can cause allergic reactions to red meat, gelatin, and cetuximab, as well as the transmission of infections [16]. The antigen that causes allergic reactions is the oligosaccharide galactose-alpha-1,3 galactose, also known as alpha-gal, which is a protein present in tick saliva. It is thus presented to B lymphocytes in the context of a Th2 response, which induces the production of anti-alpha-gal IgE. This allergy is increasing in association with climate change, as rising temperatures facilitate the survival and spread of ticks and, as a result, the risk of bites. The tick species involved in alpha-gal allergy differ according to geographical area, with *Ixodes ricinus* predominating in Europe, *A. americanum* in the United States, and *Ixodes holocyclus* in Australia.

3.2 Clinical Manifestations and Anaphylactic Reactions

Once sensitization to alpha-gal occurs, subsequent clinical responses can trigger two types of anaphylaxis [17]: (a) an immediate reaction due to exposure to intravenous cetuximab, a chimeric monoclonal antibody used in oncology that contains alpha-gal. (b) Food allergy to meat and products derived from non-primate mammals. In this case, the reaction occurs between 2 and 6 h after the food is ingested.

The delay in the onset of symptoms appears to be due to the slow digestion of oligosaccharides, such as fats, which makes a diagnosis difficult. Another distinguishing feature is that anaphylaxis due to alpha-gal is not usually accompanied by symptoms of asthma. The clinical manifestations are very broad, with digestive symptoms such as abdominal pain, reflux, vomiting, or diarrhea, or skin symptoms such as pruritus, urticaria, or angioedema; in exceptional cases, urological or neurological symptoms have also been reported. A recent study of forestry workers with high exposure to tick bites revealed sensitization to alpha-gal in 35% of them, but alpha-gal allergy was diagnosed in only 5% of sensitized subjects. These different clinical presentations suggest that the dose of the allergen and/or cofactors may affect the likelihood and severity of the reaction. In addition, several patients with alpha-gal allergy showed a complete disappearance of clinical symptoms, even after reexposure to the allergen, indicating the idiosyncratic nature of the response to alpha-gal. Thus, the wide range of symptoms and the variability of each case, together with the lack of knowledge among doctors and the general population, lead to the frequent underdiagnosis of the disease [18].

Another characteristic of alpha-gal allergy is that alpha-gal exhibits cross-reactivity with other similar allergens because it is a carbohydrate, which is known as CCD syndrome. Patients with blood Group B show greater protection against this pathology and a propensity to become sensitized to Hymenoptera venom or to experience symptoms from the bites of other arthropods, such as the *Glossina* (tsetse) fly [19].

The diagnosis of alpha-gal allergy to tick bites is based on a detailed medical history, with the onset of compatible symptoms 2–6 h after eating meat from certain mammals (beef, lamb, or pork). The delay in the onset of systemic reactions is distinct from that of other IgE-mediated food allergy reactions, which occur immediately. Furthermore, the reaction depends not only on the allergen but also on the amount, and if this threshold is not reached, the allergic response is not triggered. Similarly, cofactors such as exercise, alcohol consumption, or intake of nonsteroidal anti-inflammatory drugs have been described as determinants of the individual susceptibility of each patient. Skin tests are not valid for this type of allergy, and thus sensitization is determined by quantifying specific IgE against alpha-gal. The determination of sIgE antibodies provides essential information on the etiology of the disease because these antibodies are produced after the exposure of a susceptible individual to an allergen, but sensitization is not the same as clinical allergy, and the presence of IgE does not confirm the diagnosis of alpha-gal allergy. In addition, patients with alpha-gal anaphylaxis and indolent systemic mastocytosis may have lower sIgE values.

3.3 Recommendations

Once the diagnosis of alpha-gal allergy has been made, the patient should avoid consuming mammalian meat, its derivatives, and medications such as cetuximab and infliximab that contain alpha-gal. They should also avoid spending time in rural areas with a risk of exposure to ticks, but can avoid these risks by wearing light-colored clothing, long-sleeved shirts, and long trousers, walking in the middle of paths in risk areas and using repellents such as permethrin. If, despite taking precautions, an individual is bitten by a tick, the parasite should be immediately removed with fine-tipped tweezers by grasping it as close to the skin as possible and pulling upwards firmly, without twisting or crushing it. If an individual has a history of anaphylaxis, he or she should carry adrenaline autoinjectors for immediate treatment.

Since most dishes in food culture contain red meat, the elimination of red meat from the diet can negatively affect patients' quality of life. Thus, the option of specific immunotherapy treatment with alpha-gal, the most important treatment in allergology, has been explored. Alpha-Gal glycoconjugates based on poly-L-lysine or biodegradable nanoparticles encapsulating alpha-Gal have been proposed to improve immunological tolerance [20]. Similarly, Ünal et al. conducted a study with oral immunotherapy (OIT) in a total of 12 patients (five rapid and eight slow). Once the maximum dosage of 100 g of red meat was reached, its administration was repeated daily for 6 months and every other day for the following 6 months. After 1 year, consumption of red meat two to three times per week was recommended, which is equivalent to normal consumption of food. The authors concluded that all patients tolerated red meat, confirming good tolerance at 5 years, accompanied by a decrease in the levels of specific IgE against alpha-gal [21].

4 Methods

4.1 In Vitro Tests for the Diagnosis of Hymenoptera Venom Allergy

In vitro tests, together with the clinical history and skin tests, play an important role in the diagnosis of Hymenoptera sting allergies. Hypersensitivity to Hymenoptera venom is mediated by the reactivity of specific IgE (sIgE) in patients to one or more of the components present in the venom injected by the sting. This IgE is produced after contact with the venom from a previous sting. Therefore, the basis for the in vitro diagnosis of an allergy to Hymenoptera venom is the detection of reactive IgE against one or more of the components present in the venom injected into the patient's blood. In many cases, the Hymenoptera species responsible for the sting cannot be identified, and the venoms of all the species must be studied. Therefore, in addition to the diagnosis of a potential allergy, the detection of specific IgE will allow the identification of the species responsible for the allergy. Tests to detect the presence of sIgE against venom components are the initial step in the in vitro diagnosis of allergy to Hymenoptera venoms and allow us to confirm the clinical diagnosis and identify the responsible species.

1. IgE is bound to the surface of two preferred cell types, basophils and mast cells. These cells produce high-affinity IgE receptors on their surface, known as FcεRI. The high affinity of these receptors, together with the molecular characteristics of IgE, means that most of the IgE produced remains anchored to the cell surface, with low levels circulating in the blood. Their analysis requires the development of detection techniques, unlike analyses of other types of immunoglobulins, such as IgG, IgM, or IgA, which have higher blood concentrations. Different generations of assays have been developed, and currently, the most widely used is ImmunoCAP.
2. After the specific antigen binds to the IgE anchored in the cell membrane, mast cells and basophils generate intracellular signals that induce the degranulation of their cytoplasmic granules, releasing their components into the extracellular medium. The action of these released components is responsible for the activity in the tissues involved, ultimately causing clinical symptoms mediated by the sting. The basophil activation test (BAT) allows the effects of venoms on patient blood basophils to be reproduced in vitro and reveals whether basophils can activate and degranulate their products in the presence of the venoms studied. This analysis detects the presence of a specific IgE-mediated response of these cells to antigen exposure. This technique can complement sIgE detection and is very useful when IgE detection results and/or skin tests are inconclusive.

3. Most of the components released by basophils and tissue mast cells are highly unstable and cannot be detected in the blood. However, tryptase is stable and quantifiable in blood. Increased blood tryptase levels are highly indicative of an IgE-mediated reaction.
4. The different Hymenoptera species are phylogenetically closely related, indicating that the components of their venom are very similar and favor the occurrence of frequent cross-reactions. In addition, the venoms of Hymenoptera, especially bees and wasps, contain large amounts of cross-reactive carbohydrate determinants (CCDs) that are recognized by IgE. These CCDs induce cross-reactions in tests and cause multiple positive results, even though CCDs are not capable of generating an IgE-mediated response. Due to the presence of CCDs and cross-reactions due to homology, tests can commonly produce multiple positive results for the venoms of different Hymenoptera species. This property greatly complicates the identification of the species responsible for the sting reaction to plan appropriate treatment. Avoiding administering immunotherapy against species that are not responsible for the allergy reduces costs and prevents adverse effects and new sensitizations (*see* **Note 1**).

4.2 Determination of IgE Response to Hymenoptera

1. sIgE is defined as the detection and quantification of circulating levels of IgE with the ability to react to specific components of the venom of a particular Hymenoptera species in the blood of patients. In the event of a significant reaction after a sting, the presence of sIgE against antigens from the venoms of various species of Hymenoptera is studied. The detection of one or more reactivity responses allows, on the one hand, confirmation of the patient's sensitization and, on the other, in many cases, the identification of the species responsible for the reaction.
2. The determination is based on the classic sandwich between a solid phase containing the allergen under study and an anti-IgE antibody conjugated with molecules that will produce coloration or fluorescence. Anti-IgE will bind to IgE if it is present in the patient's sample, and depending on the concentration of IgE present, coloration or fluorescence will be emitted and quantified. The low concentration of circulating sIgE in blood has necessitated the development of highly sensitive systems. The latest generation offers high sensitivity through the ImmunoCAP system. The main component of the CAP system is its solid phase, which consists of an activated cellulose polymer inside a capsule (ImmunoCAP). This solid phase can bind a greater amount of antigen, and in addition, the binding conditions seem to make the system more sensitive when

antibodies against certain antigens are detected. Currently, sIgE detection systems are fully automated and standardized, making them the first choice for in vitro testing.

3. Although sIgE is less sensitive than skin tests in studies of Hymenoptera venom allergy, sIgE is positive in 5%–10% of patients with negative skin tests. Currently, sIgE tests are available for venom extracts from *A. mellifera* (bee), *Vespula* spp., *Polistes* spp., *P. dominulus, bumblebees, Vespa crabro, Vespa velutina, Dolichovespula maculata,* and *Dolichovespula arenaria*. An important aspect to consider when evaluating the results is that positive results indicate only sensitization, and the presence of compatible clinical symptoms after a sting is necessary to reach a diagnosis of Hymenoptera venom allergy. As with skin tests, >30% of patients present negative results after the sting because they are in a refractory period due to IgE consumption, and the test should be repeated after a certain period of time.
4. Two different types of allergens are used in sIgE assays that significantly affect the interpretation of results [22]. The first are crude extracts obtained from allergenic sources. These extracts contain a mixture of molecules, some of which are allergenic. Furthermore, molecules with an allergenic capacity do not behave similarly in patients; each patient may react to different molecules. Positive results identify the responsible species but not the molecule or allergen. The second type of allergen is molecular components. In this case, the test is performed against a single allergen obtained by purification or by obtaining recombinant proteins. In studies with molecular components, the responsible allergen is identified, and we know the producing species; however, negative results are obtained if the responsible component is another antigen. Molecular components are very informative about the molecules involved in the allergy and are very useful when immunotherapy treatments are prescribed. Table 1 shows the Hymenoptera venom allergens used for sIgE detection.
5. In many cases, the presence of CCDs causes multiple positive reactions to various venoms, and is common among bees and wasps, where 75% of multiple reactions are due to this effect. IgE antibodies against CCDs have high affinity and yield positive sIgE results, but their clinical relevance is very low. sIgE determinations using recombinant molecular components partially resolve this situation, since recombinant proteins do not undergo glycosylation and therefore do not cause cross-reactivity. Other options, such as determining the reactivity to MUXF3, a structure rich in the CCDs present in Bromelia, are available.

Table 1
Allergens available for the determination of sIgE against Hymenoptera venoms

Allergens	*Apis*	*Vespula*	*Polistes*
Whole extract	i1	i3	i77
Components	rApi m1 Phospholipase A2	rVes v1 Phospholipase A1	rPol d1 Phospholipase A1
	rApi m2 Hyaluronidase	rVes v5 Antigen 5	rPol d5 Antigen 5
	rApi m3 Acidic Phosphatase		
	rApi m5 Dipeptidylpeptidase IV		
	rApi m10 Icarapin		

6. Another possibility that can be considered and is common in cross-reactivity between *Vespula* and *Polistes* is that multiple positive reactions are due to structural homologues, given the structural similarity of the venom components. Double *Vespula–Polistes* sensitization is very common in southern Europe, affecting between 50% and 60% of patients. Polistes venom is free of CCDs; thus, the origin of this cross-reactivity is structural homology. In these cases, determining whether double sensitization or a cross-reaction in the sIgE test is present is difficult. Resolving this situation is very important for planning specific treatments. One approach to resolving this issue is to compare sIgE levels. When sIgE levels against molecular components of one species are much higher, more than double, than those of the other, this result strongly suggests monosensitization to the component with higher values. In many circumstances, sIgE levels are quite similar for the venoms of two or more species. In these cases, so-called CAP inhibition is very useful [23]. It consists of preincubating the problem serum with the venom extract of the species involved in multiple positive reactions. Preincubation is performed for several hours at 4 °C so that the IgE with the highest affinity will bind more strongly to the venom and its components, resulting in significant inhibition of subsequent sIgE detection compared with controls without preincubation with the venom. For venoms not responsible for allergies due to lower sIgE affinity, binding to the venom will be lower, and therefore, less inhibition will be observed (Fig. 2). When the inhibition of venom detection is greater than 75%, it is considered genuine reactivity. This modification in sIgE detection greatly resolves the identification of reactivity in cases of multiple reactivities.

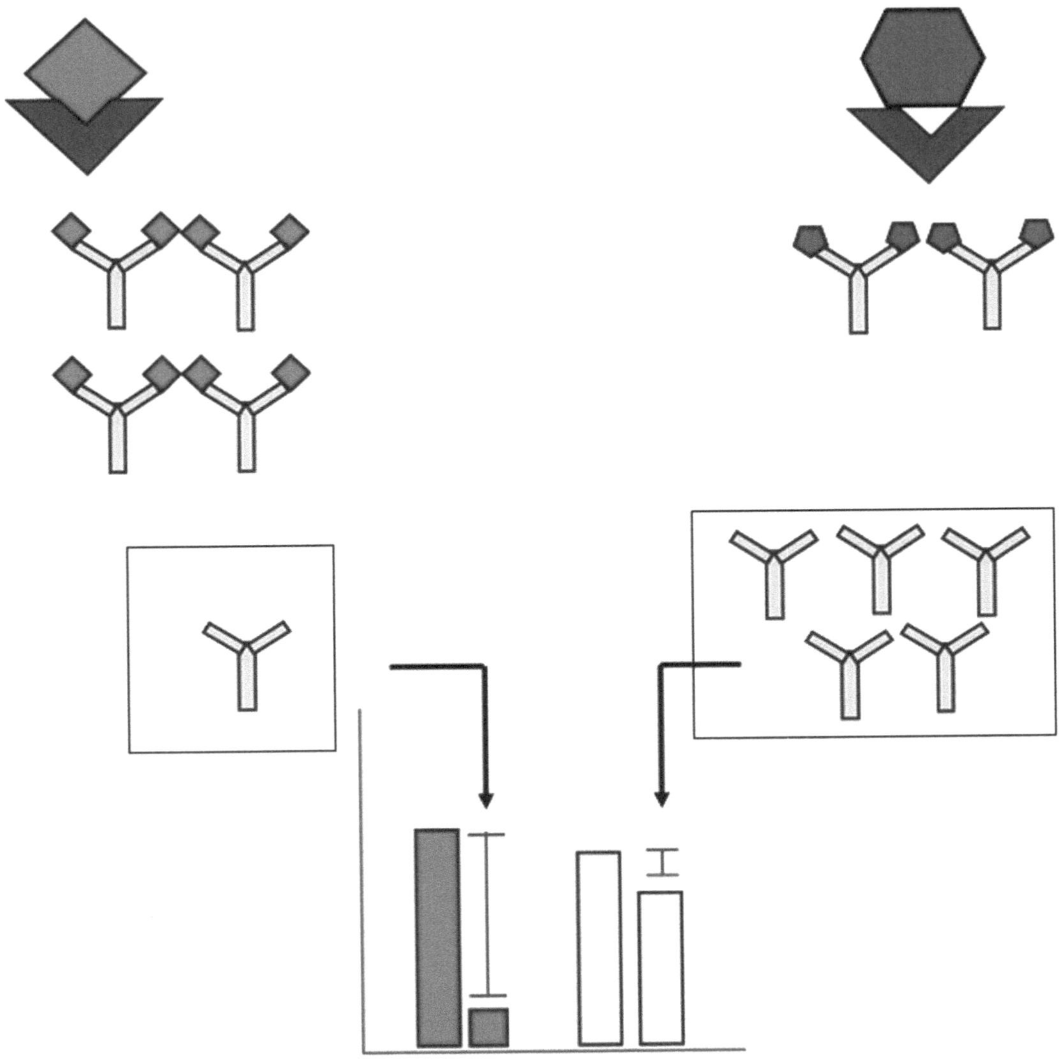

Fig. 2 Effect of preincubation with *Vespula* and *Polistes* venoms on CAP inhibition to select the sIgE responsible for the reaction. The sIgE with the highest affinity corresponding to the venom responsible for the allergy will bind more strongly to the venom components, leaving a smaller amount free. When sIgE is determined, high inhibition occurs, which differs from the result observed with the venom that causes a false-positive due to cross-reactivity

7. The sIgE levels detected are related to the amount of sIgE present in the blood and not to the sIgE produced by B lymphocytes or to that found on the surface of mast cells and basophils. This property explains the weak relationship between sIgE levels and the clinical presentation. Similar to skin tests, sIgE levels may be very low or even negative in patients with systemic mastocytosis who are highly prone to allergy to Hymenoptera venom. In patients with mastocytosis, evaluating

low sIgE detection results as potentially positive is important. The results should be evaluated starting at 0.1 KU/L instead of 0.35 KU/L.

4.3 Basophil Activation Test

1. Tests can be performed on circulating blood basophils to identify specific IgE-mediated reactions to allergens. The basophil activation test (BAT) is based on assessing the percentage of blood basophils that undergo activation and degranulation after in vitro exposure to an allergen. The BAT consists of two phases.
2. In the first phase, basophils are stimulated by incubating the blood with different concentrations of the allergen under study for a short period in a suitable medium containing Interleukin-3 (IL-3).
3. In the second phase, flow cytometry is used to identify the basophils and assess the percentage of activated cells. Fluorochrome-labeled monoclonal antibodies that recognize surface antigens characteristic of basophils are used to identify basophils.
4. Anti-IgE or anti-CCR3 antibodies are preferred. Both molecules are located on the membrane of basophils and mast cells; the former binds to the high-affinity IgE receptor (FcεRI), and the latter is a receptor for the chemokine eotaxin.
5. Activated basophils are characterized by the production of two types of markers, CD203c or CD63. CD203c is produced in basophils and mast cells at low levels that increase greatly after IgE-mediated activation. CD63 is produced on the surface of intracellular granules and not on the cell membrane. As shown in Fig. 3, after activation, degranulation occurs through the fusion of granules with the cell membrane. This process causes CD63 to be expressed on the cell membrane and can be identified by quantifying activated basophils using flow cytometry. If, as shown in Fig. 4, the percentage of activated basophils increases by 5%–15% compared with the negative control (without exposure to the antigen), the BAT is considered positive [24]. In tests with Hymenoptera venoms, namely, *Apis*, *Vespula*, or *Polistes* venoms, the BAT yields a very high sensitivity and specificity of 85%–100%, comparable to sIgE detection. As with specific IgE detection, a positive result indicates sensitization to the allergen used in the test. The BAT is more complex to perform, is not automated and is less standardized. Therefore, the BAT test is not the initial choice, but it is useful under certain situations (*see* **Note 2**).
6. Between 4% and 30% of patients who have experienced a systemic reaction after a sting report negative result for specific IgE determination. This negative result may be due, first, to

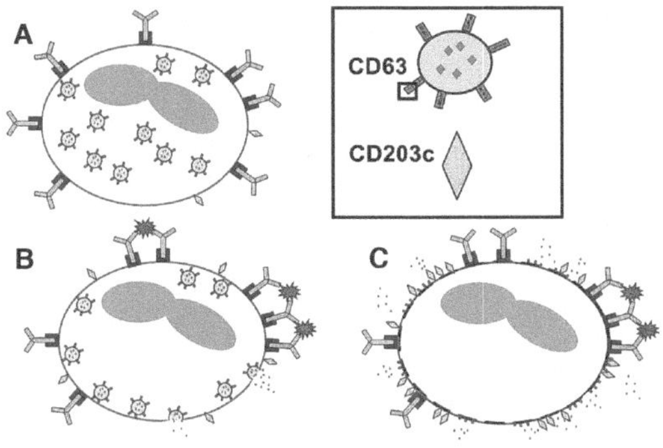

Fig. 3 Representation of CD63 and CD203c in basophils. **A**, In inactive basophils, CD63 is not expressed on the cell membrane but on the surface of cytoplasmic granules, and CD203c is expressed at low levels on the cell membrane. **B**, When the allergen interacts with IgE, the expression of CD203c increases, and the granules begin to fuse with the cell membrane to undergo exocytosis. **C**, Finally, basophils activated by allergens express large amounts of CD203c on the membrane, and through exocytosis, CD63 is expressed on the cell surface

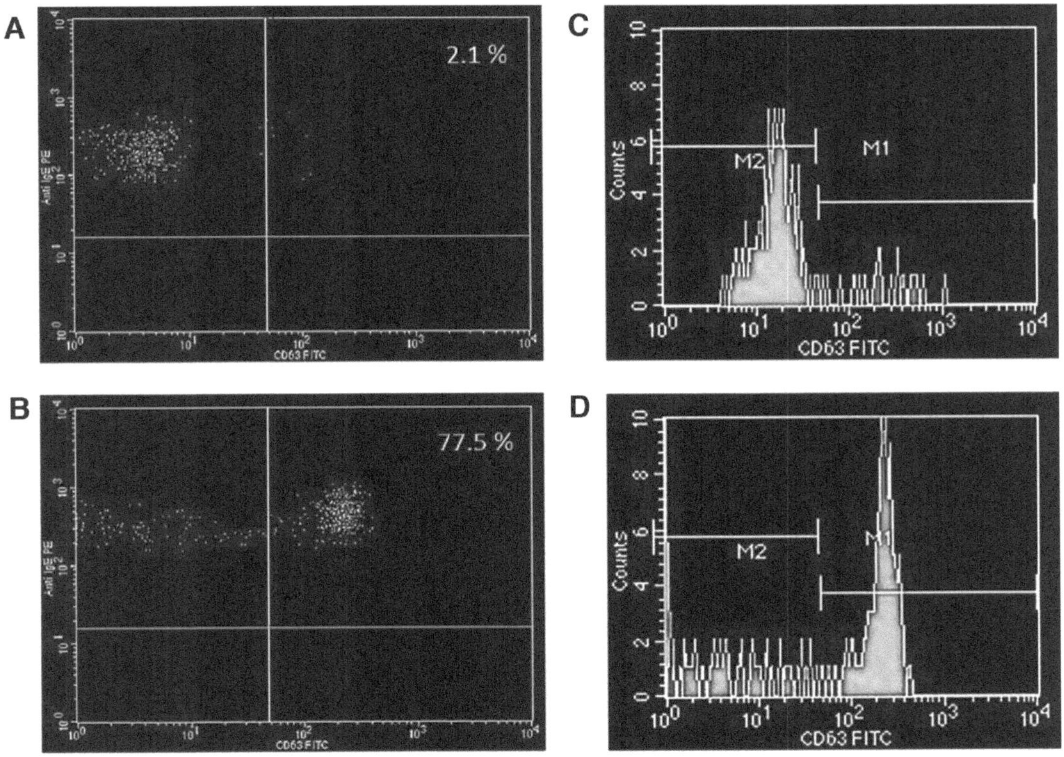

Fig. 4 Positive BAT result for *Polistes* venom. **A**, **C**, CD63 production in a negative control is shown (**A**) in an anti-IgE/CD63 dot plot and (**C**) in a cell number/CD63 histogram. **B**, **D**, Positive results obtained under the same conditions after an incubation with *Polistes* venom

the patient being in a refractory period after the sting. After a fatal reaction to a Hymenoptera sting, negative results have been observed in up to 30% of patients. Furthermore, if a long time has elapsed between the sting and the test, the patient may have developed spontaneous desensitization, and finally, up to 15% of patients with systemic mastocytosis who are highly prone to reactions to Hymenoptera venom test negative. In these situations, the characteristics of the BAT allow more than two-thirds of cases to be resolved with positive results.

7. In European studies, the rate of sensitization to bee or wasp venom varies between 25% and 44%, while only 3% of the population develops reactions after a sting. The high rate of sensitization without any reaction after stings can be partly explained by the presence of IgE against cross-reactive carbohydrate determinants (CCDs). The CCDs are epitopes present in glycoproteins that are present almost ubiquitously in plants, pollen and Hymenoptera venom. As previously mentioned, they are responsible for cross-reactivity between potential allergens, which has a negative effect on the correct diagnosis of allergies. The BAT has the ability to discriminate between specific responses and CCD-mediated cross-reactions and select sensitized patients, but it does not discriminate between clinically relevant and nonrelevant sensitizations.
8. The BAT has proven to be a very valuable test for assessing the course of immunotherapy. In studies where BAT results were compared with those of the controlled sting challenge test (repicadura) at the end of treatment, a significant and lasting decrease in the BAT result was observed in all patients who tolerated repicadura. Furthermore, in short assessment periods of 3–6 months of treatment, the BAT score significantly decreased and was maintained at that level until the end of treatment. As shown in Fig. 5, after 6 months of immunotherapy, the decrease in BAT score was more evident than that of sIgE levels.

4.4 Tryptase

1. Tryptase is a protease present inside mast cell granules and is released externally by IgE-mediated degranulation. Under normal conditions, constant levels of tryptase are maintained in the blood because of the release of a monomeric preform, and its level depends on the number of mast cells in the patient. Following an IgE-mediated reaction to a hymenopteran sting, especially if it has systemic effects, the active tetrameric form is released, increasing the blood levels. The increase in tryptase levels occurs between 15 min and 3 h after the reaction and remains stable for 24/48 h. After this period, tryptase levels return to the prereaction levels. Therefore, if an IgE-mediated

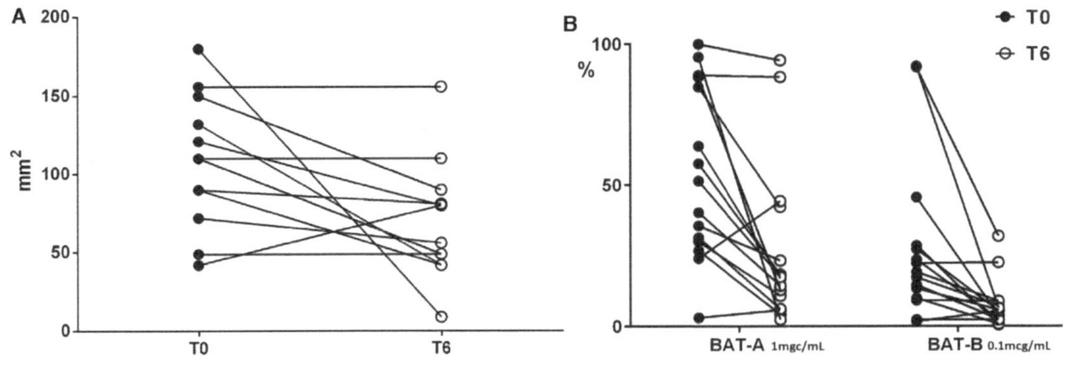

Fig. 5 The changes observed in patients with Hymenoptera venom allergy after 6 months of immunotherapy. **A**, Results of the skin tests. **B**, Results of the BAT following exposure to two venom doses

reaction is suspected, blood tryptase levels should be measured. The test should be repeated after a few days to determine the baseline blood tryptase levels. An increase of more than 20% compared with baseline values is highly suggestive of an IgE-mediated reaction [25].

2. Given the significant association between systemic mastocytosis and reactions to Hymenoptera venom, determining patients' baseline tryptase levels is essential. These levels reflect the number of mast cells in patients. Levels above 20 μg/L indicate the presence of a mast cell disorder and are a criterion for the diagnosis of mastocytosis. Patients with these characteristics are at high risk of anaphylaxis from Hymenoptera venom and those in vitro tests play important roles in the diagnosis of Hymenoptera venom allergy and provide essential information for identifying the venom responsible, which will enable appropriate treatment of the disease.

4.5 Diagnosis

1. Two fundamental pillars to the diagnosis of Hymenoptera venom allergy exist. The first is a detailed medical history, and the second is complementary tests that help us to objectively determine the patient's sensitization to the suspected Hymenoptera venom [26].
2. Regarding the clinical history, knowledge of the latency between the sting and the onset of symptoms, the characteristics and intensity of the symptoms presented, and the evolution of symptoms with the treatment administered is important. The patient should be asked about where they were when they were stung, whether they were engaged in any activity, the characteristics of the environment, whether there were wasp nests or beehives nearby, and the presence of food or water.

3. The presence of a stinger in the lesion points to a bee sting, but it is not a pathognomonic finding. For example, if the patient is stung while eating meat or seafood, the probability that it is derived from *Vespula* is 89% [27], whereas if the sting occurs in the swimming pool or while performing an activity such as cycling, it is more likely to be caused by *Polistes*.
4. Including the patient's place of residence in the medical history, as people living in rural areas are at greater risk, as well as their profession or leisure activities are important, since beekeepers, those working in agriculture and gardening, people who work outdoors and those who enjoy the countryside are at higher risk of further stings.
5. The medical history should include information on whether the patient suffers from any severe underlying disease (e.g., cardiovascular disease, respiratory disease, mastocytosis) that could aggravate the reaction after a Hymenoptera sting or whether they take any medication such as beta-blockers that could interfere with the response to adrenaline treatment.
6. Regarding the second pillar of the diagnosis, complementary tests are useful for detecting specific IgE against Hymenoptera venoms, which are necessary to establish a definitive diagnosis since, as we will see later, some patients react to stings without showing any signs of sensitization.
7. Once we have a suggestive medical history, we will perform skin tests, which are the fastest, most sensitive and least expensive methods for detecting specific IgEs against venoms [26]. We currently have four purified venom extracts corresponding to the Hymenoptera species that cause the most problems in our environment. Three extracts are from wasp venoms, *Vespula* (a mixture of *V. germanica* and *V. vulgaris)*, *P. dominulus and Polistes* spp. (a mixture of American *Polistes* species), and the fourth is purified bee venom. The tests should be performed 4–6 weeks after the reaction to avoid false negatives.
8. The recommended diagnostic test to confirm sensitization is the intradermal (ID) test, as the intraepidermal test is often negative. In a study of 24 patients in whom the usual skin tests were negative, Golden observed positive ID tests with dialyzed *Vespula* venom at concentrations of 10 μg/mL in 79% of these patients [28]. They are usually used at concentrations of 100 μg/mL for intraepidermal testing and up to 1 μg/mL (0.001–1 μg/mL) for intradermal testing because higher concentrations are irritating [26].
9. The specificity and sensitivity of skin tests vary greatly (54%–100% and 26%–89%, respectively) depending on the extract and concentration used [29]. A positive test indicates sensitization, and any sting can induce the temporary formation of

specific IgE. In other words, a positive test is not synonymous with allergy; we must always correlate it with the clinical presentation. Conversely, negative tests do not rule out the diagnosis of allergy; up to approximately one-third of people allergic to venom may have negative skin tests [30].

10. Once skin tests have been performed, we will complete the study with in vitro *tests* to diagnose as many patients as possible. Currently, the most widely used test for the detection of IgE against total venom is ImmunoCAP. Additionally, IgE against isolated allergens can be detected through molecular diagnosis or component diagnosis, which allows us to perform a more detailed and targeted study of each patient. As mentioned elsewhere in this chapter, CAP inhibition, a variant of ImmunoCAP, and the BAT can also be performed.

11. Personalized medicine has substantially reshaped the management of Hymenoptera allergy, particularly through the use of component-resolved diagnosis (CRD). This approach allows clinicians to move beyond conventional diagnostic tests and to precisely determine the specific venom molecules to which each patient is sensitized [31]. The principal allergens of honeybee (*A. mellifera)* venom are presented in Table 2 [32]. Api m 1, Api m 3, Api m 4, and Api m 10 are marker allergens indicative of primary sensitization to honeybee venom and are therefore species-specific, whereas Api m 2, Api m 5, and

Table 2
Main allergens of *A. mellifera*

Allergens	Proteins	MW (kDa)
Api ml, Api c 1, Api d 1	Phospholipase A2	16
Api m 2	Hyaluronidase	39
Api m 3	Acid phosphatase	43
Api m 4	Melittin	3
Api m 5	Dipeptidyl peptidase IV	100
Api m 6	Protease inhibitor	8
Api m 7	CUB serine protease	39
Api m 8	Carboxylesterase	70
Api m 9	Serine carboxypeptidase	60
Api m 10	Icarapin	45
Api m 11.0101	Major royal jelly protein 8	45
Api m 11.0201	Major royal jelly protein 9	46
Api m 12	Vitellogenin	200

Abbreviation: MW, molecular weight. Additional information in reference [32]

Api m 12 are considered cross-reactive allergens. Predominant sensitization to Api m 1 has been reported in 57%–97% of patients with a history of *Apis* allergy.

4.6 Evaluation of Allergen Markers

1. The allergens of *Polistes dominula* venom are shown in Table 3 [33]. Pol d 1 and Pol d 5 serve as marker allergens for identifying primary sensitization to *Polistes* venom and are species-specific. In contrast, Pol d 2 and Pol d 3 are cross-reactive allergens. Predominant sensitization to Pol d 1 has been documented in 87% of patients with a history of *Polistes* allergy.
2. The principal allergens of *Vespula vulgaris* venom are summarized in Table 4 [33]. Ves v 1 and Ves v 5 function as marker allergens for primary sensitization to *Vespula* venom and are species-specific, while Ves v 2, Ves v 3, and Ves v 12 correspond to cross-reactive allergens. Predominant sensitization to Ves v 5 has been observed in 82%–98% of patients with a history of *Vespula* allergy.
3. A diagnostic problem arises when double sensitization is observed or when sensitization is due to cross-reactive markers. The usual diagnostic techniques used in the study of Hymenoptera venom allergy show double positivity for *Vespula* and bees in up to 59% of patients [34]. Among wasps, this percentage increases, reaching 70.9%. Therefore, Sturm recently updated the diagnostic algorithm (Fig. 6) proposed by the European Academy of Allergy and Clinical Immunology in 2024 [35], indicating that, in the case of double positivity, reciprocal IgE inhibition tests (CAP inhibition), TAB or molecular diagnosis can be used to reach a definitive diagnosis (*see* **Note 3**).

4.7 Repeat Sting Test

1. Since it was first applied in the 1950s by Fackler and Loveless, the sting challenge test has been used to evaluate the efficacy of Hymenoptera venom immunotherapy (ITV). It consists of provoking a real sting by the Hymenoptera to which the patient

Table 3
Main allergens of *P. dominula*

Allergens	Proteins	MW (kDa)
Pol d 1, Poli g 1	Phospholipase A1	34
Pol d 2	Hyaluronidase	50
Pol d 3	Dipeptidyl peptidase IV	100
Pol d 4	Serine protease	33
Pol d 5, Poli g 5	Antigen 5	23-24

Abbreviation: MW, molecular weight. Additional information in reference [33]

Table 4
Main allergens of *V. vulgaris*

Allergens	Proteins	MW (kDa)
Ves v 1, Ves m 1, Ves s 1	Phospholipase	34
Ves v 2.0101, Ves m 2	A1 Hyaluronidase	45-46
Ves v 2.0201	Hyaluronidase (inactive)	45
Ves v 3	Dipeptidyl peptidase IV	23
Ves v 5, Ves f5, Ves g 5, Ves m5, Ves p 5, Ves s 5, Ves vi 5	Antigen 5	500
Ves v 6	Vitellogenin	200

Abbreviation: MW, molecular weight. Additional information in reference [33]

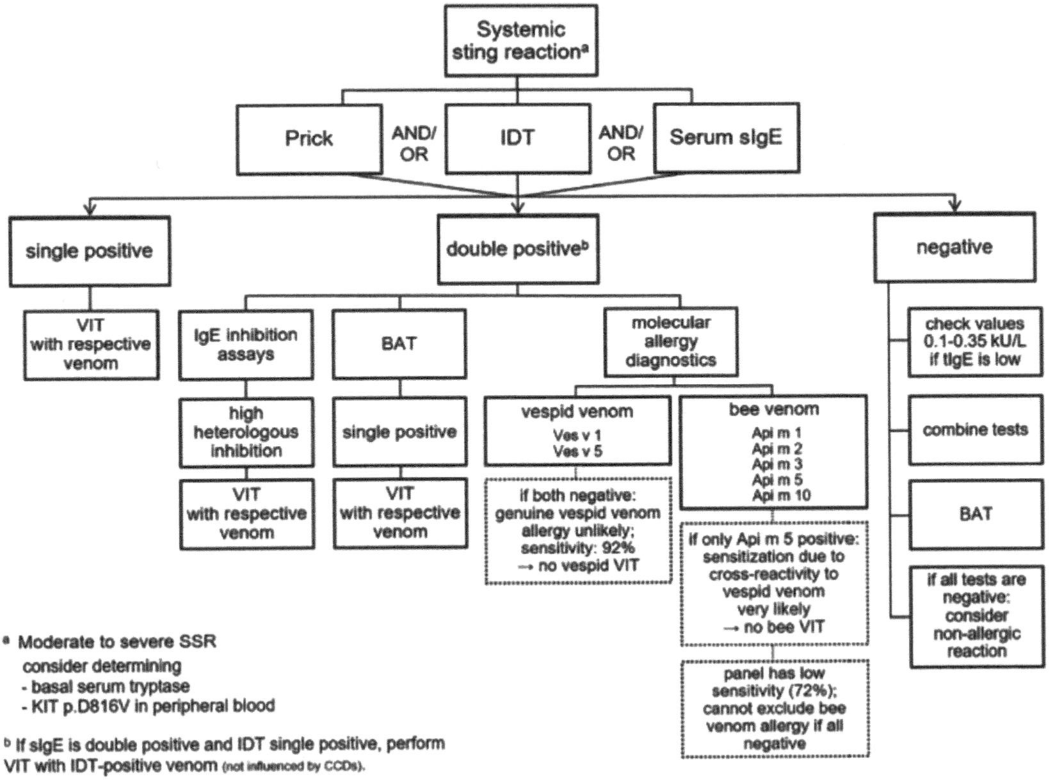

Fig. 6 Update of the diagnostic algorithm proposed by Sturm in [27]

is allergic and who is receiving specific immunotherapy treatment. Currently, it is only recommended for patients who are already receiving immunotherapy treatment. Controlled stinging could be the best method for diagnosing Hymenoptera venom allergy; however, a risk of serious, even fatal, reactions exist, which is why it is considered ethically unacceptable.

2. Given that no in vitro biomarkers are capable of predicting the effectiveness of ITV, the rechallenge test is considered the *gold standard* for assessing the risk of systemic reactions during and after specific immunotherapy and is indicated in the following situations [36]:
   (a) Patients who were allergic to Hymenoptera venom at different stages of active ITV treatment had at least 2 months of good tolerance to it.
   (b) To assist in the decision to discontinue treatment (if the patient has not had spontaneous stings) and to assess the degree of protection achieved after years of treatment.
   (c) In the event of a decrease in quality of life due to the uncertainty of the risk of reaction to a spontaneous sting.
   (d) To verify protection in beekeepers or other highly exposed professionals before reexposure.
3. The repeat sting test is contraindicated in patients who are not receiving ITV and in patients undergoing ITV treatment who have previously experienced a systemic reaction after a spontaneous sting or who have severe or uncontrolled cardiorespiratory disease (FEV1 <70%) during pregnancy and in patients with acute inflammatory disease. The test must be performed with the species responsible for the clinical reaction, which requires correct identification by a trained entomologist, biologist, or allergist.
4. CO_2 or cold is used to temporarily immobilize the responsible Hymenoptera species, during which their wings and hind legs are removed to prevent accidents. They are then stored in ventilated plastic tubes until use. The test must be performed in a hospital setting with controlled conditions and by staff trained in the management of severe allergic reactions. A complete cardiopulmonary resuscitation kit must be available.
5. Prior to the sting test, the patient must be prepared by placing a peripheral intravenous line in the arm opposite to the one used for the sting, monitoring vital signs (heart rate, blood pressure, and oxygen saturation) and taking a baseline blood sample for assessments of markers such as tryptase.
6. For the procedure, the insect is placed on the patient's forearm and held in place with tweezers in an area marked with a circle, and its abdomen is gently pressed against the skin to cause the sting (Fig. 7). The stinger of bees remains embedded in the skin and continues to inject venom, and thus it should be left in place for 30 s and then removed. Wasps must be held in place for approximately 30 s to ensure venom inoculation. Wasps can

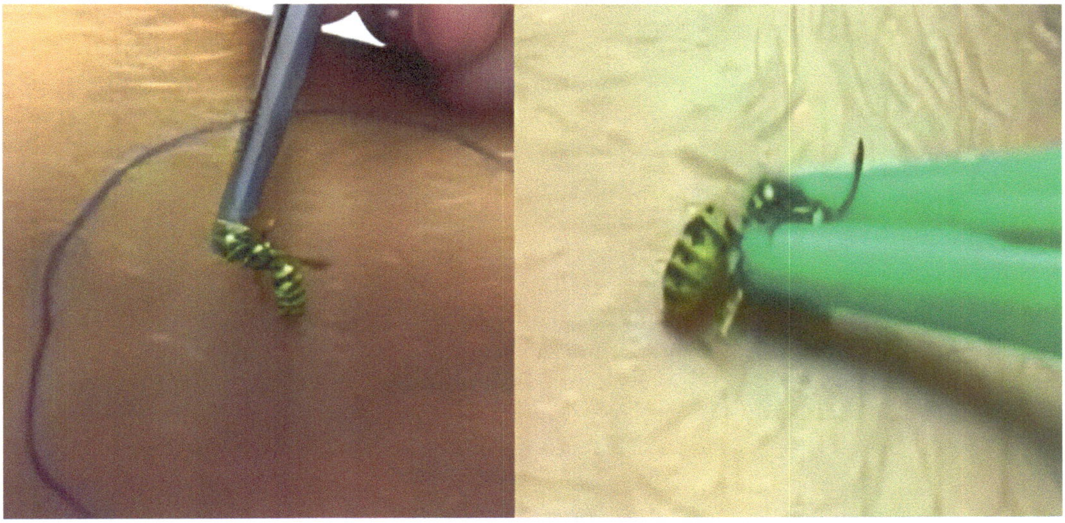

Fig. 7 Repeat sting test

sting several times and do not leave their stinger behind. After the sting, the patient must be observed for 2 h (at 15, 30, 60, and 120 min). The test is considered positive if systemic symptoms (similar to the patient's previous symptoms) appear and negative if no systemic symptoms occur. Confirmation that the venom was actually injected through an immediate local skin reaction in the area of the sting is necessary.

4.8 Treatment

1. Preventive measures: Patients who are allergic to Hymenoptera stings should be informed of the risks to which they are exposed, always carry emergency medication to treat possible reactions, and be trained in its use.
2. Acute treatment of reactions: If the reaction is localized, even if it is extensive, measuring more than 10 cm in diameter, the area should be washed thoroughly with soap and water, and a cold compress should be applied. Oral administration of antihistamines at the appropriate dose and corticosteroids at doses ranging from 0.51 mg/kg body weight can help relieve symptoms of pain and itching. If the sting has occurred in the mouth or pharynx, the patient should seek assistance at an emergency department for observation.
3. If the reaction is systemic, its severity should first be assessed. Depending on the severity, the following steps will be performed:
   (a) If only the skin is affected (urticaria or angioedema), antihistamines and corticosteroids are good choices, especially when angioedema is present. With regard to

corticosteroids, a single dose should not be administered, and a descending regimen of oral corticosteroids should be continued to reduce the rebound effect. The patient should be kept under observation to assess the progression of symptoms.

(b) If the patient has multiple organ involvement (skin, cardiovascular, digestive, and respiratory systems), anaphylaxis should be suspected. In this case, intramuscular administration of adrenaline at a dilution of 1:1000 (1 mg/mL) and a dosage of 0.01 mg/kg body weight is recommended, with a maximum of 0.3 mg for children weighing less than 30 kg, or if an autoinjector is available, a fixed dose of 150 mcg should be administered in subjects weighing less than 30 kg, 300 mcg in those weighing more than 30 kg, or 500 mcg in those weighing more than 50 kg, in addition to antihistamines and corticosteroids. If the symptoms do not subside or are severe, the dose may be administered every 5–15 min if necessary. Intravenous adrenaline should be administered only in cases of severe hypotension or cardiorespiratory arrest, which are always monitored.

(c) All patients with a severe anaphylactic reaction should be monitored (electrocardiogram, blood pressure, and pulse oximetry), and the necessary supportive measures should be added to stabilize the patient (oxygen therapy, serum therapy, bronchodilators in patients with bronchospasm, atropine, and vasopressor drugs in patients with prolonged bradycardia and glucagon in patients on beta-blocker therapy who do not respond to adrenaline).

4.9 Immunotherapy

1. Indications: Immunotherapy is administered to both adults and children with high exposure, such as in a beekeeping environment, who have suffered an anaphylactic reaction and urticaria. It is also indicated in patients with extensive local reactions with high exposure to Hymenoptera, who receive numerous treatments, are far from emergency health centers and/or have a poor quality of life because of anxiety about the possibility of stings [37].
2. Contraindications: No absolute contraindications for its administration exist, although each case must always be assessed. Notably, immunotherapy with venoms is not recommended in patients who have not experienced reactions to stings or in those with unusual or toxic reactions.
3. Extracts: The first immunotherapies for Hymenoptera were developed using extracts from the entire body of the insect. Extracts derived from Hymenoptera venom were subsequently shown to be superior in terms of efficacy and safety. Currently,

the two most widely used and commercially available venom extracts are pure aqueous venom and depot venom, which are adsorbed onto aluminum hydroxide. Both have similar efficacy profiles, although depot venom is better tolerated than aqueous venom [38].

4. Choice of venom for immunotherapy: When choosing the venom, the insect should be identified whenever possible, but in most cases, prior testing with skin tests using Hymenoptera venom extract and laboratory tests such as measurement of specific IgE to the complete extract and to the proteins identified to date, as well as basophil activation tests, are needed. If positive results are obtained for more than one genus (commonly *Polistes* and *Vespa*), immunotherapy targeting both genera should be administered. The efficacy of immunotherapy with *P. dominulus* venom is greater than that with venom from American species (*Polistes* spp.) used on the Iberian Peninsula. In patients allergic to *Vespa* or *Dolichovespula*, immunotherapy with *Vespula* venom can protect the majority of patients. In patients allergic to *Bombus* venom, the use of this venom is recommended, as protection with bee venom immunotherapy is insufficient [39].
5. Guidelines: Numerous guidelines have been developed, and although their comparison can be complicated, they are divided, in simplified terms, into ultra-rush, rush, cluster, and conventional. Treatments should always be administered in a hospital by trained personnel. In all cases, the aim is to obtain the optimal dose, which has been set at 100 μg, with some exceptions. Currently, cluster protocols are considered the most cost-effective, making them more than adequate and desirable for doctors and patients [39].
6. Treatment begins on Day 1 with a dose of 5 μg of venom, followed by 10 μg after 20 min, 20 μg after another 20 min, and another 20 μg after another 20 min. The patient will be seen again on Day 7 for the administration of 50 μg of venom, followed by another 50 μg after 30 min. Treatment will continue on Day 22 with 100 μg, thus reaching the maintenance dose. Thereafter, the maintenance dose will be administered every 4 weeks in the first year, every 6 weeks in the second year, and every 8 weeks from the third year onwards. Patients with underlying diseases such as indolent mastocytosis will require maintenance treatment for more than 5 years, for whom case administration every 12 weeks may be considered [4].
7. If two types of immunotherapy need to be administered because the patient is allergic to two different Hymenoptera species, the same regimen described above will be started, but during maintenance treatment, the doses of each immunotherapy will

be alternated every 4 weeks in the first year, every 6 weeks in the second year, and every 8 weeks from the third year onwards.

8. Evaluation of immunotherapy efficacy: Intradermal skin tests, as well as various in vitro tests such as venom-specific IgE and IgG4, are used in specialized hospitals to monitor the response to treatment, with the aim of achieving a reduction or even negativization (in 20% of patients) of these tests after 5 years of immunotherapy [40].
9. Controlled reexposure to live insects is the best test for evaluating the efficacy of immunotherapy and for improving a patient's quality of life. The risk of a systemic reaction after a negative rechallenge ranges from 10% to 20% [4].

4.10 Quality of Life

1. For most patients and their families, any allergic reaction (regardless of its severity) is a distressing experience. Given the effort required to avoid accidental exposure and the inherent uncertainty of its occurrence, living with a Hymenoptera venom allergy negatively affects quality of life, especially in patients with frequent exposure who engage in outdoor activities [41].
2. Currently, different quality-of-life questionnaires are available for assessing the social and occupational impacts of allergic diseases on patients' lives. These questionnaires assess the deterioration caused by these diseases in the daily lives of those who suffer from them. The VQLQ (Vespid Allergy Quality of Life Questionnaire), developed in 2002 by Elberink and colleagues (Oude Elberink and Dubois 2003) and the HRQLHA (Health Related Quality of Life Questionnaire for Hymenoptera Venom Allergy), developed by the Allergy Committee of the Spanish Society of Allergology and Clinical Immunology (SEAIC), adapted to Spanish customs and valid for patients allergic to *Apis, Vespula,* and *Polistes* are available [42]. This questionnaire is valid for patients aged more than 14 years and contains 14 questions with scores ranging from one to seven points, with lower scores indicating poorer quality of life.

5 Notes

1. Total IgE determination is of little value in the diagnosis of Hymenoptera venom allergy. In fact, many patients have normal total IgE levels in their blood, and this test does not identify the Hymenoptera species responsible for the allergy.

2. Despite the high sensitivity and specificity of the BAT in the diagnosis of Hymenoptera venom allergy, given its complexity and lack of standardization, it is not used as a first-line in vitro diagnostic test for specific IgE. However, the BAT is very useful for resolving discrepancies between clinical findings and skin and sIgE tests and has a high predictive value for immunotherapy desensitization.
3. A small percentage of patients suffer from systemic or intense local reactions after a Hymenoptera sting, and an adequate diagnostic study fails to demonstrate the presence of specific IgE against Hymenoptera venom in these patients either in vivo or in vitro.

Competing Interests

The authors have no conflicts of interest to declare that are relevant to the content of this chapter.

References

1. Pacheco SE, Guidos-Fogelbach G, Annesi-Maesano I et al (2021) American Academy of Allergy, Asthma & Immunology Environmental Exposures and Respiratory Health Committee. Climate change and global issues in allergy and immunology. J Allergy Clin Immunol 148(6):1366–1377. https://doi.org/10.1016/j.jaci.2021.10.011
2. Agache I, Canelo-Aybar C, Annesi-Maesano I et al (2024) The impact of outdoor pollution and extreme temperatures on asthma-related outcomes: a systematic review for the EAACIelines on environmental science for allergic diseases and asthma. Allergy 79(7):1725–1760. https://doi.org/10.1111/all.16041
3. Losol P, Sokolowska M, Hwang YK et al (2023) Epithelial barrier theory: the role of exposome, microbiome, and barrier function in allergic diseases. Allergy Asthma Immunol Res 15(6):705–724. https://doi.org/10.4168/aair.2023.15.6.705
4. Sturm GJ, Varga EM, Roberts G et al (2018) EAACI guidelines on allergen immunotherapy: Hymenoptera venom allergy. Allergy. 73(4):744–764. https://doi.org/10.1111/all.13262
5. Soriano Gomis V, Guspi Bori R (2016) Introducción y epidemiología de la alergia a veneno de himenópteros. Tratado de Alergología 2ª Edición. In: Dávila IJ, Jáuregui I, Olaguibel JM, Zubeldia JM (eds) Ergon Ediciones, pp 1231–1241. ISBN: 978-84-16270-40-8.
6. Martini M, Mauro M, Bignardi D et al (2024) Quality of life in hymenoptera venom allergy: minimal clinically important difference for the vespid quality of life questionnaire. Clin Exp Allergy 54(12):996–998. https://doi.org/10.1111/cea.14540
7. Ortega Carnicer J, Feo Brito F, Ruiz de Alegria JR, et al (1989) Muerte por picadura de avispa. Medicina Intensiva 13:303–304
8. Feás X, Vidal C, Remesar S (2022) What we know about sting-related deaths? Human fatalities caused by hornet, wasp and bee stings in Europe (1994–2016). Biology (Basel) 11(2):282. https://doi.org/10.3390/biology11020282
9. Blank S, Korošec P, Slusarenko BO et al (2025) Venom component allergen IgE measurement in the diagnosis and management of insect sting allergy. J Allergy Clin Immunol Pract 13(1):1–14. https://doi.org/10.1016/j.jaip.2024.07.023
10. Malling HJ, Hansen KS, Halken S et al (2025) Insektgiftsallergi [Hymenoptera venom allergy]. Ugeskr Laeger 187(25):V09240593. https://doi.org/10.61409/V09240593
11. Vaz-Rodrigues R, Feo Brito F, Guzmán Rodríguez R et al (2023) Allergic reactions associated with medically relevant arthropods.

Ann Med 55(2):2286531. https://doi.org/10.1080/07853890.2023.2286531
12. de la Fuente J, Urra JM, Contreras M et al (2020) A dataset for the analysis of antibody response to glycan alpha-Gal in individuals with immune-mediated disorders. F1000Res 9:1366. https://doi.org/10.12688/f1000research.27495.2
13. Nieto-Nieto A, Tejedor-Alonso MA, Farias-Aquino E, Moro-Moro M, Rosado Ingelmo A, Gonzalez-Moreno A, Gil de Miguel A (2017) Clinical profile of patients with severe anaphylaxis hospitalized in the Spanish hospital system: 1997–2011. J Investig Allergol Clin Immunol 27(2):111–126. https://doi.org/10.18176/jiaci.0146
14. Bilò MB, Pravettoni V, Bignardi D et al (2019) Hymenoptera venom allergy: management of children and adults in clinical practice. J Investig Allergol Clin Immunol 29(3):180–205. https://doi.org/10.18176/jiaci.0310
15. Marone G, Genovese A, Varricchi G et al (2014) Human heart as a shock organ in anaphylaxis. Allergo J Int 23(2):60–66. https://doi.org/10.1007/s40629-014-0007-3
16. de la Fuente J, Kocan KM (2003) Advances in the identification and characterization of protective antigens for development of recombinant vaccines against tick infestations. Expert Rev Vaccines 2:583–593
17. Gutiérrez Fernández D, Sánchez Morillas L, Moreno Ancillo A (2016) Alergia a otros insectos y artrópodos. Tratado de Alergología 2ª Edición. Dávila IJ, Jáuregui I, Olaguibel JM, Zubeldia JM. Ergon Ediciones, pp 1267–1277. ISBN: 978-84-16270-40-8
18. Germán-Sánchez A, Alonso-Llamazares A, García-González F, Matala-Ahmed B, Melgar-Reyes CS, Antepara-Ercoreca I (2023) Diagnostic validity of specific immunoglobulin E levels to alpha-gal in alpha-gal syndrome: a cross-sectional analysis. Allergy Asthma Clin Immunol 19(1):102. https://doi.org/10.1186/s13223-023-00856-6
19. Cabezas-Cruz A, Hodžić A, Román-Carrasco P, Mateos-Hernández L, Duscher GG, Sinha DK, Hemmer W, Swoboda I, Estrada-Peña A, de la Fuente J (2019) Environmental and molecular drivers of the α-Gal syndrome. Front Immunol 10:1210. https://doi.org/10.3389/fimmu.2019.01210
20. Olivera-Ardid S, Bello-Gil D, Tuzikov A, Araujo RN, Ferrero-Alves Y, García Figueroa BE, Labrador-Horrillo M, García-Pérez AL, Bovin N, Mañez R (2022) Poly-L- lysine-based αGal-glycoconjugates for treating anti-α Gal IgE-mediated diseases. Front Immunol 13:873019. https://doi.org/10.3389/fimmu.2022.873019
21. Ünal D, Eyice-Karabacak D, Kutlu A, Demir S, Tüzer C, Arslan AF, Işık SR, Gelincik A (2023) Oral immunotherapy in alpha-gal red meat allergy: could specific IgE be a potential biomarker in monitoring management?. Allergy 78(12):3241–3251. https://doi.org/10.1111/all.15840
22. Jakob T, Müller U, Helbling A et al (2017) Component resolved diagnostics for hymenoptera venom allergy. Curr Opin Allergy Clin Immunol 17(5):363–372. https://doi.org/10.1097/ACI.0000000000000390
23. Cabrera CM, Palacios-Cañas A, Joyanes-Romo JB et al (2022) Basophil activation test as alternative method to CAP-inhibition in patients with double sensitization to vespid venoms. Mol Immunol 149:59–65. https://doi.org/j.molimm.2022.06.002
24. Urra JM, Pérez-Lucendo I, Extremera A et al (2020) The method for selecting basophils might be determinant in the basophil activation test in patients with mastocytosis. J Investig Allergol Clin Immunol 30(1):65–67. https://doi.org/10.18176/jiaci.0442
25. Bonadonna P, Scaffidi L, Boni E (2019) Tryptase values in anaphylaxis and insect allergy. Curr Opin Allergy Clin Immunol 19(5):462–467. https://doi.org/10.1097/ACI.0000000000000569
26. Vega Castro A, Antolín Amérigo D, Ruiz León B (2016) Diagnóstico y tratamiento de la alergia al veneno de los himenópteros. Tratado de Alergología 2ª Ed. Dávila IJ, Jáuregui I, Olaguibel JM, Zubeldia JM. Ergon Ediciones, pp 1249–1265
27. Vega-Castro A, Castro L, Carballada F et al (2023) Hymenoptera allergy diagnosis through their presence on human food. Toxins (Basel) 15(12):680. https://doi.org/10.3390/toxins15120680
28. Golden DB, Kelly D, Hamilton RG, Wang N, et al (2009) Dialyzed venom skin tests for identifying yellow jacket-allergic patients not detected using standard venom. Ann Allergy Asthma Immunol 102(1):47–50. https://doi.org/10.1016/S1081-1206(10)60107-X
29. Jeep S, Reiprich G, Kunkel G (1992) Yellow jacket allergy. Comparison of skin prick tests and intradermal tests with three different yellow jacket venom extracts. Allergy 47(1):35–40. https://doi.org/10.1111/j.1398-9995.1992.tb02247.x
30. Golden DB, Kagey-Sobotka A, Norman PS et al (2001) Insect sting allergy with negative venom skin test responses. J Allergy Clin Immunol

107(5):897–901. https://doi.org/10.1067/mai.2001.114706
31. Blank S, Bilò MB, Ollert M (2018) Component-resolved diagnostics to direct in venom immunotherapy: Important steps towards precision medicine. Clin Exp Allergy 48(4):354–364. https://doi.org/10.1111/cea.13090
32. Blank S, Bilò MB, Jakob T et al (2022) Bee venom allergy. Molecular Allergology User's Guide 2.0. Zurich: European Academy of Allergy and Clinical Immunology (EAACI), pp 407–418
33. Bilò MB, Jakob T, Ollert M et al (2022) Vespid venom allergy. Molecular Allergology User's Guide 2.0. Zurich: European Academy of Allergy and Clinical Immunology (EAACI), pp 419–434
34. Müller UR, Johansen N, Petersen AB, Fromberg-Nielsen J, Haeberli G (2009) Hymenoptera venom allergy: analysis of double positivity to honeybee and *Vespula venom* by estimation of IgE antibodies to species-specific major allergens Api m1 and Ves v5. Allergy 64(4):543–548. https://doi.org/10.1111/j.1398-9995.2008.01794.x
35. Sturm GJ, Arzt-Gradwohl L (2024) An algorithm for the diagnosis and treatment of Hymenoptera venom. Allergy: 2024 update. Allergy 79(8):2298–2301
36. Ruiz-León B, Martínez San Ireneo M, de la Roca F, Arenas L, Alfaya Arias T, Cordobés C, Marqués L, Vega A, Moreno-Aguilar C (2022) The lights and the shadows of controlled sting challenge with Hymenoptera. J Investig Allergol Clin Immunol 32(5):357–366. https://doi.org/10.18176/jiaci.0838
37. Ruëff F, Wolf H, Schnitker J, Ring J, Przybilla B (2004) Specific immunotherapy in honeybee venom allergy: a comparative study using aqueous and aluminium hydroxide adsorbed preparations. Allergy. 59(6):589–595. https://doi.org/10.1111/j.1398-9995.2004.00505.x
38. Alfaya Arias T, Soriano Gómis V, Soto Mera T et al (2017) Hymenoptera allergy committee of the SEAIC. Key issues in Hymenoptera venom allergy: an update. J Investig Allergol Clin Immunol 27(1):19–31. https://doi.org/10.18176/jiaci.0123
39. Sánchez-Machín I, Moreno C, González R, Iglesias-Souto J, Pérez E, Matheu V (2010) Safety of a 2-visit cluster schedule of venom immunotherapy in outpatients at risk of life-threatening anaphylaxis. J Investig Allergol Clin Immunol 20(1):91–92
40. Ruiz León B, Alfaya Arias T, Alonso Llamazares A, Carballada Gónzalez FJ, Cordobés Durán C, De La Roca Pinzón F, et al (2021). Alergia a Himenopteros. SEAIC. https://www.seaic.org/wp-content/uploads/2021/05/Alergia_Himenopteros_SEAIC_20_04.pdf
41. Carballada F, Boquete M, Núñez R, Lombardero M, de la Torre F (2010) Follow-up of venom immunotherapy (VIT) based on conventional techniques and monitoring of immunoglobulin E to individual venom allergens. J Investig Allergol Clin Immunol 20(6):506–513
42. Alfaya T, Vega A, Domínguez-Noche C, Ruiz B, Marqués L, Sánchez-Morillas L (2015) Longitudinal validation of the Spanish version of the health-related quality of life questionnaire for Hymenoptera Venom Allergy (Hrqlha). J Investig Allergol Clin Immunol 25(6):426–430

Check for updates

Chapter 22

Genotyping Protocol for Identifying Genetic Risk Markers Associated with Alpha-Gal Syndrome

Jasmina Grujić and Marko Lilić

Abstract

Alpha-gal syndrome (AGS) is an IgE-mediated allergy to galactose-α-1,3-galactose (α-Gal), typically triggered by tick bites. Genetic factors, including human leucocyte antigen (HLA) class II polymorphisms and selected single nucleotide polymorphisms (SNPs) in immune and glycan-related genes, are hypothesized to modulate susceptibility and clinical expression. Here we provide a step-by-step protocol to genotype risk markers from ethylenediaminetetetraacetic acid (EDTA) whole blood sample, combining (i) manual DNA extraction (Qiagen), (ii) SNP genotyping by a real-time/quantitative PCR (qPCR) method (FluoQube system), and (iii) HLA class I/II genotyping by sequence-specific oligonucleotide (SSO method using a Luminex® xMAP system). The protocol details reagent preparation, thermal cycling and hybridization conditions, data analysis, quality control (QC), and reporting. We also include troubleshooting notes, acceptance criteria, and guidance for clinical interpretation (e.g., integrating genotypes with α-Gal sIgE and clinical phenotype). This protocol enables reproducible implementation of genetic-risk workflows for AGS in molecular diagnostics or translational-research settings.

Keywords Alpha-gal syndrome, Tick-bite allergy, HLA class II, SNP genotyping, Luminex, SSO, Real-time PCR, qPCR, Immunogenetics, Translational protocol

1 Introduction

Alpha-gal syndrome (AGS) is an IgE-mediated delayed hypersensitivity to galactose-α-1,3-galactose (α-Gal), typically triggered by tick bites and characterized by allergic reactions to mammalian-derived products [1–3]. Although environmental exposure is essential for sensitization, marked interindividual variability in clinical presentation suggests that host genetic factors may contribute to disease susceptibility and expression. Human leukocyte antigen (HLA) class II polymorphisms, particularly within the HLA-DRB1 and HLA-DQB1 loci, are hypothesized to influence antigen presentation and downstream Th2 polarization, thereby modulating IgE responses [4, 5]. In addition, functional single nucleotide polymorphisms (SNPs) in genes involved in glycan biology and IgE signaling

José de la Fuente and Alejandro Cabezas-Cruz (eds.), *Allergic Disease Models*, Methods in Molecular Biology, vol. 3092, https://doi.org/10.1007/978-1-0716-5635-8_22,

pathways (e.g., FUT2, ABO, IL4R, FCER1A) may further shape immune regulation and sensitization thresholds [4].

Targeted SNP genotyping using hydrolysis probe-based real-time PCR (qPCR) provides a rapid and robust method for analyzing predefined genetic variants relevant to immune response modulation [6–11]. In parallel, sequence-specific oligonucleotide (SSO)-based HLA typing enables multiplex detection of HLA alleles through hybridization of locus-specific amplicons to microsphere-coupled probes, with allele assignment supported by curated reference databases such as IPD-IMGT/HLA [12–17]. These established molecular approaches allow reproducible genotyping in both diagnostic and research laboratories.

This chapter describes a practical laboratory workflow for genotyping selected AGS-associated genetic risk markers from ethylenediaminetetraacetic acid (EDTA) whole blood samples (Fig. 1). The protocol integrates: (i) manual genomic DNA extraction, (ii) SNP genotyping by real-time PCR with allelic discrimination, and (iii) HLA class I/II typing using SSO hybridization on a Luminex® xMAP platform. The workflow is suitable for implementation in molecular diagnostics and translational research settings and can be applied in cohort or case–control studies investigating genetic susceptibility to AGS, particularly when integrating genotype data with α-Gal-specific IgE levels and clinical phenotype.

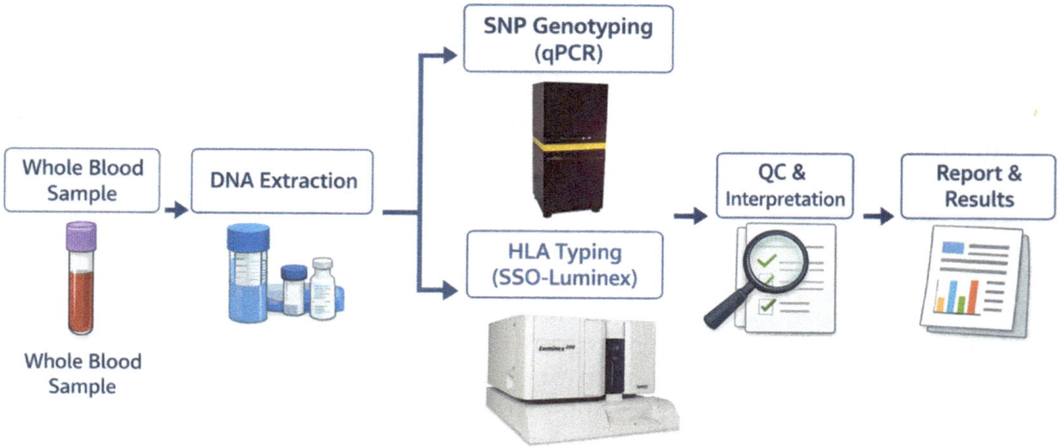

Fig. 1 Genetic testing workflow from blood collection to final result interpretation. Schematic overview of the laboratory workflow starting from peripheral EDTA whole blood collection and genomic DNA extraction. Following DNA isolation, the workflow branches into two parallel genotyping approaches: (i) SNP genotyping by real-time PCR (qPCR) using allele-specific hydrolysis probes for selected immune- and glycan-related variants, and (ii) HLA class I/II typing by sequence-specific oligonucleotide (SSO) hybridization on a Luminex® xMAP platform. After data acquisition, results undergo quality control (QC), genotype validation, and allele assignment using dedicated analysis software. The final step includes integrated interpretation and reporting of genetic findings in the context of clinical and serological data relevant to alpha-Gal syndrome (AGS)

2 Materials

2.1 Samples and Consumables

1. Fresh venous whole blood sample collected in ethylenediaminetetetraacetic acid (EDTA) tubes (3–5 mL).
2. DNase/RNase-free microtubes (1.5 and 0.2 mL PCR tubes or 96-well plates).
3. Filtered pipette tips and calibrated micropipettes, ranges 0.5–20, 10–100, and 100–1000 μL.

2.2 Manual DNA Extraction

The protocol used in this chapter describes a manual total DNA extraction (genomic DNA, but also mitochondrial and viral, if present) from a whole blood EDTA sample using a microcentrifuge.

1. QIAamp® DNA Blood Mini Kit (Qiagen) for 50 or 250 preparations.
2. Ethanol (96%–100%) (*see* **Note 1**).
3. 1.5-mL microcentrifuge tubes.
4. Pipet tips with aerosol barrier.
5. Microcentrifuge (with rotor for 1.5 or 2-mL tubes).
6. Vortex mixer.
7. Water bath or heating block at 56 °C.
8. Phosphate-buffered saline (PBS) may be required for some samples.
9. Tube racks.
10. Nuclease-free water or kit-contained elution buffer (AE buffer) for elution (50–100 μL).

2.3 Real-Time PCR (SNP Genotyping—qPCR Instrument)

1. FluoQube qPCR instrument with allelic-discrimination software (qPCRsoft).
2. White optical 96-well qPCR plates, validated for use with the FluoQube instrument.
3. Optical adhesive film/plate sealers.
4. Microcentrifuge tubes.
5. Plate spinner/mini centrifuge.
6. Probe-based master mix [e.g., TaqMan Genotyping Master Mix (2×), or ampther master mix compatible with TaqMan SNP genotyping).
7. Custom/validated 20× TaqMan SNP Genotyping assay (examples listed in Table 1).
8. Nuclease-free water.
9. Genomic DNA extracted from whole blood.

Table 1
Candidate risk markers for AGS studies

Gene/ locus	dbSNP ID (Marker)	Nucleotide/ amino-acid variant	Functional note	Rationale/ reported relevance
FUT2	rs601338	c.428 G>A (Trp143Ter)	Defines *non-secretor* status; loss of α -1,2-fucosyltransferase activity and absence of ABH antigens in mucosa	Reduced oral tolerance and altered exposure to α-Gal epitopes
ABO	rs8176746	c.803 C>A (Leu268Met)	Determines B antigen expression on glycoconjugates	Structural similarity of B antigen to α-Gal glycan may influence sensitization
IL4R	rs1801275	c.1727 A>G (Gln576Arg)	Alters IL-4 receptor α signaling; enhances STAT6 activation	Promotes Th2 polarization and IgE class switching
FCER1A	rs2252226	T>C (c.-315T>C in promoter)	Modulates FcεRIα transcription and IgE receptor density	Affects baseline and allergen-specific IgE levels
FUT3	rs812936	c.299 T>C (Leu100Ser)	Modifies Lewis antigen synthesis	Impacts glycan composition of mucosal epithelia relevant to α-Gal exposure
TLR2	rs5743708	c.2258 G>A (Arg753Gln)	Decreases TLR2 signaling capacity	Alters innate recognition of tick salivary components
CD14	rs2569190	C>T (c.-159C>T promoter)	Regulates CD14 expression and Th1/Th2 balance	Linked with IgE production and allergy susceptibility

2.4 HLA Genotyping (HLA-SSO; Microsphere-Based)

1. Luminex® xMAP analyzer—Luminex 200 or FLEXMAP 3D (DiaSorin/Luminex Corporation) with xPONENT software.
2. Sheath fluid plus—system liquid for Luminex 200 or FLEXMAP 3D instruments.
3. Validated thermal cycler by the assay manufacturer or by the user.
4. Water bath or heating block capable to reach 56°C.
5. Ultrasonic cleaner.

6. LIFECODES® HLA SSO Typing Kits (e.g., Immucor GTI Diagnostics), listed in Sect 3.3.
7. Streptavidin–PE conjugate.
8. Nuclease-free water.
9. Corning® Thermowell® 96-well Polycarbonate PCR Microplate, labeled as Costar 96-well plates later in this chapter.
10. Adequate plate sealers (e.g., MicroAmp™ Clear Adhesive Film from Applied Biosystems/Thermo Fisher Scientific, cat. no. 4306311).
11. HLA analysis software: MATCH IT® DNA for LIFECODES HLA-SSO kits
12. Alternative assays for HLA typing: LABType® SSO (One Lambda/Thermo Fisher).
13. Alternative HLA analysis software: HLA Fusion® for LABType SSO kits.

2.5 Quality Control and Ancillary Reagents

1. For optimal results, DNA concentration should be measured after the extraction by any appropriate method accessible in the laboratory.
2. Positive and negative controls should be used in HLA-SSO testing per local and international guidelines.
3. 70% ethanol.
4. 20% bleach.

3 Methods

All methods presented in this section are based on the laboratory standard and operating procedures (SOP) developed in accordance with the manufacturers' instructions.

Perform all steps under nuclease-free conditions.

3.1 Manual DNA Extraction from Whole Blood (QIAamp® Column-Based)

1. Equilibrate samples to room temperature. Preheat a water bath or heating block to 56°C.
2. Prepare QIAGEN Protease according to the manufacturer's instructions by adding 1.2 mL (50-prep kit) or 5.5 mL (250-prep kit) protease solvent to the lyophilized vial (*see* **Note 2**).
3. Ensure that Buffer AL is thoroughly mixed before use. Add ethanol (96%–100%) to Buffers AW1 and AW2 prior to first use as indicated by the manufacturer (*see* **Notes 3–5**).
4. Use 200 μL EDTA whole blood per extraction. If the sample volume is <200 μL, adjust to 200 μL with PBS. For volumes >200 μL, proportionally increase the volumes of protease,

Buffer AL, and ethanol. Wash buffer volumes remain unchanged (*see* **Notes 7–10**).

5. Pipette 20 μL QIAGEN Protease into the bottom of a 1.5-mL microcentrifuge tube.
6. Add 200 μL blood sample and mix briefly.
7. Add 200 μL Buffer AL and mix thoroughly by pulse-vortexing for 15 s to ensure complete lysis (*see* **Notes 11 and 12**).
8. Incubate at 56°C for 10 min (*see* **Note 13**). Briefly centrifuge to remove drops from the tube lid.
9. Add 200 μL ethanol (96%–100%) and mix thoroughly by pulse-vortexing for 15 s. Briefly centrifuge again to remove droplets (*see* **Note 14**).
10. Carefully apply the mixture to a QIAamp Mini spin column placed in a 2-mL collection tube, avoiding contact with the rim.
11. Close the cap and centrifuge at 6000 ×*g* for 1 min. Discard the filtrate and place the column into a clean collection tube (*see* **Notes 15–17**).
12. Add 500 μL Buffer AW1 without wetting the rim. Centrifuge at 6000 ×*g* for 1 min. Discard the filtrate and transfer the column to a new collection tube (*see* **Notes 15, 18, and 19**).
13. Add 500 μL Buffer AW2. Centrifuge at full speed (20,000 ×*g*) for 3 min.
14. Transfer the column to a new collection tube and centrifuge at full speed for 1 min to eliminate residual wash buffer (*see* **Note 20**).
15. Place the column in a clean 1.5-mL microcentrifuge tube. Add 200 μL Buffer AE or nuclease-free water directly onto the membrane.
16. Incubate at room temperature for at least 1 min (*see* **Note 21**).
17. Centrifuge at 6000 ×*g* for 1 min to elute DNA.
18. Measure DNA concentration if required and store DNA at −30 °C to −15 °C for long-term storage (*see* **Notes 22–25**).
19. During column handling, avoid wetting the rim of the column, change pipette tips between liquid transfers, and keep columns closed during centrifugation to minimize aerosol formation (*see* **Note 6**).

3.2 SNP Genotyping by qPCR (FluoQube)

The protocol in this section is illustrated by the TaqMan SNP Genotyping Assay (20×) for FUT2 rs601338 with detected SNPs A/G.

All the steps can be appropriately applied to any TaqMan Assay (20×) for genes and SNPs listed in Table 1.

1. Make sure there is 1–20 ng of extracted genomic DNA per reaction. In this protocol we will use 5 ng DNA, by pipetting 5 μL of pre-diluted DNA to the concentration 1 ng/μL.
2. Begin a PCR run setup in the qPCRsoft software by creating an empty project.
3. Within Settings/Thermocycler project window set up a standard (non-fast) two-step PCR program (cycling conditions) for TaqMan SNP genotyping (*see* **Note 26**).
4. 95 °C 10 min (initial denaturation and enzyme activation).
5. 40 cycles: 95 °C 15 s; 60 °C 60 s (data acquisition in this step).
6. Select two dyes corresponding to the assay in the Settings/Scan project window. For FUT2 rs601338 assay these are: FAM (allele G) and JOE (allele A). Make sure to set the measurement for both dyes (*see* **Note 27**).
7. Steps 3 and 4 can be saved as a template which can be opened instead in the future runs.
8. For a manual sample selection, select the Define scan region manually option on the Settings/Scan tab. A graphical representation of the sample block is opened.
9. Use the mouse to select the columns. Active columns are highlighted blue.
10. If necessary, edits to the sample entries can be made on the Samples tab after pressing the Edit layout button.
11. Thaw, mix, and quickly spin all the reagents necessary for the working master mix (reagents listed in Table 2 minus DNA). The volume of the working master mix should be 15 μL (*see* **Note 28**).
12. Make sure to estimate 5%–10% higher volume than necessary for the number of samples to be run to compensate for the pipetting volume loss.

Table 2
SNP genotpying PCR reaction setup (working mastermix + sample DNA)

Component	Volume per sample(per PCR well / single SNP)
2× Genotyping Master Mix	10 μL
20× TaqMan SNP Genotyping Assay Here: for FUT2 rs601338	1 μL
Genomic DNA	5 μL, concentration 1 ng/μL (total 5 ng)
Nuclease-free water	4 μL to the final volume 20 μL

13. Dispense 15 μL of working master mix to the wells of the PCR plate.
14. Add 5 μL DNA to the sample wells. If control samples are run, add these to their designated wells.
15. Add 5 μL water or elution buffer to the NTC wells instead of DNA.
16. Seal the PCR plate and briefly spin (*see* **Note 29**).
17. All functions required for starting and monitoring a qPCR run are combined on the Monitoring project tab in the qPCRsoft.
18. The toolbar displays the Play icons for starting the PCR protocol (qPCR run).
19. After the qPCR run is finished save the project (*.rtp file).
20. Open the completed run/project (File/Open project) or continue from previous step if analysis is to be done immediately after the qPCR cycling finishes.
21. Go to Analysis/Genotyping tab and click on the Add genotyping (new evaluation).
22. Enter an evaluation name (e.g., "FUT2_rs601338_*date*").
23. qPCRsoft then shows:
24. Parameter settings (for defining threshold and fluorescent dye assignment).
25. Plots (fluorescence curves, scatter plot or bar graph).
26. Results table.
27. Assign the two reporter channels (dyes) correctly to the two genotypes (alleles). In the Genotyping parameters set genotype1 = dye for allele 1 (e.g., FAM) and genotype2 = dye for allele 2 (e.g., JOE/HEX/VIC) (*see* **Note 30**). For the FUT2 rs601338 (A/G) assay on the FluoQube instrument: G allele → FAM, A allele → JOE.
28. As a quick quality control (QC) step, view and check fluorescence/amplification curves for both dyes (*see* **Note 31**). Confirm NTC stays at baseline (no true amplification). Confirm samples show clear signal separation in the two channels.
29. For genotype calls use the Scatter (allele discrimination) plot tab (Fig. 2, *see* **Note 32**). Allele 1 is on one axis, allele 2 on the other.
30. There are four clusters formed—verify (*see* **Notes 33–35**):
    - high Ct (low fluorescence) FAM, low Ct (high fluorescence) JOE—homozygous allele 1 (e.g., AA).

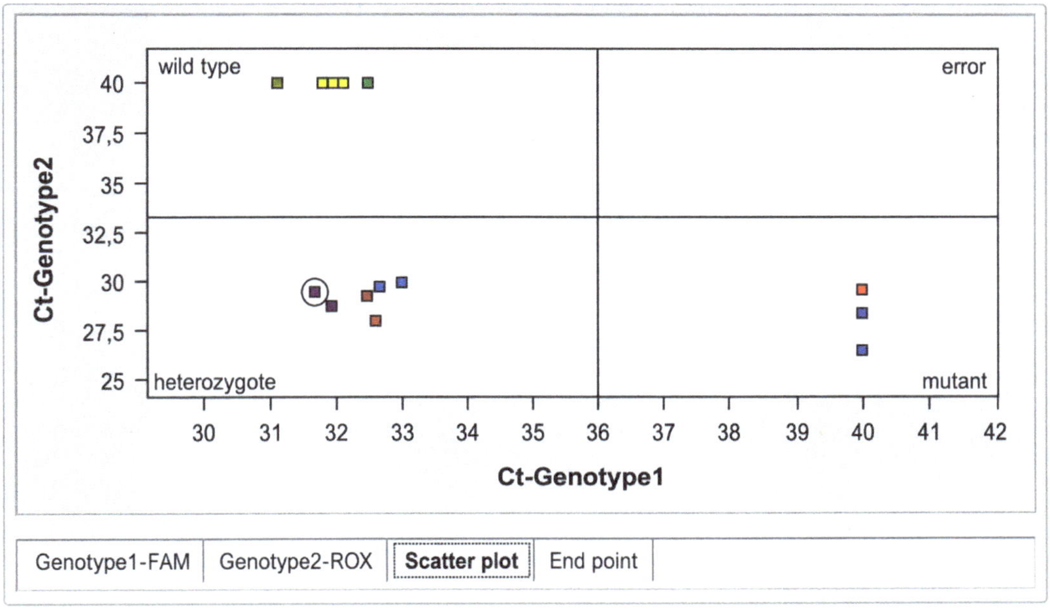

Fig. 2 Scatter plot tab view of the SNP genotyping results in qPCRsoft software. Representative allelic discrimination scatter plot generated after real-time PCR amplification using a TaqMan SNP Genotyping Assay. The *x*-axis (Ct-Genotype1) represents the threshold cycle (Ct) value associated with reporter dye assigned to genotype1, while the *y*-axis (Ct-Genotype2) represents the Ct value for the reporter dye assigned to genotype2. Each dot corresponds to an individual sample. Samples clustering in the upper left quadrant correspond to wild-type homozygotes (signal predominantly in Genotype2 channel), samples in the lower right quadrant correspond to mutant homozygotes (signal predominantly in Genotype1 channel), and samples located in the lower left quadrant represent heterozygous genotypes with signal detected in both channels. The upper right quadrant represents potential outliers or assay errors. Quadrant separation lines define genotype calling thresholds used for automated or manual genotype assignment

- high Ct (low fluorescence) JOE, low Ct (high fluorescence) FAM—homozygous allele 2 (e.g., GG).
- intermediate/low Ct (intermediate/high fluorescence) in both (diagonal cluster)—heterozygous (e.g., AG).
- near baseline (no amplification): NTC.

31. Confirm cluster identity with a known-genotype control at least the first time the genotyping assay is performed.
32. Interpret the genotypes obtained according to the instructions for use of the utilized SNP genotyping assay.
33. Assign genotype calls in the results table manually or automatically (*see* **Note 36**).
34. Export the results table (and optionally the scatter plot image) for documentation (*see* **Note 37**).

35. For the FUT2 rs601338 (A/G) assay (Fig. 3): AA = homozygous allele 1; AG = heterozygous; GG = homozygous allele 2.

3.3 HLA Class I/II Genotyping by HLA-SSO

The procedure described below is appropriate for performing the HLA-SSO typing by using the following kits listed in Table 3.

1. Make sure that the respective lot-specific templates for the used HLA kits are uploaded to both xPONENT software (DiaSorin/Luminex Corporation) for the Luminex instrument and MATCH IT! DNA Software (Immucor GTI

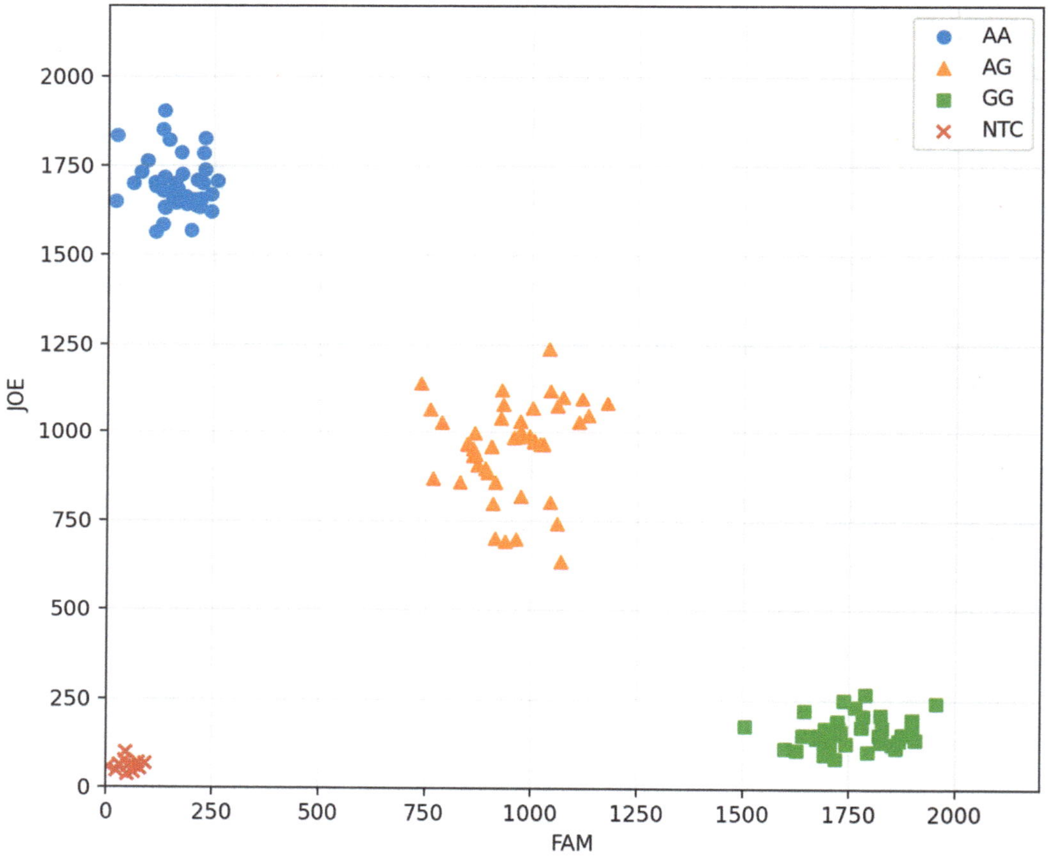

Fig. 3 Schematic representation of the scatter plot clusters in the FUT2 genotyping assay. Illustrative scatter plot showing endpoint fluorescence distribution obtained from the TaqMan SNP genotyping assay for FUT2 rs601338 (A/G). The *x*-axis represents fluorescence intensity of the FAM reporter dye (allele G) and the *y*-axis represents fluorescence intensity of the JOE reporter dye (allele A). Each symbol corresponds to an individual sample. Samples with high JOE and low FAM fluorescence cluster in the upper left region and correspond to homozygous AA genotypes. Samples with high FAM and low JOE fluorescence cluster in the lower right region and correspond to homozygous GG genotypes. Samples with intermediate fluorescence intensity in both channels form a central diagonal cluster representing heterozygous AG genotypes. No-template controls (NTC) are located near the origin, indicating absence of amplification. This schematic facilitates interpretation of genotype assignment based on fluorescence signal separation

Table 3
LIFECODES® HLA SSO typing kits by Immucor GTI diagnostics

HLA Locus	Kit Name	Catalog No.	Tests	HLA allele resolution
HLA-A	LIFECODES® HLA-A eRES	628913	50	Low to enhanced resolution (medium)
HLA-B	LIFECODES® HLA-B eRES	628917	50	Low to enhanced resolution (medium)
HLA-C	LIFECODES® HLA-C eRES	628921	50	Low to enhanced resolution (medium)
HLA-DRB1	LIFECODES® HLA-DRB1 eRES	628925	50	Low to enhanced resolution (medium)
HLA-DRB3/4/5	LIFECODES® HLA-DRB3/4/5	628927	50	Multi-DRB-locus coverage
HLA-DQA1 and HLA-DQB1	LIFECODES® HLA-DQA1/B1	628930	50	Low to medium resolution
HLA-DPA1 and HLA-DPB1	LIFECODES® HLA-DPB1	628931	50	Low resolution

Diagnostics) for the HLA typing results evaluation from the LIFECODES HLA SSO kits.

2. The templates for both software applications are readily available on the manufacturer's website of the HLA kits, in this instance, Immucor GTI Diagnostics.
3. Make sure that the instrument is in its calibrated state and perform calibration and verification steps when necessary. Please refer to the IFUs of the used Luminex instrument appropriate Calibration and Performance Verification kits.
4. It is recommended that one negative and one positive control sample be run with each test. The negative control sample can be a water blank or elution buffer from the DNA extraction kit used. The positive control sample may be a previously typed sample; its DNA should follow this procedure.
5. The HLA locus-specific Master Mix needs to warm to room temperature (18–30 °C) before use.
6. When ready, gently vortex for approximately 10 s, which will keep the salts homogenized in the solution.
7. Spin briefly in microcentrifuge (5–10 s) to bring the liquid to the bottom of the tube (*see* **Note 38**).
8. Prepare HLA locus-specific PCR sample reaction mix according to Table 4 (*see* **Notes 39 and 40**). Do not add the DNA at this moment. Always prepare sufficient surplus to compensate for pipetting volume loss (e.g. for $n+1$ samples).
9. Aliquot the amplification mix in the PCR tubes or wells of the PCR plate. The volume pipetted in each well in this step should be (20 μL—calculated DNA volume per well).
10. Pipette the calculated volume of genomic DNA into the wells (*see* **Note 41**).
11. Cap the PCR tubes or seal the PCR plate to prevent evaporation during PCR.

Table 4
PCR reaction components for LIFECODES HLA-SSO amplification

Component	Amount per sample (per PCR well)
LIFECODES Master Mix	6 μL
Genomic DNA concentration 10–200 ng/μL	Total of 80 ng per PCR reaction (tolerated range 40–120 ng)
LIFECODES Taq Polymerase	0.2 μL (1 U)
Nuclease-free water	Add, depending on the DNA volume, to the final volume 20 μL

12. Place the PCR tubes or plate in the thermal cycler and run the program (*see* Table 5) (*see* **Note 42**).
13. If the amplicons are processed further immediately and not stored at 2–8 °C (for a few days) or −20 °C (longer), turn on the Luminex 200 Instrument and its XY platform or FLEXMAP 3D instrument to allow for a 30-min laser warm-up. This step can also be performed concurrently while the PCR is running.
14. After the 30-min laser warm-up perform the daily start-up procedure from the xPONENT software (*see* **Note 43**).
15. Turn on the heating block or water bath to allow it to reach 55–60 °C. Warm the HLA locus-specific probe mix in the heating block for at least 5–10 min, to thoroughly solubilize all components in the probe mixture (*see* **Notes 44 and 45**).
16. After that, sonicate the probe mix briefly (at least 15 s), then vortex probe mix for at least 15 s, so that xMAP microspheres are thoroughly suspended (*see* **Note 45**).
17. Take one 96-well Costar plate from its packaging (*see* **Notes 46 and 47**).
18. Combine 15 μL of the appropriate HLA locus-specific probe mix with 5 μL of all PCR products from the same HLA locus into each well of a 96-well Costar plate (*see* **Note 48**).
19. Seal Costar plate with a clear adhesive tape. MicroAmp™ Clear Adhesive Film (Applied Biosystems cat. no. 4306311) was validated for use with LIFECODES HLA SSO kits.
20. Place silicone compression mat (pad) on top of plate prior to hybridization.

Table 5
PCR thermal cycler conditions for LIFECODES HLA-SSO amplification

Step	Temperature and Incubation Time	Number of Cycles
1	95 °C for 3 min	1
2	95 °C for 15 s 60 °C for 30 s 72 °C for 30 s	12
3	95 °C for 10 s 63 °C for 30 s 72 °C for 30 s	28
4	72 °C for 2 min	1
5	4 °C forever	1

21. The following incubation conditions for the 20-min hybridization protocol should be used (Table 6).
22. During the hybridization step, prepare a 200:1 mixture of 170 μL Dilution Solution (DS) and 0.85 μL streptavidin–phycoerythrin (SA–PE) conjugate (1 mg/mL). The DS/SA–PE mixture must be kept in the dark at room temperature, because SA–PE is light-sensitive. Make sure to prepare enough mixture to account for any pipetting losses, so at least $n+1$ for n hybridizing samples (*see* **Note 49**).
23. After the 20-min incubation program, at the 56 °C HOLD step, while the tray remains on the thermal cycler, dilute each sample with 170 μL of the prepared DS/SA–PE mixture. It is critical that this step lasts for a maximum of 5 min, during which all the samples must be diluted (after the 8-min 56 °C step of the hybridization protocol).
24. Remove the sample tray from the cycler and place it in the Luminex instrument for microsphere acquisition.
25. Click the *Eject* icon in xPONENT software to eject the plate holder.
26. Place the Costar plate containing the samples in the XYP heater block present on the plate holder.
27. Click the *Retract* icon. The samples are now ready to be analyzed.
28. Set up a Batch Run within xPONENT software.
29. Open the *Batches* page.
30. Click *Create a New Batch from an Existing Protocol.*
31. Type the batch name in the *Batch Name* box. When naming the batch, do not include commas in the name because information after a comma will be lost upon data export.
32. Click the protocol you want to use in the *Select a Protocol* list. For example, when performing an HLA-DRB1 testing, select the lot-specific protocol for the used HLA-DRB1 kit.
33. Click Next. The Plate layout tab appears (Fig. 4).

Table 6
Thermal cycler conditions for LIFECODES HLA-SSO hybridization

97 °C for 2 min
47 °C for 10 min
56 °C for 8 min
56 °C HOLD

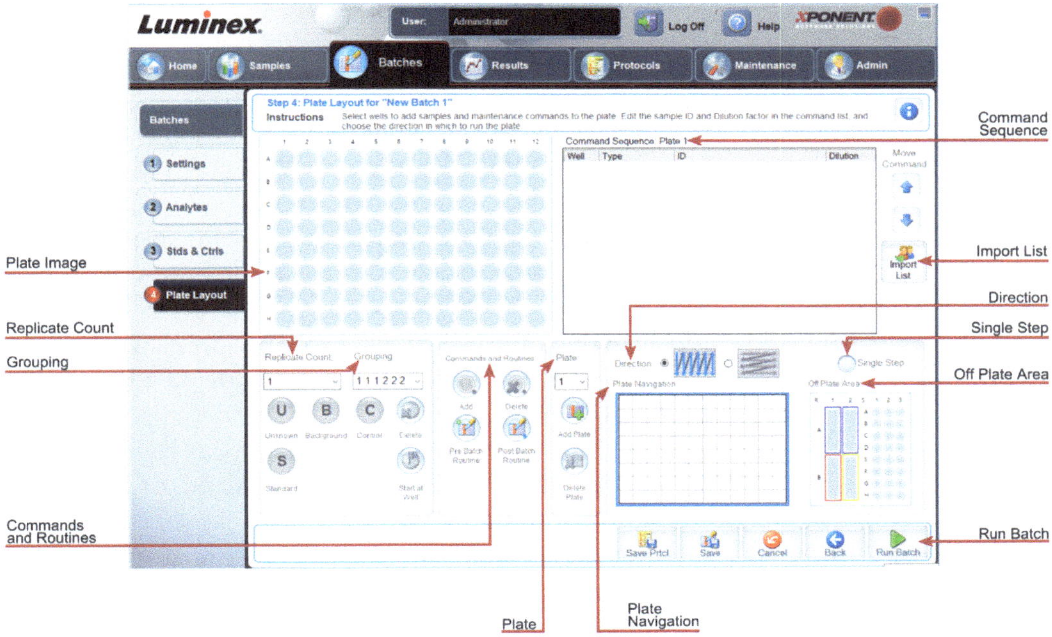

Fig. 4 xPONENT software plate layout interface when creating a batch. Screenshot of the plate layout tab in xPONENT software during batch setup for HLA class I/II genotyping using LIFECODES® HLA-SSO kits. The interface allows selection and assignment of wells on a 96-well plate, definition of replicate count and grouping, and designation of sample types (e.g., unknown samples, controls, standards, or background). The command sequence panel enables entry or import of sample identifiers and definition of dilution factors. Plate navigation and directional options determine the order of well acquisition during microsphere reading. After configuration of the plate layout and verification of sample positioning, the batch is initiated by selecting "Run Batch," followed by microsphere acquisition on the Luminex instrument and subsequent data export for allele assignment in dedicated HLA typing software

34. On the Plate layout tab, assign wells used in the Costar plate for the selected HLA locus (batch).
35. Label the samples by clicking *U* (*unknown*).
36. Insert the sample IDs in the Command Sequence. Alternatively, import the previously created patient list with sample IDs, saved as a text (.txt) file. The patient list can be created in Windows' Notepad application (even before starting the test, to save time at this point in the protocol).
37. Click *Run Batch* to begin batch acquisition.
38. After the samples have been run through the instrument, a daily shutdown procedure with 20% household bleach should be performed, followed by two wash steps.
39. The Luminex instrument can be turned off at this point if it is not going to be used for the remainder of the day.

40. After a batch is complete, the data is exported from the xPONENT software as a comma-separated values (.csv) file. These files are saved in a xPONENT Output folder with the Batch Name. The folder is available via Desktop shortcut. The saved data is then available for making HLA typing assignments, preferably by using the appropriate evaluation software as described below.
41. The MATCH IT! DNA Software is recommended to be used for data analysis. The.csv files are to be imported into the MATCH IT! DNA Software.
42. Provided the software settings are set to automatically analyze the samples, the HLA typing will be listed in the Final Typing Assignment table.
43. The allele assignments for each HLA locus should be carefully reviewed, all ambiguities resolved, in order that the assignment would be Exact.
44. After the analysis, the report(s) can be generated and printed or saved digitally.
45. If No Match was able to be made (hence, that assignment was provided by the software) or if there are any apprehensions about the final typing, the sample should be escalated to an alternative method (e.g., next-generation sequencing or other appropriate method), if required.
46. Always document a lot of the kits and the software version used in the testing.
47. The LIFECODES Probe Mix(es) contain one or more Consensus SSO probes per HLA locus. They are identified in the typing kit worksheets. These consensus probes hybridize to all alleles of that HLA locus and act as internal controls to verify amplification and to confirm that hybridizations occurred. If the minimum value is not obtained for these SSOs, the sample may not produce the correct typing, and the sample test should be repeated.
48. Values obtained with Consensus SSOs from positive controls should exceed the threshold value for the SSO as set forth in the Threshold Table Worksheet.
49. Values obtained with Consensus SSOs from negative controls should be below the threshold value for the SSO as set forth in the Threshold Table Worksheet.
50. To obtain reliable results, there must be sufficient data gathered by the Luminex instruments. For LIFECODES HLA SSO assays this requirement is at least 40 events per each microsphere in the Bead Mix.

51. If more than two probes are indeterminate, the sample should be re-assayed. If a probe hit pattern does not produce an HLA type, the sample should be re-amplified and re-assayed. It may also be necessary to re-isolate DNA from the sample and re-amplify and re-assay.
52. The observed two-field HLA allele concordance with other methods of the LIFECODES kits is higher than 98% across all HLA loci in experiments with more than 120 well-characterized samples, while the manufacturer's acceptance criteria were higher than 95%.

4 Notes

1. Denatured alchohol should not be used as it contains methanol.
2. Dissolved QIAGEN Protease is stable up to 12 months when stored at 2 –8 °C.
3. QIAGEN Protease must not be added directly to Buffer AL.
4. Buffer AL is stable for 1 year when stored at room temperature (15–25 °C).
5. Buffers AW1 and AW2 are stable for 1 year when stored closed at room temperature (15–25 °C).
6. It is recommended that the aerosol-barrier pipette tips are utilized for the described DNA extraction protocol.
7. Centrifugation of QIAamp Mini columns is performed at 6000 × *g* to reduce centrifuge noise from larger speeds. DNA yield will not be affected if centrifuging QIAamp Mini columns at full speed. It is acceptable to use lower speeds for centrifugation, provided that almost the whole volume of each solution passes through the QIAamp membrane.
8. If the sample volume is larger than 200 μL, increase the amount of QIAGEN Protease (or proteinase K) and Buffer AL proportionally; for example, a 400 μL sample will require 40 μL QIAGEN Protease (or proteinase K) and 400 μL Buffer AL. If sample volumes larger than 400 μL are required, there are other DNA extraction kits available that can process up to 2 mL or even up to 10 mL of sample.
9. If the sample volume is less than 200 μL, add the appropriate volume of PBS up to 200 μL. QIAamp Mini spin columns copurify RNA and DNA when both are present in the sample. RNA may inhibit some downstream enzymatic reactions, but it does not inhibit PCR. If RNA-free genomic DNA is required,

4 μL of an RNase A stock solution (100 mg/mL) should be pipetted to the sample before addition of Buffer AL.

10. It is possible to add QIAGEN Protease (or proteinase K) to samples that have already been dispensed into microcentrifuge tubes. If that is the case, it is important to ensure proper mixing after adding the enzyme.
11. If a precipitate has formed in Buffer AL, dissolve by incubating at 56 °C.
12. Do not add QIAGEN Protease or proteinase K directly to Buffer AL. The enzyme should be added to the sample before Buffer AL.
13. DNA yield reaches a maximum after lysis for 10 min at 56 °C. Longer incubation times have no noticeable effect on yield or quality of the purified DNA.
14. If the sample volume is greater than 200 μL, the amount of added ethanol should be increased proportionally; for example, a 400 μL sample will require 400 μL of ethanol.
15. The filtrate in this and subsequent centrifugation steps may contain hazardous waste and should be disposed of appropriately.
16. Close each spin column to avoid aerosol formation during centrifugation. Centrifugation is performed at 6000 x*g* (8000 rpm) to reduce noise. Centrifugation at full speed will not affect the yield or purity of the DNA. In this step it is important that the lysate completely passes through the column after centrifugation. If that is not the case, you may centrifuge again at higher speed until the QIAamp Mini spin column is entirely empty.
17. When preparing DNA from buffy coat or lymphocytes, centrifugation at full speed is recommended to avoid clogging.
18. It is not necessary to increase the volume of Buffer AW1 if the original sample volume is larger than 200 μL.
19. Flow through in this step contains Buffer AL or Buffer AW1 which are not compatible with bleach.
20. These steps help to eliminate the chance of possible Buffer AW2 carryover.
21. Incubating the QIAamp Mini spin column loaded with Buffer AE or water for 5 min at room temperature before centrifugation generally increases DNA yield.
22. A second elution step with a further 200 μL Buffer AE will increase yields by up to 15%. Please note that volumes of more than 200 μL should not be eluted into a 1.5-mL microcentrifuge tube because the eluate will come into contact

with the spin column, leading to possible aerosol formation during centrifugation.

23. Elution with volumes of less than 200 μL can increase the final DNA concentration in the eluate significantly, but it can slightly reduce the overall DNA yield.
24. A total of 200 μL of whole blood usually yields 3–12 μg of DNA.
25. For samples containing less than 1 μg of DNA, elution in 50 μ L Buffer AE or water is recommended. Eluting with 2 × 100 μ L instead of 1 × 200 μL does not increase elution efficiency.
26. The described cycling conditions should be appropriate for most genotyping assays. Make sure to always check the instructions for use of the utilized product. Alternatively, if in-house mixes are used in the assay, validate the protocol by adjusting the temperatures and time intervals in the cycling protocol as needed.
27. If a passive reference is used in the genotyping assay, it must be explicitly defined; otherwise, leave the passive reference off. In qPCRsoft, there is the option to include passive reference only if one is defined on the Settings/Scan project tab. If the "include passive reference" option is activated, the fluorescence of the dye that has been set as a passive reference is used for standardization.
28. During the whole procedure protect the TaqMan probes from strong light.
29. This step of PCR plate spin is necessary to remove bubbles from pipetting and collect liquid at the bottom of the well. Usually, a short spin protocol available on the device is enough. Alternatively, spin for 2 min at 1000 x *g*.
30. If applicable, select the correct Group if you ran multiple groups on one PCR plate.
31. Check curves and set the threshold only if it is required in your custom setup. Apply or adjust thresholding only if your analysis method requires it for your dataset. For many SNP calls the endpoint scatter clustering is the key output.
32. The scatter plot is divided into four quadrants for genotype1, genotype2, heterozygote, and error. The samples are assigned to one of the quadrants based on the measured relative fluorescence or the Ct values of the two dyes. The respective cutoff value for the sample assignment is represented by two black lines in the scatter plot view. To change the position of the lines and thus change the cutoff value, select the lines and keep the left mouse button pressed.

33. If clusters overlap, repeat with consistent DNA mass per well and/or confirm the correct dye mapping (which dye = nt1 vs nt2) from assay documentation.
34. NTC should sit near baseline (no clear cluster away from origin).
35. Replicates should cluster tightly. Wide scatter suggests: pipetting variability, bubbles, poor sealing (evaporation); low DNA quality or PCR inhibitors—in this case re-extract/dilute DNA.
36. Exclude wells with aberrant Ct values or noise.
37. If needed, also export the PCR plate layout for traceability.
38. Failure to do this step could decrease the number of tests available in the kit, as there could be some Master Mix volume losses.
39. The volume for PCR reactions including the LIFECODES Master Mix is the same for all LIFECODES HLA kits (HLA loci). In other words, all Master Mixes are pipetted 6 μL per sample.
40. If performing the amplification step for several HLA loci simultaneously, prepare separate PCR reaction mixes for each of the kits used (for each HLA locus).
41. Make sure to mix several times by the pipette, so that DNA is well mixed with the rest of the PCR sample reaction in this step.
42. PCR cyclers usually request that the lid temperature be defined when creating the PCR program. It is necessary that the lid temperature be higher than the maximum temperature during the PCR, which is 95 °C. The lid temperatures of 100 °C, 105 °C, or 110 °C are normally used in practice.
43. The Luminex instrument must be ready for microsphere acquisition when the hybridization step is completed, so make sure to time this step appropriately so as to avoid any downstream delays with the microsphere acquisition.
44. The microspheres in this assay are not so temperature sensitive that the temperature should be precisely set in this step. The range of 5 °C is appropriate and whichever temperature is used, it is important to remain constant for the whole 5–10 min. Usual practice is that the temperature in this step is 56 °C, because this is the temperature for hybridization.
45. When performing simultaneous assays with different LIFECODES kits, all microsphere mixes can be heated, sonicated and vortexed at the same time.
46. All hybridization reactions and forward labeling and microsphere acquisition steps are performed in the Costar plate.

47. Please make sure that the protocol is not stopped any time before the microsphere acquisition. At least, perform the hybridization and labeling steps consecutively, with a maximum of 30 min before the plate is read in the Luminex instrument. The results could not be valid if the subsequent steps are not performed immediately one after another.
48. When aliquoting probe mix to more than eight wells (one column), gently vortex probe mix after each column. This is necessary in order to keep the microsphere sets evenly suspended in the solution.
49. The DS/SA–PE mixture needs to be prepared fresh each time an assay is performed and any remaining portion must be discarded. Therefore, make sure that the pipetted volumes are not overestimated.

Competing Interests

The authors have no conflicts of interest to declare that are relevant to the content of this chapter.

References

1. Platts-Mills TAE, Li RC, Keshavarz B, Smith AR, Wilson JM (2020) Diagnosis and management of patients with the α-gal syndrome. J Allergy Clin Immunol Pract 8:15–23.
2. Wilson JM, Schuyler AJ, Workman L, Gupta M, James HR, Posthumus J et al (2019) Investigation into the α-gal syndrome: characteristics of 261 children and adults reporting red meat allergy. J Allergy Clin Immunol Pract 7:2348–2358.
3. Wilson JM, Erickson L, Levin M, Ailsworth SM, Commins SP, Platts-Mills TAE (2024) Tick bites, IgE to galactose-alpha-1,3-galactose and urticarial or anaphylactic reactions to mammalian meat: The alpha-gal syndrome. Allergy 79:1440–1454
4. Banović P, Jakimovski D, Mijatović D, Bogdan I, Simin V, Grujić J et al (2025) Genetic and immunological insights into tick-bite hypersensitivity and alpha-gal syndrome: a case study approach. Int J Mol Sci 26:680
5. Steinke JW, Platts-Mills TA, Commins SP (2015) The alpha-gal story: lessons learned from connecting the dots. J Allergy Clin Immunol 135:589–596.
6. Kubista M, Andrade JM, Bengtsson M, Forootan A, Jonák J, Lind K et al (2006) The real-time polymerase chain reaction. Mol Aspects Med 27:95–125
7. Arya M, Shergill IS, Williamson M, Gommersall L, Arya N, Patel HR (2005) Basic principles of real-time quantitative PCR. Expert Rev Mol Diagn 5:209–219
8. Heid CA, Stevens J, Livak KJ, Williams PM (1996) Real time quantitative PCR. Genome Res 6:986–994
9. Livak KJ (1999) Allelic discrimination using fluorogenic probes and the 5' nuclease assay. Genet Anal 14:143–149
10. Ranade K, Chang MS, Ting CT, Pei D, Hsiao CF, Olivier M et al (2001) High-throughput genotyping with single nucleotide polymorphisms. Genome Res 11:1262–1268
11. Bustin SA, Benes V, Garson JA, Hellemans J, Huggett J, Kubista M et al (2009) The MIQE guidelines: minimum information for publication of quantitative real-time PCR experiments. Clin Chem 55:611–622
12. Sanchez JA, Pierce KE, Rice JE, Wangh LJ (2004) Linear-after-the-exponential (LATE)-PCR: an advanced method of asymmetric PCR and its uses in quantitative real-time analysis. Proc Natl Acad Sci U S A 101: 1933–1938

13. Pierce KE, Wangh LJ (2007) Linear-after-the-exponential polymerase chain reaction and allied technologies. Real-time detection strategies for rapid, reliable diagnosis from single cells. Methods in Mol Med 132:65–85
14. Dunbar SA (2006) Applications of Luminex xMAP technology for rapid, high-throughput multiplexed nucleic acid detection. Clin Chim Acta 363:71–82
15. Erlich H (2012) HLA DNA typing: past, present, and future. Tissue Antigens 80:1–11
16. Barker DJ, Maccari G, Georgiou X, Cooper MA, Flicek P, Robinson J et al (2023) The IPD-IMGT/HLA Database. Nucleic Acids Res 51:D1053–D1060
17. Shiina T, Hosomichi K, Inoko H, Kulski JK (2009) The HLA genomic loci map: expression, interaction, diversity and disease. J Hum Genet 54:15–39

Check for updates

Chapter 23

Clinical Protocol for Evaluating Ectoparasite-Induced Allergies with Emphasis on Alpha-Gal Syndrome

Pavle Banović and Andrija Jović

Abstract

Alpha-gal syndrome (AGS) is an ectoparasite-associated IgE-mediated hypersensitivity in which reactions occur after exposure to galactose-alpha-1,3-galactose (alpha-Gal) in mammalian foods, biologics, medications, and other mammalian-derived products. The delayed onset after oral exposure and the imperfect correlation between sensitization and clinical allergy complicate diagnosis. This chapter provides a practical clinical protocol for evaluating suspected ectoparasite-induced allergy with emphasis on AGS. The protocol integrates structured history taking, tick and other ectoparasite exposure assessment, focused clinical examination, serum IgE testing, skin prick and prick-to-prick testing, basophil activation testing (BAT), and medically supervised oral challenge when needed. A consensus BAT workflow is incorporated, including whole-blood handling, negative and positive controls, stimulation, antibody staining, erythrocyte lysis, flow cytometry acquisition, gating, quality-control criteria, and reporting. The protocol is intended for implementation in specialist allergy clinics and translational research settings where diagnostic safety, reproducibility, and standardized data capture are required.

Keywords Alpha-gal syndrome, Ectoparasite allergy, Tick bite, Basophil activation, Specific IgE, Skin prick test, Oral challenge, Anaphylaxis, Mammalian meat

1 Introduction

Alpha-gal syndrome (AGS) is a clinically distinctive form of IgE-mediated allergy associated with ectoparasite exposure, most commonly tick bites. It is characterized by sensitization to the carbohydrate galactose-alpha-1,3-galactose (alpha-Gal) and by allergic reactions after exposure to mammalian meat, innards, gelatin, selected dairy products, biologics, medical devices, vaccines, antivenoms, or other mammalian-derived products [1–8]. Oral reactions are classically delayed, often appearing 2–6 h after ingestion, whereas parenteral exposure to an alpha-Gal-containing product can produce a rapid immediate-type reaction [4–7].

The clinical spectrum includes urticaria, angioedema, pruritus, flushing, gastrointestinal pain, nausea, vomiting, diarrhea, respiratory

José de la Fuente and Alejandro Cabezas-Cruz (eds.), *Allergic Disease Models*, Methods in Molecular Biology, vol. 3092, https://doi.org/10.1007/978-1-0716-5635-8_23,

symptoms, hypotension, syncope, and anaphylaxis. Some patients present primarily with gastrointestinal symptoms, and symptoms may occur intermittently despite apparently similar meals because dose, fat content, digestion kinetics, alcohol intake, exercise, nonsteroidal anti-inflammatory drugs, intercurrent illness, mast-cell burden, and other cofactors can modify the reaction threshold [1–8].

A diagnostic protocol must therefore distinguish three clinically important states: clinically relevant AGS, asymptomatic alpha-Gal sensitization, and non-AGS ectoparasite-associated allergy. Serum alpha-Gal-specific IgE (sIgE) is central to the work-up but is not sufficient as an isolated test result because sensitization can occur without symptoms [4, 9, 10]. Skin testing can support the diagnosis but may be insensitive with some commercial meat extracts, making prick-to-prick testing with clinically relevant mammalian foods useful in selected cases [11, 12].

The basophil activation test (BAT) provides an ex vivo functional readout of IgE-mediated effector-cell activation. In AGS, BAT with alpha-Gal-bearing proteins and pork kidney or other mammalian extracts has been shown to help distinguish symptomatic AGS from asymptomatic alpha-Gal sensitization and can reduce reliance on potentially risky oral provocation tests [13–15]. Because BAT results are sensitive to pre-analytical handling, stimulation conditions, gating strategy, cytometer setup, and threshold definitions, this chapter incorporates a standardized whole-blood BAT workflow adapted for AGS evaluation [14–16].

The complete protocol consists of the following major procedures: (i) pre-test safety triage and medication review; (ii) structured clinical and exposure assessment; (iii) focused examination; (iv) serum IgE testing and adjunctive laboratory testing; (v) skin prick and prick-to-prick testing; (vi) BAT with alpha-Gal and ectoparasite-associated stimuli; (vii) oral challenge only when clinically justified; and (viii) integrated diagnostic classification and patient counseling. The oral challenge component follows current food-challenge safety principles and is reserved for cases in which the result is expected to change management [17, 18].

2 Materials

All clinical procedures must be performed under local ethics, consent, infection-control, biosafety, and clinical governance requirements. Use only locally approved reagents for patient testing. For non-commercial or research ectoparasite materials, obtain appropriate institutional approval before patient sample exposure (*see* **Notes 1–4**).

2.1 Patient Assessment and Documentation

1. Standardized AGS and ectoparasite-allergy case report form or electronic template (Table 1).

Table 1
Structured clinical and exposure variables for suspected ectoparasite-induced allergy and AGS

Domain	Minimum variables to record	Interpretive purpose
Index reaction	Date, trigger, route, dose, preparation method, time to first symptom, time to peak, treatment, emergency care	Defines clinical probability and separates delayed oral AGS from immediate parenteral or venom-like reactions
Symptom phenotype	Cutaneous, gastrointestinal, respiratory, cardiovascular, neurologic, isolated gastrointestinal, anaphylaxis, objective signs	Identifies systemic allergy, isolated gastrointestinal AGS phenotype, and mimics
Mammalian foods	Beef, pork, lamb, venison, innards, pork kidney, gelatin, dairy, processed foods, tolerance of poultry/fish	Matches test stimuli to the suspected alpha-Gal source and avoids unnecessary restriction
Tick exposure	Geography, season, occupation, recreation, pets/livestock, known bites, attachment duration, prolonged bite pruritus or erythema	Supports ectoparasite-associated sensitization and prevention counseling
Other ectoparasites	Hymenoptera, mosquitoes, fleas, mites, lice, bedbugs, flies, occupational insect exposure	Supports non-AGS ectoparasite allergy differential diagnosis
Cofactors	Alcohol, exercise, NSAIDs, infection, sleep deprivation, high-fat meals, menstruation, stress	Explains intermittent reactions and modified thresholds
Risk modifiers	Asthma, cardiovascular disease, mast-cell disease, baseline tryptase, pregnancy, beta-blocker use, previous severe anaphylaxis	Determines testing and challenge safety

2. Written informed consent forms for diagnostic testing, photography where applicable, BAT, and oral challenge where performed.
3. Medication and cofactor checklist including antihistamines, leukotriene receptor antagonists, tricyclic antidepressants, systemic corticosteroids, beta-blockers, angiotensin-converting enzyme inhibitors, omalizumab or other biologics, alcohol, exercise, nonsteroidal anti-inflammatory drugs (NSAIDs), infection, menstruation, sleep deprivation, and stress.
4. Symptom diary template allowing documentation of food exposure, dose, fat content, preparation method, timing of symptom onset, objective signs, treatment used, and duration.
5. Tick and ectoparasite exposure questionnaire documenting geography, season, outdoor occupation or recreation, pets/

livestock exposure, bites, retained mouthparts, prolonged local pruritus or erythema, and previous reactions to hymenoptera, mosquito, flea, mite, or other arthropod exposures.

6. Clinical camera or secure medical image system for anonymized documentation of skin findings, when permitted by local policy.
7. Vital-sign equipment: noninvasive blood-pressure monitor, pulse oximeter, thermometer, and stethoscope.
8. Measuring device for skin lesions and skin-test wheals: transparent millimeter ruler or digital caliper.

2.2 Safety and Emergency Equipment

1. Clinical area equipped for recognition and treatment of anaphylaxis, with staff trained in immediate management.
2. Intramuscular epinephrine/adrenaline 1 mg/mL (1:1000), syringes, needles, and dosing chart according to local guidelines.
3. Oxygen source with masks or nasal cannulae, suction, bag-valve-mask device, and airway equipment according to institutional policy.
4. Intravenous cannulas, administration sets, and isotonic crystalloid fluids.
5. H1-antihistamine, H2-antihistamine when used locally, inhaled short-acting beta2-agonist, systemic corticosteroid, and antiemetic according to local emergency protocol.
6. Crash cart or immediate access to resuscitation equipment and emergency medical response.
7. Observation area capable of prolonged monitoring after oral mammalian food challenge because AGS reactions may be delayed (*see* **Notes 17–20**).

2.3 Serum IgE and Adjunctive Laboratory Testing

1. Blood collection supplies appropriate for the reference laboratory performing serum alpha-Gal sIgE (see Chap. 19), total IgE, and selected mammalian allergen tests.
2. Laboratory request forms specifying alpha-Gal sIgE, total IgE, beef, pork, lamb, cow milk or dairy components, gelatin, and drug or biologic testing where clinically indicated.
3. Serum separator tubes or other validated blood collection tubes specified by the performing laboratory.
4. Centrifuge and aliquot tubes if serum is processed locally before shipment.
5. Cold-chain shipping materials where required by the reference laboratory.
6. Serum tryptase request forms and tubes for baseline tryptase and acute tryptase after recent anaphylaxis, according to local policy (*see* **Note 21**).

2.4 Skin Prick and Prick-to-Prick Testing

1. Disposable single-use skin-prick lancets.
2. Fine-tip skin marker or standardized skin-test grid.
3. Timer set for 15–20 min readings.
4. Millimeter ruler or digital caliper.
5. Absorbent tissue or gauze for removing excess allergen without cross-contamination.
6. Sharps container and personal protective equipment.
7. Negative control: 0.9% sodium chloride or the glycerinated diluent used for allergen extracts.
8. Histamine positive control: histamine dihydrochloride at a locally approved concentration, commonly 10 mg/mL histamine or equivalent.
9. Commercial mammalian meat extracts, gelatin extract, or other validated extracts where available.
10. Fresh mammalian foods for prick-to-prick testing when clinically indicated and locally permitted: beef, pork, lamb, pork kidney or other innards, gelatin-containing food, or cooked and raw preparations selected according to the clinical history.
11. Separate clean surfaces, forceps, and lancets for each food to avoid cross-contamination.

2.5 Basophil Activation Test (BAT)

1. Sodium or lithium heparin whole-blood collection tubes for BAT; process blood within 4 h of phlebotomy whenever possible.
2. Flow cytometry tubes labeled for negative control, positive control, and each allergen or ectoparasite stimulus concentration.
3. Calibrated micropipettes and sterile filtered tips suitable for 2–1000 microliter volumes.
4. Phosphate-buffered saline (PBS): sterile, calcium- and magnesium-free, pH 7.2–7.4.
5. Optional stimulation buffer: Hanks balanced salt solution containing calcium and magnesium, used only if validated locally and applied consistently across the run.
6. Positive-control stimulus: anti-IgE antibody or anti-Fc epsilon RI antibody at a locally validated working dilution.
7. Negative-control stimulus: PBS, stimulation buffer, or the vehicle/protein matrix matching the allergen preparation.
8. Activation and identification antibody panel: anti-CD63 conjugated to allophycocyanin or equivalent fluorochrome, anti-CD193/CCR3 conjugated to fluorescein isothiocyanate or equivalent, and anti-CD203c conjugated to phycoerythrin or equivalent.

9. Red blood cell lysis solution: FACS lysing solution or locally validated equivalent.
10. Wash buffer: PBS stored at 2–8 °C for post-lysis washing.
11. Flow cytometry calibration beads compatible with the instrument and fluorochromes used.
12. Flow cytometer capable of detecting the chosen fluorochromes, with forward scatter, side scatter, and compensation controls available.
13. Flow cytometry analysis software allowing sequential gating, histogram thresholding, and export of percent CD63-positive basophils and CD63 median fluorescence intensity (MFI).
14. Alpha-Gal-bearing stimuli: locally validated alpha-Gal-human serum albumin, alpha-Gal-bovine serum albumin, alpha-Gal-containing mammalian tissue extract such as pork kidney extract, mammalian meat extract, gelatin, or relevant medication/biologic excipient preparation (*see* **Notes 10–16**).
15. Ectoparasite-associated stimuli for research or translational protocols: sterile, non-cytotoxic, validated extracts or purified salivary molecules from ticks or other ectoparasites; use only after biosafety and ethics approval (*see* **Note 13**).

2.6 Oral Mammalian Food or Product Challenge

1. Written challenge protocol approved by the clinical service, including stopping rules, treatment algorithm, observation duration, and discharge criteria.
2. Challenge consent form describing the risk of delayed anaphylaxis and prolonged observation.
3. Pre-weighed challenge food or product selected from the clinical history, such as cooked mammalian meat, pork kidney, gelatin-containing product, or other mammalian-derived material; dose and matrix must be locally approved.
4. Placebo food matrix if a blinded or placebo-controlled challenge is used.
5. Dose-escalation record sheet documenting dose, time administered, symptoms, signs, vital signs, and treatments.
6. Emergency medications and equipment listed in Sect. 2.2.
7. Observation bed or chair and capacity for at least 6 h of monitoring after the final dose, with longer observation or admission for high-risk patients.

3 Methods

3.1 Eligibility, Safety Triage, and Pre-test Preparation

The workflow is summarized in Table 2. Perform the steps in the order below unless a safety concern requires immediate deviation.

1. Confirm that the referral question is evaluation of suspected ectoparasite-induced allergy, AGS, or both. Record whether the primary concern is food-associated reactions, medication or biologic reactions, tick-bite reactions, other arthropod reactions, or unexplained anaphylaxis.
2. Screen for current instability. Do not proceed with elective skin testing, BAT phlebotomy, or oral challenge during acute anaphylaxis, uncontrolled asthma, significant intercurrent infection, hemodynamic instability, or other clinical instability (*see* **Note 1**).

Table 2
Integrated diagnostic workflow and decision points

Step	Procedure	Proceed when	Output
1	Safety triage and consent	Patient is clinically stable, and testing is justified	Testing plan and emergency precautions
2	Structured history and exposure assessment	All reactions and exposures have been recorded	Pre-test clinical probability
3	Focused clinical examination	Skin, mucosa, respiratory, cardiovascular, and gastrointestinal signs have been assessed	Objective baseline and differential diagnosis
4	Serum alpha-Gal sIgE and adjunct tests	History suggests AGS, unexplained anaphylaxis, or alpha-Gal product reaction	Sensitization profile
5	Skin prick or prick-to-prick testing	Skin testing will change probability or guide challenge/ product decisions	Immediate skin reactivity profile
6	BAT	Clinical history and sensitization are discordant, the challenge is risky, or functional confirmation is needed	Ex vivo basophil activation to alpha-Gal or ectoparasite stimuli
7	Oral challenge	Diagnosis remains uncertain, and benefit outweighs risk	Confirmed tolerance or objective reaction under supervision
8	Diagnostic classification and counseling	All data have been reviewed together	Confirmed, probable, possible, unlikely AGS, asymptomatic sensitization, or non-AGS ectoparasite allergy

3. Document pregnancy status where relevant and follow local policy for allergy testing and challenge in pregnancy (*see* **Note 22**).
4. Review medications that can suppress skin testing or alter reaction treatment. H1-antihistamines should be withheld for the locally required interval before skin testing and oral challenge. Do not stop beta-blockers, anticoagulants, or other high-risk medications without prescriber review (*see* **Notes 2 and 3**).
5. Ask the patient to avoid avoidable cofactors before elective testing, especially alcohol, strenuous exercise, NSAIDs, and unusually high-fat meals, unless a cofactor challenge is intentionally planned under specialist supervision (*see* **Note 4**).
6. Record baseline vital signs and assess respiratory, cardiovascular, cutaneous, and gastrointestinal status.
7. Obtain written informed consent for clinical assessment and each planned test. For BAT using research-grade ectoparasite materials, obtain specific research consent and approval before blood collection (*see* **Note 13**).
8. Create a single diagnostic record containing the clinical history, exposure history, test results, interpretation, and final diagnostic category.

3.2 Structured Clinical and Exposure Assessment

1. For each reaction, document the date, clock time of exposure, suspected trigger, dose, preparation method, fat content if food-related, and whether the food was muscle meat, innards, gelatin, dairy, or a processed product with unclear mammalian ingredients.
2. Document the latency from exposure to first symptom. Classify oral reactions as delayed when symptoms begin at least 2 h after ingestion and typically within 2–6 h. Record earlier reactions separately, particularly when parenteral exposure or major cofactors are present.
3. Record the symptom phenotype: cutaneous, gastrointestinal, respiratory, cardiovascular, neurologic, isolated gastrointestinal, or mixed systemic symptoms.
4. Document objective signs when available: urticaria, angioedema, flushing, wheeze, hypoxemia, hypotension, syncope, vomiting, diarrhea, or emergency department findings.
5. Record all treatments used, including epinephrine, antihistamines, corticosteroids, bronchodilators, intravenous fluids, and emergency department or intensive care attendance.
6. Document reactions to mammalian-derived medications and products: cetuximab or other monoclonal antibodies, gelatin-containing vaccines or capsules, antivenoms, heparin, colloid volume expanders, pancreatic enzymes, thyroid

preparations, heart valves, hemostatic agents, and other biologics or excipients.

7. Document tick exposure in detail: geographic location, season, outdoor activities, occupational risk, pets or livestock, number of known bites, attachment duration, local bite reaction, prolonged pruritus or erythema, and whether bites preceded symptom onset.
8. Document other ectoparasite or arthropod exposures: hymenoptera, mosquitoes, fleas, mites, lice, bedbugs, flies, or occupational insect exposure. Record whether reactions were local, large local, systemic immediate, delayed, or dermatitis-like.
9. Record known atopy, asthma, chronic urticaria, mast-cell disease, baseline tryptase if available, childhood cow-milk allergy, atopic dermatitis, blood group if known, and previous food or venom allergy.
10. Enter the findings into the structured data table (Table 1) before ordering additional tests so that laboratory results are interpreted against a pre-defined clinical probability rather than in isolation (*see* **Notes 5 and 6**).

3.3 Focused Examination and Reaction Documentation

1. Perform the examination in a well-lit room at a comfortable temperature, with privacy and appropriate gowning to expose affected areas.
2. Inspect the skin, scalp, lips, eyelids, oral mucosa, trunk, extremities, palms, soles, and previous bite sites when relevant.
3. Describe each lesion by distribution, symmetry, color, morphology, elevation, borders, size, blanching, tenderness, warmth, scale, vesiculation, crusting, excoriation, or residual hyperpigmentation.
4. Ask specifically about pruritus, pain, burning, and duration of lesions. Urticarial lesions that persist in one location for more than 24 h, bruise, or are painful should prompt evaluation for alternative diagnoses.
5. Assess the respiratory and cardiovascular systems when symptoms suggest systemic reactions. Record wheeze, stridor, oxygen saturation, blood pressure, heart rate, and perfusion.
6. Assess abdominal tenderness, hydration status, and ongoing vomiting or diarrhea when gastrointestinal symptoms are prominent.
7. Photograph lesions only with consent and according to institutional anonymization policy. Do not include identifiable facial features unless clinically essential and permitted.
8. Update the diagnostic record with the examination findings before performing skin tests or challenges.

3.4 Serum Testing and Initial Interpretation

1. Order serum alpha-Gal sIgE (see Chap. 19) in all patients with compatible delayed reactions to mammalian foods, reactions to mammalian-derived biologics or medications, unexplained nocturnal anaphylaxis, or a strong tick-exposure history.
2. Order total IgE when local interpretation uses the relationship between alpha-Gal sIgE and total IgE, or when very high total IgE may influence interpretation.
3. Order additional sIgE tests according to the history. Useful adjuncts can include beef, pork, lamb, cow milk, gelatin, cat serum albumin, dog serum albumin, or specific medications/excipients where validated tests are available.
4. If the patient recently had anaphylaxis, obtain acute serum tryptase as soon as feasible and a baseline tryptase after recovery according to local practice. Elevated baseline tryptase may indicate mast-cell disease and increased reaction risk (*see* **Note 21**).
5. Interpret a positive alpha-Gal sIgE result as evidence of sensitization, not as a standalone diagnosis. Confirm clinical AGS only when the history includes compatible symptoms after alpha-Gal exposure or when a supervised challenge is positive (*see* **Note 5**).
6. If alpha-Gal sIgE is negative but the history is strongly suggestive, review timing of testing, laboratory method, alternative diagnoses, and whether repeat testing or BAT is justified (*see* **Note 6**).
7. Proceed to skin testing, BAT, or oral challenge only when the result will change diagnostic classification or management.

3.5 Skin Prick and Prick-to-Prick Testing

1. Confirm that antihistamine washout is adequate and that there is no uncontrolled asthma, active generalized urticaria, extensive dermatitis at the test site, dermatographism that prevents interpretation, or other contraindication.
2. Seat the patient comfortably. Use a supine position for children, anxious patients, needle-phobic patients, or patients with previous vasovagal episodes.
3. Expose the volar forearm or upper back. Select intact skin free from lesions, heavy hair, lotion, moisturizer, or topical medication.
4. Do not vigorously rub the skin. Avoid alcohol or antiseptic immediately before testing unless local infection-control policy requires it; if used, allow the skin to dry completely before allergen placement (*see* **Note 7**).
5. Mark test sites with a pen or grid. Space sites approximately 2.5 cm apart, avoiding skin creases and keeping at least 2–3 cm from the wrist and antecubital fossa when the forearm is used.

6. Place one drop of negative control at the first site and one drop of histamine positive control at the final site. Place one drop of each commercial mammalian extract or gelatin extract at the corresponding marked sites.
7. For prick-to-prick testing, insert a fresh lancet into the selected mammalian food or product and then prick the marked skin site with gentle pressure. Use a new lancet and clean material for each food. Test cooked and raw forms separately only when clinically justified and locally approved.
8. Prick through each commercial extract drop with a separate lancet using consistent pressure sufficient to penetrate the superficial epidermis without causing bleeding.
9. Remove excess allergen or food material with tissue after all sites are pricked, taking care not to drag liquid across sites.
10. Read the test at 15–20 min. Measure the largest wheal diameter and the orthogonal diameter, then record the mean wheal diameter. Record flare size separately if required locally.
11. A test is positive when the wheal is at least 3 mm greater than the negative control, and the negative control is not reactive. The histamine control must be positive for a valid test.
12. Record photographs or diagrams when clinically useful. Enter all results into the diagnostic record with the extract type, food form, lot or source, and reading time (*see* **Notes 8 and 9**).

3.6 BAT Pre-analytical Setup

1. Schedule BAT blood collection so that stimulation can begin within 4 h of phlebotomy. Record the exact time of phlebotomy and the exact time of stimulation start (*see* **Note 10**).
2. Collect peripheral venous blood into sodium or lithium heparin tubes. Mix gently by inversion. Do not clot, centrifuge, freeze, or refrigerate the sample before stimulation unless a locally validated alternative protocol is being used.
3. Prepare a tube map before handling blood. At minimum, include an unstimulated negative control, anti-IgE or anti-Fc epsilon RI positive control, a vehicle/protein control if relevant, and each alpha-Gal or ectoparasite stimulus concentration (Table 3).
4. Prepare allergen and ectoparasite stimuli at locally validated concentrations. For diagnostic AGS BAT, include an alpha-Gal-conjugated carrier protein and a mammalian tissue extract such as pork kidney extract when available. For research protocols, include tick saliva, purified tick salivary molecules, or other ectoparasite materials only after validation for sterility, endotoxin, cytotoxicity, and nonspecific activation (*see* **Notes 11–13**).

Table 3
Example BAT tube layout. Stimulus concentrations must be locally validated before diagnostic use

Tube	Stimulus	Purpose	Result expected for valid run
1	PBS or matched vehicle/ protein control	Measures spontaneous activation and nonspecific vehicle effect	Low background; CD63 threshold set so 2.5% of resting basophils are positive
2	Anti-IgE or anti-Fc epsilon RI	Positive control for basophil responsiveness	Clear activation above the laboratory validity threshold
3	Unconjugated carrier protein, if alpha-Gal conjugate is used	Controls for carrier-protein activation	No clinically meaningful activation
4-6	Alpha-Gal-conjugated carrier protein concentration series	Assesses alpha-Gal-specific functional activation	Dose-related activation in compatible AGS; absent or weak in controls
7-9	Pork kidney or mammalian tissue extract concentration series	Assesses clinically relevant mammalian tissue response	Dose-related activation in compatible AGS; inspect for nonspecific high-dose effects
Optional	Gelatin, medication excipient, biologic, tick saliva, or ectoparasite stimulus	Addresses patient-specific or research question	Interpret only with matched controls and local validation

5. Prepare all tubes, pipettes, antibodies, lysis solution, cold PBS, and cytometer settings before opening the blood tube. BAT is time-sensitive and should be run without avoidable interruption.
6. Run daily cytometer quality control and fluorescence calibration beads according to local flow cytometry procedures before acquiring patient samples (*see* **Note 14**).

3.7 BAT Stimulation, Staining, Lysis, and Washing

Use one of the two reagent formats below consistently within a run. The PBS format is often the simplest for clinical laboratories. The HBSS format maintains divalent cation concentration and may be preferred by laboratories that have validated it. Do not compare patient results generated with different formats unless equivalence has been established locally.

1. Label flow cytometry tubes for each patient: negative control, positive control, vehicle/protein control if required, and all allergen or ectoparasite stimulus concentrations.

2. Add 100 μLs of well-mixed heparinized whole blood to each tube.
3. PBS format: pre-warm the blood-containing tubes for 15 min at 37 °C.
4. PBS format: add 2 μL of PBS to the negative-control tube, 2 μL of anti-IgE or anti-Fc epsilon RI to the positive-control tube, and 2 μL of each allergen, ectoparasite stimulus, or vehicle control to the appropriate tubes.
5. PBS format: prepare an antibody master mix containing 5 μL each of anti-CD203c-PE, anti-CD193/CCR3-FITC, and anti-CD63-APC plus 10 μL of PBS. Add 5 μL of this master mix to each tube.
6. HBSS format: dilute 2 microliters of anti-IgE or anti-Fc epsilon RI in 100 microliters of PBS for the positive control. Use 100 μL PBS or matching vehicle as the negative control and prepare each allergen stimulus in 100 microliters.
7. HBSS format: pre-warm blood and stimuli separately for 15 min at 37 °C.
8. HBSS format: mix 100 μL of blood with 100 μL of the appropriate stimulus in each tube and add 2 μL each of anti-CD203c-PE, anti-CD193/CCR3-FITC, and anti-CD63-APC to each tube.
9. Incubate all blood–stimulus–antibody mixtures for 15 min at 37 °C. Keep the incubation time identical for all tubes in the run.
10. Add 2 mL red blood cell lysis solution to each tube. Mix gently and incubate for 10 min at room temperature in the dark.
11. Centrifuge at 300 ×*g* for 5 min at 4 °C.
12. Carefully remove the supernatant without disturbing the cell pellet.
13. Add 2 mL cold PBS to each tube and resuspend gently.
14. Centrifuge at 300 ×*g* for 5 min at 4 °C.
15. Remove the supernatant and resuspend the cell pellet in sufficient PBS for acquisition on the flow cytometer.
16. Acquire samples as soon as possible after staining and washing. Protect samples from light until acquisition.

3.8 Flow Cytometry Acquisition and Gating

1. Run calibration beads and confirm that fluorescence detection, compensation, and scatter settings are acceptable for the instrument and antibody panel.
2. Acquire the entire basophil-containing fraction from the equivalent of 100 μL of whole blood when feasible. Aim to acquire more than 200 gated basophils per tube; results

below this number should be interpreted cautiously (*see* **Note 15**).

3. On the forward scatter/side scatter plot, create a lymphocyte–monocyte region that includes basophils and excludes debris, granulocyte clusters, and obvious dead-cell events.
4. When the instrument and software allow it, perform doublet exclusion using forward-scatter height versus area and side-scatter height versus area.
5. Gate basophils using simultaneous CD203c and CD193/CCR3 positivity, together with low side scatter. Confirm that side scatter distribution allows discrimination of basophils from eosinophils (*see* **Note 16**).
6. Create a CD63 histogram for the basophil gate in the negative-control tube.
7. Set the CD63-positive threshold so that 2.5% of resting basophils in the negative-control tube fall above the threshold.
8. Apply the same CD63 threshold to all stimulated tubes from the same patient and run.
9. Record percent CD63-positive basophils for each tube. Record CD63 MFI for resting basophils and for activated basophils when the software permits this export.
10. Calculate net activation as percent CD63-positive basophils in the stimulus tube minus percent CD63-positive basophils in the negative-control tube.
11. Calculate the response normalized to the positive control when useful: percent CD63-positive basophils after allergen stimulation divided by percent CD63-positive basophils after anti-IgE or anti-Fc epsilon RI stimulation. Do not use this ratio when the positive control is invalid (*see* **Note 17**).
12. Export the gated event count, percent CD63-positive basophils, CD63 MFI, net activation, positive-control response, spontaneous activation, allergen lot/source, stimulus concentrations, and acquisition time.

3.9 BAT Quality Control, Interpretation, and Reporting

1. Confirm sample acceptability: heparinized whole blood, stimulation begun within the accepted time window, no clotting, no hemolysis interfering with acquisition, and adequate patient identifiers.
2. Confirm negative-control acceptability. High spontaneous CD63 expression can obscure weak allergen responses and should prompt cautious interpretation or repeat testing (*see* **Note 18**).
3. Confirm positive-control acceptability. If anti-IgE or anti-Fc epsilon RI fails to produce a valid activation response, classify

the patient as a BAT non-responder for that run and do not interpret allergen tubes as negative (*see* **Note 17**).

4. For the consensus analytical approach, consider net activation of at least 5% CD63-positive basophils above the resting threshold as a positive response, provided that negative and positive controls are valid.
5. Assess dose response. A reproducible dose–response curve across alpha-Gal-bearing stimuli strengthens interpretation. Isolated activation at a single high concentration should be checked for cytotoxicity, endotoxin, nonspecific lectin effects, or carrier-protein effects (*see* **Notes 11 and 19**).
6. Classify a positive alpha-Gal BAT as supportive of clinically relevant AGS only when the clinical history is compatible. BAT does not replace the clinical history, serum sIgE, or supervised challenge when challenge is required (*see* **Note 20**).
7. Classify a negative alpha-Gal BAT as lowering the probability of clinically relevant AGS only if controls are valid and the tested stimuli match the suspected trigger. A negative BAT does not exclude AGS in a patient with a compelling history (*see* **Note 20**).
8. Report the result in a structured format: specimen type and time to stimulation; antibody panel; gating strategy; number of basophils acquired; spontaneous activation; positive-control activation; stimulus names and concentrations; percent CD63-positive basophils; CD63 MFI if available; net activation; interpretation; and limitations.
9. Archive representative plots for the negative control, positive control, and strongest allergen response to support audit and inter-laboratory comparison (Table 4).

3.10 Oral Mammalian Food or Product Challenge

1. Use oral challenge only when the diagnosis remains uncertain after history, sIgE, skin testing, and BAT, and when the expected benefit outweighs the risk. Do not perform a challenge solely to prove a diagnosis that is already clinically clear or in a patient with recent severe anaphylaxis unless there is a compelling specialist indication (*see* **Notes 23 and 24**).
2. Perform the challenge in a specialist allergy setting with immediate access to anaphylaxis treatment and prolonged observation capacity.
3. Select the challenge material according to the clinical question. Examples include cooked mammalian meat, pork kidney or another innard, gelatin-containing product, or the clinically relevant mammalian-derived product. Use a standardized, locally approved recipe and dose schedule.

Table 4
BAT quality-control and reporting checklist

Item	Acceptance target	Action if not met
Blood sample	Heparinized whole blood; no clots; stimulation within validated time window, ideally less than 4 h	Reject, repeat, or report limitation.
Cytometer setup	Daily QC passed; bead performance and compensation acceptable	Do not acquire clinical samples until corrected
Basophil gate	CD203c/CD193-positive, low side scatter population clearly separated from eosinophils and debris	Review gating, antibody staining, and scatter settings
Basophil events	More than 200 gated basophils per tube, where feasible	Repeat acquisition or report low-event limitation
Negative control	Low spontaneous CD63 activation; threshold set at 2.5% resting basophils	Repeat sample if high background prevents interpretation
Positive control	Valid activation to anti-IgE or anti-Fc epsilon RI by local criteria	Classify as non-responder or invalid run; do not call allergens negative
Allergen response	Net activation at least 5% above negative control, preferably with dose response	If isolated high-dose response, assess cytotoxicity, endotoxin, or nonspecific activation
Report	Include event count, percent CD63-positive basophils, CD63 MFI if available, net activation, controls, stimuli, concentrations, and limitations	Amend report before clinical release

4. Confirm absence of uncontrolled asthma, active urticaria, intercurrent infection, or other conditions that would obscure interpretation or increase risk.
5. Confirm medication washout where required and document baseline symptoms, vital signs, peak flow or spirometry if indicated, and physical examination findings.
6. Administer escalating doses according to the approved local schedule. A common structure is four to six incremental portions separated by 15–30 min, but the final cumulative dose and escalation steps must be approved locally for the specific challenge material and patient risk level.
7. Record symptoms, objective signs, vital signs, and treatment after each dose. Stop the challenge for objective allergic symptoms, persistent or progressive gastrointestinal symptoms, respiratory symptoms, cardiovascular symptoms, or clinician concern.

8. Observe for at least 6 h after the final dose because AGS reactions can be delayed. Extend observation or arrange inpatient monitoring for high-risk patients, innard challenges, previous severe reactions, mast-cell disease, or challenges involving cofactors (*see* **Note 25**).
9. Treat reactions according to the anaphylaxis protocol. Use intramuscular epinephrine promptly for anaphylaxis or progressive systemic symptoms.
10. Before discharge, provide written instructions for delayed or biphasic symptoms, emergency medication use, tick-bite prevention, dietary and medication avoidance until the final report, and follow-up contact.
11. Record the challenge as positive only when objective allergic signs occur or when reproducible symptoms meet the local positive-challenge criteria. Record subjective symptoms without objective findings as equivocal unless the protocol defines them otherwise.

3.11 Integrated Diagnostic Classification

1. Review the structured history, exposure timing, objective symptoms, tick/ectoparasite exposure, serum sIgE, skin testing, BAT, and challenge results together.
2. Classify as confirmed AGS when the patient has compatible reactions to alpha-Gal-containing exposure plus alpha-Gal sensitization, and either a convincing history with improvement on avoidance or a positive supervised challenge.
3. Classify as probable AGS when the history is highly compatible and alpha-Gal sensitization is present but confirmatory challenge is not performed because risk or patient preference precludes it.
4. Classify as asymptomatic alpha-Gal sensitization when alpha-Gal sIgE is present without compatible clinical reactions despite relevant exposure.
5. Classify as possible AGS when the history is suggestive, but laboratory results are equivocal or discordant. Define a follow-up plan, which may include repeat sIgE, BAT using the suspected trigger, supervised reintroduction, or challenge in a specialist setting.
6. Classify as unlikely AGS when reactions are temporally inconsistent, alpha-Gal testing is negative, BAT is negative with valid controls, and another diagnosis better explains the symptoms.
7. Classify non-AGS ectoparasite-induced allergy separately, specifying the ectoparasite, reaction type, evidence for IgE-mediated allergy, and relationship to alpha-Gal testing.

8. Document the final classification, management implications, emergency plan, and recommended follow-up interval in the clinical report.

3.12 Post-diagnostic Counseling and Follow-up

1. For confirmed or probable AGS, provide individualized avoidance guidance for mammalian meat, innards, gelatin, dairy or high-fat dairy where relevant, and mammalian-derived medical products according to reaction history and risk profile.
2. Review all current medications, supplements, vaccines, biologics, and planned procedures for mammalian-derived ingredients when clinically relevant. Coordinate with pharmacy, anesthesia, surgery, dentistry, and primary care when needed.
3. Prescribe and train in epinephrine autoinjector use for patients with systemic reactions or risk of anaphylaxis according to local guidelines.
4. Provide tick-bite prevention counseling: protective clothing, repellents, prompt tick checks, safe tick removal, pet tick control, and avoidance of high-risk habitats when possible.
5. Advise the patient to document future bites and reactions, including photographs of bite reactions and a time-stamped symptom diary.
6. Plan follow-up to reassess reaction control, accidental exposures, epinephrine use, new tick bites, and changes in alpha-Gal sIgE or BAT only when repeat testing will influence management.
7. For asymptomatic sensitization, explain that sensitization does not equal clinical allergy. Avoid unnecessary dietary restriction unless clinically indicated, but counsel on symptom recognition and tick-bite prevention.

4 Notes

1. This protocol is not a substitute for emergency care. Patients with ongoing anaphylaxis, hypotension, respiratory compromise, or rapidly progressive symptoms must be treated immediately before elective diagnostic evaluation is continued.
2. Medication washout should follow local allergy service policy. H1-antihistamines are the main cause of false-negative skin tests. Some antidepressants and sedating medications also suppress skin responses. BAT is less affected by antihistamines but can be affected by systemic corticosteroids, biologics, sample age, and cell viability.

3. Beta-blockers and angiotensin-converting enzyme inhibitors can complicate treatment of anaphylaxis. They should not be stopped automatically; discuss the risk with the prescribing clinician before challenge or high-risk testing.
4. Cofactors are common in AGS and can lower the reaction threshold. The absence of symptoms after one mammalian meal does not exclude AGS if previous reactions occurred with alcohol, exercise, NSAIDs, infection, or high-fat mammalian foods.
5. A positive alpha-Gal sIgE result alone is not diagnostic of AGS. The result must be interpreted with the history, timing, and exposure pattern. This is particularly important in endemic tick regions where asymptomatic sensitization may be detected.
6. A negative alpha-Gal sIgE result lowers the probability of AGS but does not absolutely exclude it in all circumstances. Consider assay timing, sample identity, laboratory method, recent avoidance, immunomodulatory therapy, and whether the suspected trigger is non-alpha-Gal ectoparasite allergy.
7. Vigorous rubbing, alcohol residue, dermatographism, and closely spaced test sites can cause false-positive or uninterpretable skin tests. The negative control is essential for interpretation.
8. Commercial meat extracts can be less sensitive than clinically relevant prick-to-prick materials in some AGS patients. If prick-to-prick testing is used, document the exact food, preparation, cooking status, and source because reproducibility depends on the material.
9. Prick-to-prick testing with raw animal products carries infection-control and cross-contamination risks. Use single-use lancets, gloves, clean surfaces, and disposal procedures approved by the institution.
10. BAT is highly time-dependent. Fresh heparinized blood should ideally be stimulated within 4 h. Delayed processing can reduce responsiveness or increase background activation.
11. Allergen preparations used in BAT must be checked for nonspecific activation, cytotoxicity, endotoxin contamination, and carrier-protein effects. Include vehicle controls and unconjugated carrier-protein controls when using alpha-Gal-conjugated proteins.
12. Pork kidney and other innards may be potent alpha-Gal sources, but extracts are not standardized across laboratories. Determine protein concentration, prepare aliquots, avoid repeated freeze-thaw cycles, and validate each lot against positive and negative controls before diagnostic use.

13. Tick saliva, purified tick salivary proteins, mite extracts, flea extracts, mosquito extracts, and other ectoparasite preparations should be used in BAT only in a research or validated translational setting. Sterility, endotoxin, protease activity, and nonspecific basophil activation must be assessed before patient testing.
14. Instrument setup is a major source of inter-laboratory variability. Run beads, compensation controls, and antibody titration according to a written flow cytometry quality system. Side-scatter resolution must be adequate to separate basophils from eosinophils.
15. A basophil count below 200 gated events per tube reduces precision. Repeat acquisition if sample remains available; otherwise report the result as low-event and interpret cautiously.
16. CD203c/CD193 gating is robust, but local panels may use CD123, CRTH2, Fc epsilon RI, or exclusion markers. If a different panel is used, validate it locally and do not mix gating strategies within a study without equivalence testing.
17. Some patients are BAT non-responders to anti-IgE or anti-Fc epsilon RI. An absent allergen response in a non-responder must not be reported as a negative allergen BAT.
18. High spontaneous activation can arise from delayed processing, sample agitation, infection, active allergic disease, or technical problems. Repeat the sample if background prevents interpretation.
19. The 2.5% resting-basophil threshold and at least 5% net activation criterion provide a standardized starting point. Diagnostic laboratories should refine thresholds using local receiver-operating characteristic analysis when sufficient validated clinical samples are available.
20. BAT is supportive rather than definitive. A positive BAT with a non-compatible history can represent sensitization, nonspecific activation, or laboratory artifact. A negative BAT with a compelling history may occur if the relevant allergen matrix was not tested.
21. Serum tryptase is useful when mast-cell disease is suspected or when an acute reaction has occurred. Obtain paired acute and baseline samples according to local timing policies; do not delay emergency treatment to collect tryptase.
22. Skin testing and oral challenge should be modified or avoided in pregnancy, severe cardiovascular disease, uncontrolled asthma, or other high-risk settings unless the benefit is compelling and specialist safeguards are in place.
23. Oral challenge in AGS differs from many immediate food allergy challenges because observation must account for

delayed reactions. A short observation period can miss clinically important reactions.

24. Do not use an oral challenge when the diagnosis is already clear and the result will not change management. The risk is higher for patients with previous severe reactions, innard-associated reactions, mast-cell disease, or inability to use epinephrine promptly.
25. Cofactor challenge should be reserved for specialist centers and research protocols. Introducing exercise, alcohol, NSAIDs, or high-fat meals can increase diagnostic yield but also increases risk.
26. Patient instructions should include hidden mammalian exposures such as gelatin capsules, excipients, medical devices, biologics, vaccines, and perioperative products. The relevance varies between patients; avoid unnecessarily broad restrictions when the history indicates tolerance.

Conflict of Interest

The authors have no conflicts of interest to declare that are relevant to the content of this chapter.

References

1. Commins SP (2020) Diagnosis and management of alpha-gal syndrome: lessons from 2,500 patients. Expert Rev Clin Immunol 16:667–677. https://doi.org/10.1080/1744666X.2020.1782745
2. Reddy S, Yi L, Shields B, Platts-Mills T, Wilson J, Flowers RH (2023) Alpha-gal syndrome: a review for the dermatologist. J Am Acad Dermatol 89:750–757. https://doi.org/10.1016/j.jaad.2023.04.054
3. Rao S, Rathore SS, Goyal RM et al. (2026) Understanding alpha-gal syndrome: a systematic review and meta-analysis of clinical manifestations and outcomes. Neurogastroenterol Motil 38:e70245. https://doi.org/10.1111/nmo.70245
4. Darsow U, Gelincik A, Jappe U, Platts-Mills TAE, Unal D, Biedermann T (2024) Algorithms in allergy: an algorithm for alpha-Gal syndrome diagnosis and treatment, 2024 update. Allergy 79:3169–3172. https://doi.org/10.1111/all.16291
5. Fischer J, Yazdi AS, Biedermann T (2016) Clinical spectrum of alpha-Gal syndrome: from immediate-type to delayed immediate-type reactions to mammalian innards and meat. Allergo J Int 25:55–62. https://doi.org/10.1007/s40629-016-0099-z
6. Weins AB, Eberlein B, Biedermann T (2019) Diagnostik des alpha-gal-syndroms: aktuelle Standards, Fallstricke und Perspektiven. Hautarzt 70:36–43. https://doi.org/10.1007/s00105-018-4288-1
7. Rutkowski K, Wagner A, Rutkowski R, Sowa P, Pancewicz S, Moniuszko-Malinowska A (2020) Alpha-gal syndrome: an emerging cause of food and drug allergy. Clin Exp Allergy 50:894–903. https://doi.org/10.1111/cea.13683
8. Perusko M, Grundstrom J, Eldh M et al. (2025) Allergenic potency of various foods of mammalian origin in patients with alpha-gal syndrome. Allergy 80:181–192
9. Brestoff JR, Zaydman MA, Scott MG, Gronowski AM (2017) Diagnosis of red meat allergy with antigen-specific IgE tests in serum. J Allergy Clin Immunol 140:608–610. https://doi.org/10.1016/j.jaci.2017.01.032
10. Centers for Disease Control and Prevention (2026) Clinical diagnosis and testing: alpha-gal syndrome. https://www.cdc.gov/alpha-gal-syndrome/hcp/diagnosis-testing/index.html. Accessed 4 May 2026

11. Bousquet J, Heinzerling L, Bachert C et al. (2012) Practical guide to skin prick tests in allergy to aeroallergens. Allergy 67:18–24. https://doi.org/10.1111/j.1398-9995.2011.02728.x
12. Heinzerling L, Mari A, Bergmann KC et al. (2013) The skin prick test—European standards. Clin Transl Allergy 3: 3. https://doi.org/10.1186/2045-7022-3-3
13. Mehlich J, Fischer J, Hilger C et al. (2019) The basophil activation test differentiates between patients with alpha-gal syndrome and asymptomatic alpha-gal sensitization. J Allergy Clin Immunol 143:182–189. https://doi.org/10.1016/j.jaci.2018.06.049
14. Pascal M, Edelman SM, Nopp A et al. (2024) EAACI Task Force report: a consensus protocol for the basophil activation test for collaboration and external quality assurance. Allergy 79:290–293. https://doi.org/10.1111/all.15907
15. Santos AF, Alpan O, Hoffmann HJ (2021) Basophil activation test: mechanisms and considerations for use in clinical trials and clinical practice. Allergy 76:2420–2432
16. Ebo DG, Elst J, Van Gasse AL, De Puysseleyr L, Faber MA, Hagendorens MM et al. (2020) Basophil activation experiments in immediate drug hypersensitivity: more than a diagnostic aid. Methods Mol Biol 2163:197–211
17. Lanser BJ, Wright BL, Orgel KA et al. (2020) Conducting an oral food challenge: an update to the 2009 adverse reactions to foods committee work group report. J Allergy Clin Immunol Pract 8:75–90. https://doi.org/10.1016/j.jaip.2019.09.029
18. Sampson HA, Gerth van Wijk R, Bindslev-Jensen C et al. (2024) AAAAI-EAACI PRACTALL: standardizing oral food challenges—2024 update. Pediatr Allergy Immunol 35:e14276. https://doi.org/10.1111/pai.14276

Chapter 24

Protocol for the Manipulation of Fossil Amber to Reduce Allergenic Effects

José de la Fuente, Margarita Villar, Agustín Estrada-Peña, Almudena González-García and David Fernández-Castellanos

Abstract

Amber has a significant role in science, traditional medicine, art, education, and communication. Fossil amber inclusions are relevant for the study of the history and evolution of life on Earth. Amber has also been associated with allergic reactions in some individuals, and thus protocols should be implemented to reduce these risks when working with amber. In this chapter, we describe the experimental approaches to process fossil amber and measures to reduce the risks associated with allergic reactions to this fossilized tree resin. These methods allow the processing of amber pieces for the study of fossil organism inclusions.

Keywords Allergy, Amber, BSL-2, Fossil, Laminar flow hood, Paleoproteomics

1 Introduction

Fossils are relevant for the study of the history and evolution of life on Earth, with a significant role in traditional medicine, art, education, and communication [1]. Fossil amber inclusions are used for analysis using microscopic and tomographic morphological characterization to innovative paleoproteomics for peptide/protein identification by mass spectrometry, phylogeny and protein structure [2] (Fig. 1). Amber has been studied for its potential anti-inflammatory and other medicinal properties through inhibition of signaling pathways (e.g., MAPK and NF-κB) that reduce cytokine production [3], and has been used for centuries as a diuretic to enhance mental stability, stop bleeding, aid in wound healing, promoting anti-allergenic, anti-neurotoxic, and anti-inflammatory effects, to suppress melanin production, promote collagen production, reduce fat accumulation, and many other uses in folk medicine [4, 5] but has also been associated with allergic reactions [6, 7].

Fossil amber is generally safe and is not considered a clinical problem in personal ornaments, but it can cause reactions in individuals

José de la Fuente and Alejandro Cabezas-Cruz (eds.), *Allergic Disease Models*, Methods in Molecular Biology, vol. 3092,
https://doi.org/10.1007/978-1-0716-5635-8_24,

Fig. 1 Preparation of amber samples. (**A**), Epoxy resin embedding, trimming, polishing, and cleaning. (**B**), Morphological analysis: (1) Tick larva, Baltic amber, Eocene (38–54 Mya). (2) Crane fly, Diptera: Tupilidae, Dominican amber, Oligocene (24.5–38 Mya). (3) Spider with insects, Burma, Myanmar, Cretaceous (ca. 99 Mya). (4) Hymenopteran Chalcidoidea Chalsid wasp and Mantodea nymph, Burma, Myanmar, Cretaceous (ca. 99 Mya). (5) Moth larval case and plant fragments, silk-based larval cases built and carried by casemaking Lepidoptera (casebearer/bagworm larvae) with layered, fibrous construction, and embedded plant fragments, not a single organic body, Burma, Myanmar, Cretaceous (ca. 99 Mya)

allergic to rosin [6]. Fossil amber from Baltic origin (40–60 million years ago; Mya), came from *Pinus succinifera, an* extinct pine tree species and contains many compounds such as resin acids (e.g., abietic acid) that can produce a type IV hypersensitivity contact allergic reaction [6, 7]. Additionally, amber inclusions may contain biomolecules such as the highly conserved Actin protein [2, 8] (Fig. 2) and the Pacific coast of Asia oyster, *Magallana gigas* [2, 9]. Therefore, when processing amber inclusions, it is important to implement measures to reduce the risk of allergic reactions.

As described by Luo et al. [10], amber extracts for biomedical interventions are usually prepared by crushing, powdering, and extracting in 50% ethanol at 40°C for 1 h while stirring, and then the filtered solution is depressurized and freeze-dried to form a powder dissolved in dimethyl sulfoxide for use. In paleontological studies,

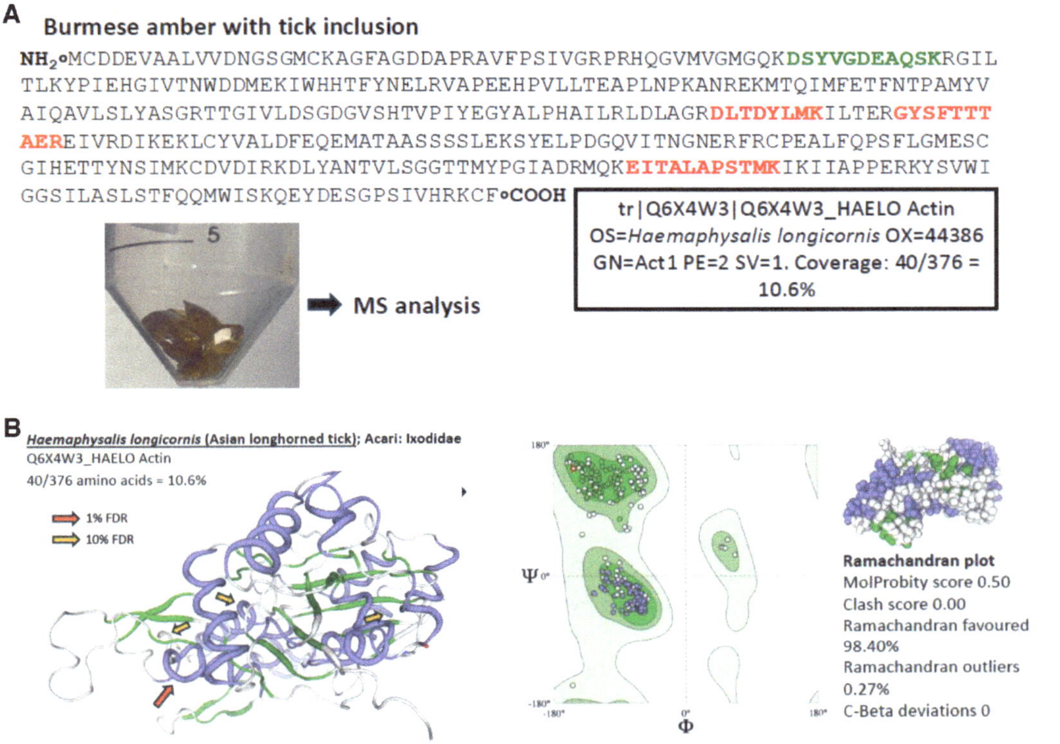

Fig. 2 Processing amber pieces for paleoproteomics analysis. Example of amber with a fossil tick inclusion (Fig. 1B1). (**A**), Only one protein was identified matching the Asian longhorn tick, *Haemaphysalis longicornis* Actin (UniProt entry Q6X4W3), with 10.6% coverage by searching against the UniProt database Tick (Ixodidae) (accessed on 23 May 2024, UniProtKB 289,360 entries; https://www.uniprot.org/uniprotkb?query=ixodidae). (**B**), Swiss-Model supports a highly stable structure for this protein. Additionally, most of the identified peptides are in highly stable (purple color) regions of the protein

amber fossil resin is trimmed to the minimum and then surfaces polished flat to visualize the fossil inclusions from all sides [11] and then processed to powder if used for paleoproteomics analyses [2].

2 Materials

2.1 Fossil Amber with Inclusions

Amber pieces are certified as authentic, as tested with UV light, saltwater floating, sinking in fresh water, acetone resistant, and heat—smell of pine resin and dated to the fossil epoch by radiometric analysis [2, 12] (*see* **Note 1**).

2.2 Epoxy Resins

Commercial EpoThnin resins (e.g., Epo-thin Resin 0.951, Epo-thin Hardener 0.471; https://www.agarscientific.com/media/import/AGB9187_&_AGB9188_Jan14.pdf) are used for epoxy resin embedding.

2.3 Holders for Trimming

For large amber pieces, pieces of PVC and then acrylic Plexiglas with rubber lining and screw fixation are applied. For large pieces, it is better held by a hand for moderate precision cuts.

2.4 Polishing Machines

Commercial polishing machines with corresponding holders are used (e.g., https://www.inovatecmachinery.com/amber-polishing).

3 Methods

Methods for preparation of amber samples are based on Sidorchuk and Vorontsov [11] and de la Fuente et al. [2] (Fig. 1). Amber pieces should be handled with sterile nitrile gloves. Processing of amber pieces should be conducted at level 2 laboratory (BSL-2) under laminar flow hood conditions.

3.1 Epoxy Resin Embedding

The epoxy resin embedding procedure is required for resin reinforcement. A vacuum chamber with a medical syringe, with its tip blocked with a plug or two pieces of plastic, screwed together to enclose a small pot of epoxy resin, with the same syringe serving to remove air with a needle poked through a sealing ring, can be used. According to the epoxy resin used, conditions should be established for curing times, proportioning, molds, cleaning, and mixing (https://www.agarscientific.com/media/import/AGB9187_&_AGB9188_Jan14.pdf).

3.2 Trimming of Amber

One-speed, water-fed circular saw with a 0.75 × 125 mm diamond disk for large amber pieces and a hand-held Dremel 300 rotary tool with a flexible drive joined it for finer operations. The tool uses dental diamond disks (thinner than 0.5 mm) and allows for speed regulation, but the flexible drive is rather heavy, and its flexibility is limited.

3.3 Polishing

The protocol should be adapted to the requirements of the selected polishing machine.

3.4 Handling and Cleaning of Polished Amber Pieces

Fine-tipped steel forceps covered with soft material are used for cleaning. Cleaning is initially done with Kimwipes. A wipe is put flat under a dissecting microscope, and then another is wrapped around a finger, both slightly moistened, and the amber piece lying on the bottom wipe is gently stroked by the finger. It usually adheres to the finger, the bottom wipe thus serving for the actual cleaning.

3.5 Morphological Analysis

Low-resolution images are captured with a routine stereo microscope using an objective with 2×–6× zoom and a stereomicroscope using the scope software. Microscope images are analyzed using Image J program (https://imagej.net/ij/) and pencil sketches of images using IOimageonline.co (https://pencilsketch.imageonline.

co/index.php). High-resolution images are captured with confocal microscopy and CT scan tomography.

3.6 Conditions for Processing Amber Pieces for Paleoproteomics Analysis

Amber pieces are selected (*see* **Note 2**) and scrubbed in 5% SDS with a brush and abundantly rinsed first in water and then in 100% methanol. Then, amber is covered with liquid nitrogen and fractured with a sterile ceramic pestle. Amber fragments without inclusions are used as a control and, as with arthropod inclusions, are triturated to the appearance of powder. Triturates are extracted in 50 μl Laemmli sample buffer by applying ten cycles of sonication followed by vortexing. Samples are then heated to 90 °C for 5 min, centrifuged at 12,000 x*g* for 5 min, and supernatants are collected, concentrated on-gel, and trypsin digested. The resulting tryptic peptides are desalted onto pipette tips, dried down, and then used for paleoproteomics analysis [2] (Fig. 2) (*see* **Note 3**).

3.7 Storage for Preservation

Amber pieces should be stored under conditions protected from light and at room temperature. Samples for mass spectrometry analysis are stored at −20 °C.

4 Notes

1. Fossils should come from collections such as KGJ Colección (Ciudad Real, Spain) for the amber pieces disclosed in Fig. 1, and the provider ensures that "Seller knows that in line with the laws and regulations in his country it is allowed to sell/export this object."
2. Paleoproteomics approach complements morphological analyses when multiple specimens of a given taxon/species are available, and so destructive sampling is affordable.
3. For paleoproteomics analysis, samples are resuspended in 10 μl of 2% acetonitrile—5% acetic acid in water and analyzed by reverse-phase liquid chromatography coupled online to mass spectrometry (RP-LC–MS/MS) using an Ekspert™ nLC 415 system coupled with a 6600 TripleTOF mass spectrometer through Information-Dependent Acquisition (IDA). The peptides are concentrated in a 0.1 × 20 mm C18 RP precolumn with a flow rate of 5 μl/min for 10 min in solvent A. Then, peptides are separated in a 0.075 × 150 mm C18 RP column with a flow rate of 300 nl/min. Peptide elution is done in a 60-min gradient from 5% to 30% solvent B, followed by a 10-min gradient from 30% to 60% solvent B (Solvent A: 0.1% formic acid in water, solvent B: 0.1% formic acid in acetonitrile), and directly injected into the mass spectrometer for analysis. Two microliters of each independent sample are analyzed in duplicate. The mass spectrometer is set to scan full spectra

from 350 to 1400 m/z (250 ms accumulation time), followed by up to 50 MS/MS scans (100–1500 m/z). Candidate ions with a charge state between +2 and +5 and counts per second above a minimum threshold of 100 are isolated for fragmentation. One MS/MS spectrum is collected for 100 ms before adding those precursor ions to the exclusion list during 15 s (mass spectrometer operated by Analyst R TF 1.7, AB Sciex). Dynamic background subtraction is turned off. Data are acquired in high-sensitivity mode with rolling collision energy on and a collision energy spread of 5. Control amber without inclusions was processed in the same manner. For data analysis, the search parameters are iodoacetamide cysteine alkylation, trypsin digestion, and gel-based ID as special factor, identification focused on biological modification and thorough ID as the search effort. The detected protein threshold is set at 0.05. An independent False Discovery Rate (FDR) analysis with the target-decoy approach provided by ProteinPilot™, is used to assess the quality of identifications. Positive identifications are considered when identified proteins reach a 1% global FDR. Protein Blast (Blastp), conserved domains, phylogenetic and taxonomic analyses are conducted at the National Center for Biotechnological Information (NCBI) (https://blast.ncbi.nlm.nih.gov/Blast.cgi?PROGRAM=blastp&PAGE_TYPE=BlastSearch&LINK_LOC=blasthome); https://www.ncbi.nlm.nih.gov/Taxonomy/Browser/wwwtax.cgi?id=2762511) and UniProt (https://www.uniprot.org/blast). Protein structure is modeled using Swiss-Model (https://swissmodel.expasy.org).

Competing Interests

The authors have no conflicts of interest to declare that are relevant to the content of this chapter.

References

1. de la Fuente J (2026) Fossils in art and science. Guidance for research, education and communication. Int J Human Soc Sci Educ (IJHSSE) 13(1):123–138. https://doi.org/10.20431/2349-0381.1301012
2. de la Fuente J, Villar M, Estrada-Peña A (2026) Paleontological approaches for the study of fossils. Am J Sci Educ Res (American J Sci Edu Re: AJSER-302) 5(1):100302. https://doi.org/10.71010/2835-6764/ajser-e302
3. Refli R, Sofyana NT, Haeiwa H, Takeda R, Okazaki K, Sekita M, Sakamoto K (2023) Amber extract suppressed mast cell-mediated allergic inflammation via the regulation of allergic mediators—an *in vitro* study. Nutraceuticals 3(1):75–90. https://doi.org/10.3390/nutraceuticals3010006
4. de la Fuente J (2025) Paleopharmaceuticals from fossil amber. Hektoen Int. A J Med Human. Winter March 24, 2025. Sections: Anthropology. https://hekint.org/2025/03/24/paleopharmaceuticals-from-fossil-amber/
5. Tian Y, Zhou S, Takeda R, Okazaki K, Sekita M, Sakamoto K (2021) Anti-inflammatory

activities of amber extract in lipopolysaccharide-induced RAW 264.7 macrophages. Biomed Pharmacother 141:111854. https://doi.org/10.1016/j.biopha.2021.111854
6. Karlberg AT, Boman A, Lidén C (1992) Studies on the allergenicity of Baltic amber. Contact Dermatitis 27(4):224–229. https://doi.org/10.1111/j.1600-0536.1992.tb03250.x
7. Mills JS, White R, Gough LJ (1984) The chemical composition of Baltic amber. Chem Geol 47(1–2):15–39. https://doi.org/10.1016/0009-2541(84)90097-4
8. Alvarez-González M, Flores-Flores A, Carbajal-Salinas V, Bazán-Perkins B (2025) Altered actin isoforms expression and enhanced airway responsiveness in asthma: the crucial role of β-cytoplasmic actin. Front Physiol 16:1627443. https://doi.org/10.3389/fphys.2025.1627443
9. Lee AJ, Thalayasingam M, Lee BW (2013) Food allergy in Asia: how does it compare. Asia Pac Allergy 3(1):3–14. https://doi.org/10.5415/apallergy.2013.3.1.3
10. Luo Y, Zhou S, Takeda R, Okazaki K, Sekita M, Sakamoto K (2022) Protective effect of amber extract on human dopaminergic cells against 6-hydroxydopamine-induced neurotoxicity. Molecules 27(6):1817. https://doi.org/10.3390/molecules27061817
11. Sidorchuk EA, Vorontsov DV (2018) Preparation of small-sized 3D amber samples: state of the technique. Palaeoentomology 001(1):080–090. https://doi.org/10.11646/palaeoentomology.1.1.10
12. Shi G, Grimaldi DA, Harlow GE, Wang J, Wang J, Yang M, Lei W, Li Q, Li X (2012) Age constraint on Burmese amber based on U-Pb dating of zircons. Cretaceous Res 37:155–163. https://doi.org/10.1016/j.cretres.2012.03.014

Kolibri 360 GmbH
Ferdinand-Jühlke-Straße 7
99095 Erfurt, Deutschland
produktsicherheit@kolibri360.de